HEAT SHOCK
From Bacteria to Man

HEAT SHOCK,
From Bacteria to Man

Edited by

Milton J. Schlesinger
Washington University School of Medicine

Michael Ashburner
University of Cambridge

Alfred Tissières
University of Geneva

Cold Spring Harbor Laboratory
1982

HEAT SHOCK
From Bacteria to Man

Front cover: E. coli DH1. (Photo courtesy of Jeffrey A. Engler, Cold Spring Harbor Laboratory.) "Man's Reach," plate from Albrecht Durer's study of the human body, *Alberti Dureri: Clarissimi pictoris et geometrae de symetria partum* (1532). (The Bettmann Archive, Inc.)

Back cover: Nucleotide sequences of hsp23 gene from *Drosophila.* (Courtesy of Richard Southgate, University of Geneva)

Library of Congress Cataloging in Publication Data

Main entry under title:

Heat shock, from bacteria to man.

   Papers presented at a meeting held at Cold Spring Harbor Laboratory, May 5-9, 1982.
   Includes index.
   1. Heat shock proteins—Congresses.   2. Gene expression—Congresses.   I. Schlesinger, Milton.
II. Ashburner, M.   III. Tissières, Alfred.
IV. Cold Spring Harbor Laboratory.
QP552.H43H4   1982   574.87'322   82-61222
ISBN 0-87969-158-1

Conference Participants

Mitchell Altschuler, *Department of Biological Sciences, State University of New York, Albany, New York*

Michael Ashburner, *Department of Genetics, University of Cambridge, Cambridge, England*

Burr G. Atkinson, *Department of Zoology, University of Western Ontario, London, Ontario, Canada*

Agnes Ayme, *Department of Molecular Biology, University of Geneva, Geneva, Switzerland*

Dennis Ballinger, *Department of Biology, Massachusetts Institute of Technology, Cambridge, Massachusetts*

Ellen Z. Baum, *Department of Biochemistry, Brandeis University, Waltham, Massachusetts*

Bernd-Joachim Benecke, *Department of Biochemistry, Ruhr University, Bochum, Federal Republic of Germany*

Edward Berger, *Department of Biology, Dartmouth College, Hanover, New Hampshire*

Mariann Bienz, *Medical Research Council Laboratory of Molecular Biology, Cambridge, England*

Harald Biessmann, *Department of Genetics, University of California, Davis, California*

J. Jose Bonner, *Department of Biology, Indiana University, Bloomington, Indiana*

Nicole Bournias-Vardiabasis, *Department of Cytogenetics, City of Hope Medical Center, Duarte, California*

v

Thomas Brady, Department of Biological Sciences, Texas Tech University, Lubbock, Texas

Peter A. Bromley, Department of Molecular Biology, University of Geneva, Geneva, Switzerland

Ian R. Brown, Department of Zoology, Scarborough College, University of Toronto, Toronto, Ontario, Canada

Joan Brugge, Department of Microbiology, State University of New York, Stony Brook, New York

Roy H. Burdon, Department of Biochemistry, University of Glasgow, Glasgow, Scotland

Carolyn Buzin, Department of Cytogenetics, City of Hope Medical Center, Duarte, California

Rino Camato, Molecular Genetics Laboratory, Centre Hôpital Université Laval, Quebec, Canada

Victor C. Corces, Department of Biochemistry and Molecular Biology, Harvard University, Cambridge, Massachusetts

Elizabeth Craig, Department of Physiological Chemistry, University of Wisconsin, Madison, Wisconsin

Amalia Dangli, Molekulare Genetik, University of Heidelberg, Federal Republic of Germany

Beth J. DiDomenico, Department of Biology, University of Chicago, Chicago, Illinois

Kevin Dybvig, Department of Microbiology and Immunology, Washington University School of Medicine, St. Louis, Missouri

Joel C. Eissenberg, Department of Zoology, University of North Carolina, Chapel Hill, North Carolina

Sarah Elgin, Department of Biology, Washington University, St. Louis, Missouri

Vivian Ernst, Department of Biochemistry, Brandeis University, Waltham, Massachusetts

Craig Findly, Department of Biology, Yale University, New Haven, Connecticut

David Finkelstein, Department of Biochemistry, University of Texas Health Science Center, Dallas, Texas

Walter J. Gehring, Department of Cell Biology, Biozentrum, University of Basel, Basel, Switzerland

Claiborne V.C. Glover, Department of Biochemistry, Stanford University School of Medicine, Stanford, California

Martin A. Gorovsky, Department of Biology, University of Rochester, Rochester, New York

Georgio Graziosi, Department of Zoology, University of Trieste, Trieste, Italy

Peter Guidon, Microbiology Section, University of Connecticut, Storrs, Connecticut

Linda Hanley-Bowdoin, Rockefeller University, New York, New York

John J. Heikkila, Department of Medical Biochemistry, University of Calgary, Alberta, Canada

Robert Heimer, Department of Medical Oncology, Yale School of Medicine, New Haven, Connecticut

Eileen Hickey, Department of Biology, University of Southern Florida, Tampa, Florida

Larry Hightower, *Microbiology Section, University of Connecticut, Storrs, Connecticut*

Hidetoshi Iida, *Tokyo Metropolitan Institute of Medical Science, Tokyo, Japan*

David Ish-Horowitz, *Imperial Cancer Research Fund Laboratories, London, England*

Robert S. Jack, *Department of Cell Biology, Biozentrum, University of Basel, Basel, Switzerland*

E. J. Kasambalides, *Department of Pathology, State University of New York, Downstate Medical Center, Brooklyn, New York*

Michael Keene, *Department of Biology, Washington University, St. Louis, Missouri*

Phillip Kelley, *Department of Genetics, University of California, Berkeley, California*

Joseph E. Key, *Department of Botany, University of Georgia, Athens, Georgia*

Robert Klevecz, *Department of Biology, City of Hope Research Institute, Duarte, California*

Jonathan Knowles, *Recombinant DNA Group, Technical Research Center of Finland, Espoo, Finland*

Stephen Kurtz, *Department of Biology, University of Chicago, Chicago, Illinois*

Karl W. Lanks, *Department of Pathology, State University of New York, Downstate Medical Center, Brooklyn, New York*

David LaRocca, *Harvard University, Cambridge, Massachusetts*

Louis Levinger, *Department of Biology, Massachusetts Institute of Technology, Cambridge, Massachusetts*

Warren Levinson, *Department of Microbiology, University of California, San Francisco, California*

Gloria C. Li, *Department of Radiation Oncology, University of California, San Francisco, California*

James Lin, *Cold Spring Harbor Laboratory, Cold Spring Harbor, New York*

Susan L. Lindquist, *Department of Biology, University of Chicago, Chicago, Illinois*

John T. Lis, *Department of Biochemistry, Department of Molecular and Cell Biology, Cornell University, Ithaca, New York*

William F. Loomis, *Department of Biology, University of California, San Diego, La Jolla, California*

N. H. Lubsen, *Department of Genetics, University of Nijmegen, Nijmegen, The Netherlands*

John C. Lucchesi, *Department of Zoology, University of North Carolina, Chapel Hill, North Carolina*

Debabrata Majumdar, *Department of Biological Sciences, Oakland University, Rochester, Michigan*

Kenneth Manly, *Department of Cell and Tumor Biology, Roswell Park Memorial Institute, Buffalo, New York*

Joseph Mascarenhas, *Department of Biological Sciences, State University of New York, Albany, New York*

Philip J. Mason, *Department of Molecular Biology, University of Geneva, Geneva, Switzerland*

Matthew Meselson, Department of Biochemistry and Molecular Biology, Harvard University, Cambridge, Massachusetts

Marc-Edouard Mirault, Department of Molecular Biology, University of Geneva, Geneva, Switzerland

R. Mitchel, Department of Radiation Biology, Atomic Energy of Canada, Ontario, Canada

Herschel Mitchell, Department of Biology, California Institute of Technology, Pasadena, California

Richard Morimoto, Fairchild Biochemistry, Harvard University, Cambridge, Massachusetts

Paul W. Morris, Department of Biological Chemistry, University of Chicago Medical Center, Chicago, Illinois

De Lill Nasser, Department of Genetic Biology, National Science Foundation, Washington, D. C.

Frederick C. Neidhardt, Department of Microbiology and Immunology, University of Michigan, Ann Arbor, Michigan

Wilma L. Neuman, Department of Biological Chemistry, University of Illinois Medical Center, Chicago, Illinois

Joseph Nevins, Rockefeller University, New York, New York

Markus Noll, Department of Cell Biology, Biozentrum, University of Basel, Basel, Switzerland

Thaddeus Nowak, Jr., National Institutes of Health, Bethesda, Maryland

Mary Lou Pardue, Department of Biology, Massachusetts Institute of Technology, Cambridge, Massachusetts

Hugh R. B. Pelham, Medical Research Council Laboratory of Molecular Biology, Cambridge, England

Jose R. Pellon, Department of Nutrition and Food Science, Massachusetts Institute of Technology, Cambridge, Massachusetts

Lawrence Pesko, Department of Biology, University of Chicago, Chicago, Illinois

Nancy S. Petersen, Department of Biology, California Institute of Technology, Pasadena, California

Judith Plesset, Department of Biological Chemistry, University of California, Irvine, California

Christopher M. Preston, Institute of Virology, Glasgow, Scotland

Ferruccio Ritossa, Institute of Genetics, University of Bari, Bari, Italy

Ira S. Rubin, Department of Pathology, University of Chicago, Chicago, Illinois

Marilyn M. Sanders, Department of Pharacology, UMDNJ-Rutgers Medical School, Piscataway, New Jersey

Milton J. Schlesinger, Department of Microbiology and Immunology, Washington University School of Medicine, St. Louis, Missiouri

Fritz Schöffl, Department of Botany, University of Georgia, Athens, Georgia

Ronald L. Seale, Department of Cellular Biology, Scripps Clinic and Research Foundation, La Jolla, California

Julie Silver, Division of Life Sciences, Scarborough College, University of Toronto, Toronto, Ontario, Canada

Ralph M. Sinibaldi, Department of Biology, University of Chicago, Chicago, Illinois

Eileen R. Sirkin, *Department of Biology, University of Chicago, Chicago, Illinois*
Karl Sirotkin, *Department of Microbiology, University of Tennessee, Knoxville, Tennesee*
Terry Snutch, *Department of Biological Sciences, Simon Fraser University, Burnaby, British Columbia, Canada*
Richard Southgate, *Department of Molecular Biology, University of Geneva, Geneva, Switzerland*
John R. Subjeck, *Department of Radiation Biology, Roswell Park Memorial Institute, Buffalo, New York*
Hewson Swift, *Department of Biology and Pathology, University of Chicago, Chicago, Illinois*
R. M. Tanguay, *Centre Hôpital Université Laval, Molecular Genetics Laboratory, Quebec, Canada*
G. Paul Thomas, *Cold Spring Harbor Laboratory, Cold Spring Harbor, New York*
Alfred Tissières, *Department of Molecular Biology, University of Geneva, Geneva, Switzerland*
István Török, *Department of Biochemistry, Hungarian Academy of Sciences, Szeged, Hungary*
Andrew A. Travers, *Medical Research Council, Cambridge, England*
Jose M. Velazquez, *Department of Biology, University of Chicago, Chicago, Illinois*
Richard Voellmy, *Department of Biochemistry, University of Miami, Miami, Florida*
Samuel Wadsworth, *Worcester Foundation for Experimental Biology, Worcester, Massachusetts*
D. B. Walden, *Department of Plant Sciences, University of Western Ontario, London, Ontario, Canada*
Lee A. Weber, *Department of Biology, University of Southern Florida, Tampa, Florida*
William J. Welch, *Cold Spring Harbor Laboratory, Cold Spring Harbor, New York*
Fredrick White, *Department of Pharmacology, University of Iowa, Iowa City, Iowa*
James Wilhelm, *Department of Microbiology, University of Rochester Medical Center, Rochester, New York*
Carl Wu, *National Cancer Institute, National Institutes of Health, Bethesda, Maryland*
Ichiro Yahara, *Tokyo Metropolitan Institute of Medical Science, Tokyo, Japan*
Wes Yonemoto, *Department of Microbiology, State University of New York, Stony Brook, New York*
Takashi Yura, *Department of Virus Research, Kyoto University, Kyoto, Japan*

First row: A. Tissières, E. Watson, M. J. Schlesinger; M. Meselson; J. S. Brugge.
Second row: M. Ashburner; Canadian contingent.
Third row: H. R. B. Pelham; F. Ritossa; M. L. Pardue.
Fourth row: K. Dybvig, M. M. Saunders; H. K. Mitchell, A. Tissières.

Preface

The first meeting to focus exclusively on heat-shock proteins was held at the Cold Spring Harbor Laboratory May 5–9, 1982. It brought together about 130 scientists — among them bacterial geneticists, plant biochemists, animal physiologists, developmental biologists, molecular biologists, and virologists. Their areas of research were equally diverse and included the cloning and sequencing of heat-shock genes, studies of in vitro RNA and protein synthesis, assays of proteins and enzymes in heated tissue culture cells, and morphological analyses of cells reacted with monoclonal antibodies made against heat-shock proteins. Results of these investigations were presented in some 90 reports — half of them as formal talks and the rest as posters. Summaries of these talks are presented in this volume; they represent our most complete current information about this important universal biological phenomenon.

We thank the following organizations for financial support of this meeting: National Science Foundation, U.S. Public Health Service, Department of Health and Human Services, and The Monsanto Company.

We wish to thank, in particular, James D. Watson, Director of the Cold Spring Harbor Laboratory, for his encouragement in arranging this meeting; Gladys Kist of the Laboratory's Meetings Office for her excellent help in organizing the conference; and Nancy Ford, Director of Publications, and Judy Cuddihy, editor, for expediting so efficiently and rapidly this monograph.

M.J.S.
M.A.
A.T.

Contents

Section 2   CHROMATIN STRUCTURE

Section 3 REGULATION

Section 5 PHYSIOLOGICAL RESPONSE

HEAT SHOCK
From Bacteria to Man

The Effects of Heat Shock and Other Stress on Gene Activity: An Introduction

Michael Ashburner
Department of Genetics
University of Cambridge
Cambridge, England

The discovery (Ritossa 1962) that specific puffs in the polytene chromosomes of *Drosophila busckii* could be induced by a brief heat shock was the curious beginning of a trail that has led to the analysis of a ubiquitous cellular response to stress. This volume, the outcome of a meeting held at Cold Spring Harbor Laboratory in May 1982, summarizes the "state of the art" in this field. In this introductory chapter, my intention is to give a bird's-eye view of the phenomena involved in this stress response, to highlight some of the major discoveries since 1962, and to identify some of the problems that remain to be solved.

A brief comment on nomenclature. It is now recognized that a variety of agents can induce changes in gene activity similar to those caused by heat shock. Because the term "heat shock" is now so well established in the literature, I will often use it as shorthand for these general stress stimuli. A difficulty in the nomenclature of heat-shock proteins will be immediately apparent to readers of this book: there is no uniformity in the description of heat-shock proteins by their apparent molecular weights. In part, this is due to differences in molecular weight of the "same" heat-shock protein in different species, in part it is due to different laboratories using different gel systems and different molecular-weight standards. It is unsafe to presume that two heat-shock proteins which are said to have the same, or different, apparent molecular weights are necessarily the "same" or different proteins, unless supporting evidence has been provided.

1

HISTORICAL

The December 1962 issue of the Swiss journal *Experientia* included a short paper titled "A new puffing pattern induced by temperature shock and DNP in *Drosophila*" by Ferruccio Ritossa, then working in Naples. Ritossa's interest was the phenomenon of puffing, i.e., the transient modifications to the banded structure of dipteran polytene chromosomes thought (Beermann 1956) to be indicative of active gene loci. It had been known since the early 1950s that the pattern of puffs active in the larval tissues of, for example, *Chironomus* or *Drosophila*, changed in a very regular manner as development proceeded. There were indications that these changes in puffing activity were controlled by the same ecdysteroid hormones that control insect development (Clever and Karlson 1960; Becker 1962). The significance of Ritossa's paper was to show that these puffing patterns could be dramatically perturbed by external environmental influences. In particular, Ritossa found that a brief heat shock, or treatment of tissues with 2,4-dinitrophenol or sodium salicylate, resulted in the appearance of three new puffs. Ritossa elaborated these observations in his second paper (Ritossa 1964), increasing the number of agents that could effect the changes in puffing and speculating that these changes result from "some chemical modification associated with the uncoupling of oxidative phosphorylation."

These two papers by Ritossa (and a brief note in the 1963 issue of *Drosophila Information Service* [Ritossa 1963]) led Hans Berendes and the author to study the effects of heat shock, and other agents, on puffing in *D. hydei* and *D. melanogaster*, respectively. Our work (e.g., Berendes and Holt 1964; Ashburner 1970) led to a fairly thorough description of the response, which was to stand subsequent molecular studies in good stead, and to the discovery of several new features of the reaction of the chromosomes to stress. Most notable was the fact that the induction of new puffs was very rapid, it occurred within minutes of the increase in temperature, but it was transient—during a typical heat shock (e.g., from 25°C to 37°C) the puffs reached their maximum size after 30 or so minutes and then regressed. Although the induction of the puffs was quite independent of protein synthesis, their regression was not. Heat shock was also found to result in the regression of nearly all puffs active before the treatment began. The induction of the heat-shock puffs in isolated organs of *Drosophila* demonstrated that the response did not require an intact organism. Moreover these studies with *Drosophila* demonstrated that the heat-shock response was neither tissue- nor developmental stage-specific.

In the period to 1970, the heat-shock puffs offered a very convenient experimental model for the study of puffing itself, but there was really very little advance in our understanding of either the mechanisms of induction or of the function of the heat-shock puffs. In fact only Hans

Berendes' laboratory really gave much attention to the problem of heat-shock puff function, and they, it has turned out, were following the wrong path. As is so often the case, it took the application of a new experimental method before any significant advance could be made. Alfred Tissières, spending a sabbatical at Caltech, took this step with Hershell Mitchell in 1973, using SDS-slab gel electrophoresis to analyze radiolabeled proteins from heat-shocked *Drosophila* salivary glands. The results (Tissières et al. 1974) were dramatic: heat shock induced the synthesis of a small number of polypeptides and repressed the synthesis of most others. The approximate numerical coincidence between the number of heat-shock puffs and newly induced polypeptides encouraged the view that the latter were coded by RNAs synthesized at the former. Within a few months, these data were confirmed in both Nijmegen and Cambridge (Lewis et al. 1975), and we then suggested a possible, but very general, hypothesis of the function of the heat-shock proteins—"we presume the response ... is homeostatic." Not, perhaps, a striking insight but one that seems to have withstood the test of time.

The parallel induction of the heat-shock puffs and proteins by heat shock and other agents encouraged the view that Beermann's hypothesis of puffs—that they were active genes—was correct. Proof of this was to await, however, the isolation of the heat-shock protein mRNAs. This was first made possible by the discovery (McKenzie et al.1975; Spradling et al. 1975) that the permanent tissue culture cell lines of *Drosophila* responded to a heat shock by the synthesis of the same heat-shock proteins as were seen in larval or adult tissues of the fly itself. At a time when biochemists were somewhat reluctant to handle flies, this was important, for it allowed the discovery of the way in which heat shock inhibited ongoing protein synthesis. Within a few minutes of heat shock, all polysomes break down, or at least cannot be recovered from cells. During the next 30 or so minutes a new polysome peak appears in the cells and this contains heat-shock protein mRNA. The isolation of this mRNA, its hybridization back to the polytene chromosomes, and, subsequently, its translation in vitro into heat-shock proteins (McKenzie and Meselson 1977; Mirault et al. 1978) gave very strong evidence that it was coded for by the induced puffs. Furthermore, at a time when the cloning of genes was limited, by and large, by the abundance of their transcripts, these observations made the heat-shock genes, and their mRNAs, good candidates for cloning.

It was the cause of some considerable confusion, therefore, that the first heat-shock-induced transcripts to be cloned (Lis et al. 1978) did not have the hallmarks of mRNAs. They did, it is true, hybridize to the 87C1 heat-shock puff but they neither appeared to be polysomal nor did they appear to code for, at least, a prominent heat-shock protein (Henikoff and Meselson 1977; McKenzie and Meselson 1977; Livak et al. 1978).

Confusion was increased by the first attempts at a genetic correlation between heat-shock puffs and heat-shock proteins. David Ish-Horowicz and Jeanette Holden, then working in Walter Gehring's laboratory in Basel, constructed embryos of *D. melanogaster* that lacked the 87C1 site. These embryos were heat-shocked and their heat-shock proteins studied with the full expectation that they would lack one heat-shock protein, i.e., that coded by 87C1. In fact they showed a normal pattern of heat-shock protein synthesis (Ish-Horowicz et al. 1977) as did embryos that lacked the 87A7 heat-shock puff (Ish-Horowicz et al. 1979). The relationship between heat-shock puff and heat-shock protein was rather more complex than first thought. In fact, we now know that 87C1 and 87A7 are duplicate loci, both containing multiple hsp70-coding sequences. The original clones isolated in Stanford were complementary to a family of sequences, called $\alpha\beta$, found at 87C1 but not 87A7. Although their transcription is under heat-shock control, they are not heat-shock protein-coding genes. It is now clear that certain middle-repetitive sequences have become subverted by those sequences required for the control of the hsp70 genes. A close relative of *D. melanogaster*, *D. simulans,* lacks $\alpha\beta$ sequences at 87C yet has a heat-shock response otherwise identical to that of the former species: the heat-induced transcription of the $\alpha\beta$ sequences in *D. melanogaster* appears to be quite gratuitous (Livak et al. 1978; Leigh Brown and Ish-Horowicz 1981.)

Since 1976 we have learnt a great deal about the molecular anatomy of the heat-shock genes in *Drosophila*. Indeed, they have proven to be very convenient and interesting models for the study of gene organization in this fly. Yet until 1978 these studies appeared to have little impact on the outside world, except for those interested in gene structure per se. In part this may be due to the fact that much of the new data from *Drosophila* took a rather long time to be seen in print, but there was also, I think, a general reluctance to believe that the response of *Drosophila* cells to heat shock would be of more general relevance, especially to homeothermic species. Whatever the reasons, when Bonner and I came to write a review on heat shock in the latter half of 1978 (Ashburner and Bonner 1979), we could devote only a short paragraph to studies on species other than *Drosophila*. This situation was to change dramatically from 1978, with the discovery of analogous stress responses in chick embryonic fibroblasts (Kelly and Schlesinger 1978), in CHO cells (Bouche et al. 1979), in *E. coli* (Lemeaux et al. 1978; Yamamori et al. 1978), in yeast (Miller et al. 1979; McAlister et al. 1979), in *Naegleria* (Walsh 1980), in *Tetrahymena* (Fink and Zeuthen 1978) and, subsequently, in many other species including plants (Barnett et al. 1978).

Quite independently of the study of heat shock in *Drosophila*, biochemists studying the effects of trauma and other agents on vertebrate cells discovered that these agents could result in dramatic changes to the patterns of protein synthesis (Hightower and Smith 1978; Hightower 1980; White 1980). The similarities between these responses and those of tissues to heat shock were soon seen (Hightower and White; 1981). The "heat-shock" response had been shown not only to be universal but also to occur under a wide variety of different stress conditions.

PROSPECTIVE

The contributions to this volume review the present state of the study of the response of cells to heat shock and other stress stimuli. Apart from a far better description of the response than is as yet available there are three major areas for future study.

The first of these concerns the extent to which the responses seen, at both the transcriptional and translational levels, in species as different as *E. coli*, yeast, *Drosophila*, and man are homologous. This problem can be investigated in several different ways: in terms of the homology of the genes themselves, in terms of the homology of the mechanisms of induction, and in terms of homology of function.

The other two broad areas for study can be succinctly stated: How? and Why? What are the mechanisms of induction? and What are the functions of the response?

The number and diversity of agents that induce the heat-shock puffs in *Drosophila*, or the heat-shock proteins in other species, are very high: they include not only heat shock and recovery from anoxia, but also amino acid analogs, sulfhydryl-reacting reagents, transition metal ions, uncouplers of oxidative phosphorylation, viral infection, and a miscellany of other agents such as ethanol, various antibiotics, certain ionophores, chelators, and (albeit indirectly) pyrogens such as LSD. It has been tempting to peruse this list to identify some common cellular target. Early studies of the heat-shock response led to the general idea that the mitochondria were involved in the induction mechanism (e.g., Leenders et al. 1974). Indeed, there is some experimental support for this view (Sin 1975). The discovery of a heat-shock response in an organism that is as happy as an anaerobe as it is as an aerobe (i.e., yeast) will allow a precise study of the role of mitochondrial functions in the heat-shock response.

It is not clear that this search for a common mechanism for induction will succeed. There is evidence that the transcriptional and translational control of the heat-shock genes is autoregulatory, that is to say, the activity of these genes is controlled by the available concentration of the heat-shock proteins themselves. An interesting consequence of autoregulation has been pointed out by Susan Lindquist (this volume): were the heat-shock protein genes to be active under control conditions (i.e., in the absence of external stimuli) then their (low) level of activity would be controlled by the effective concentration of their product(s). Then, any factor that *either* denatured the heat-shock proteins *or* bound them more effectively than their genes would result in the immediate induction of hsp gene transcription. Inducers, therefore, would fall into two classes: those that inactivated the heat-shock proteins and those that acted on some cellular target to increase the concentration or affinity, (or both) of heat-shock protein binding sites.

The functional significance of the heat-shock response is unknown. At the simplest level there is little doubt that it is homeostatic (Lewis et al. 1975), to protect the cell against the ravages of the environmental insult and ensure that the cell can continue its normal life after the crisis has passed. Perhaps the best available evidence for this is seen from experiments studying what may be called "acquired thermotolerance." Cells exposed to one heat shock, or other stimulus, are relatively protected against the effects of a second heat shock (Li and Hahn 1978; Henle and Dethlefson 1978; Mitchell et al. 1979; McAlister and Finkelstein 1980). Despite evidence that some heat-shock proteins, at least, enter nuclei after heat shock, we are almost entirely ignorant of the targets for protection. Moreover we are unlikely to understand the relationship between the heat-shock proteins and protection until we have a better idea than now of the "natural" inducer(s). Heat is certainly a ubiquitous stimulus and as little as a two-degree rise in the body temperature of an intact animal can result in the induction of heat-shock protein synthesis (e.g., Freedman et al. 1981). Yet different species may well differ in the nature of the natural inducers, according to the challenges they have evolved to meet. The study of both the mechanisms of induction of HSP synthesis and of the function of the response will be aided enormously by the discovery, in *E. coli* and other organisms, of mutations that affect the response.

It has been known since the last century that tumors are often thermosensitive. As with syphilis, the treatment of some tumors has involved deliberate fever, either by infection or by the injection of pyrogenic toxins. The effects of hyperthermia on tumors is very complex, involving such gross changes as in the vascularization of the tumor (Eddy 1980). Yet hyperthermia can be an effective therapy (e.g., Crile 1963; Suit and Shwayder 1974; Suit 1977). Despite the complexity of the response of whole tumors to heat shock, an increased sensitivity to hyperthermia has been seen in transformed cells (e.g., Kase and Hahn 1975). Although it is very far from clear whether or not the induction of the heat-shock proteins is relevant to the hypersensitivity of transformed cells to heat, the probable role of heat-shock proteins in acquired thermotolerance, at least, is clearly of clinical relevance.

REFERENCES

Ashburner, M. 1970. Patterns of puffing activity in the salivary gland chromosomes of *Drosophila*. V. Responses to environmental treatments. *Chromosoma* **31:** 356–376.

Ashburner, M. and J.J. Bonner. 1979. The induction of gene activity in *Drosophila* by heat shock. *Cell* **17:** 241–254.

Barnett, T., M. Altschuler, C.N. McDaniel, and J.P. Mascarentes. 1980. Heat shock induced proteins in plant cells. *Dev. Genet.* **1:** 331–340.

Becker, H-J. 1962. Die Puffs der Speicheldrusenchromosomen von *Drosophila melanogaster*. II. Die Auslosung der Puffbildung, ihre Spezifitat und ihre Beziehung zur Funktion der Ringdruse. *Chromosoma* **13:** 341–384.

Beermann, W. 1956. Nuclear differentiation and functional morphology of chromosomes. *Cold Spring Harbor Symp. Quant. Biol.* **21:** 217–232.

Berendes, H.D. and T.K.H. Holt. 1964. The induction of chromosomal activities by temperature shocks. *Genen. Phaenen* **9:** 1–7.

Bouche, G., F. Amalric, M. Caizergues-Ferres, and J.P. Zalta. 1979. Effects of heat shock on gene expression and subcellular protein distribution in Chinese hamster ovary cells. *Nucleic Acid Res.* **7:** 1739–1747.

Clever, U. and P. Karlson. 1960. Induktion von Puffveranderungen in den Speicheld rusenchromosomen von *Chironomus tentans* durch Ecdyson. *Exp. Cell Res.* **20:** 623–626.

Crile, G. 1963. The effects of heat and radiation on cancer implanted on the feet of mice. *Cancer Res.* **23:** 372–380.

Eddy, H.A. 1980. Alterations in tumor microvascalature during hyperthermia. *Radiobiology* **137:** 515–521.

Fink, K. and E. Zeuthen. 1978. Heat shock proteins in *Tetrahymena*. *ICN-UCLA Sympos. Mol. Cell. Biol.* **12:** 103–115.

Freedman, M.S., B.D. Clarke, T.F. Cruz, J.W. Gurd, and I.R. Brown. 1981. Selective effects of LSD and hyperthermia on the synthesis of synaptic proteins and glycoproteins. *Brain Res.* **207:** 129–145.

Henikoff, S. and M. Meselson. 1977. Transcription at two heat shock loci in *Drosophila melanogaster*. *Cell* **12:** 594–600.

Henle, K.J. and L.A. Dethlefsen. 1978. Heat fractionation and thermotolerance: A review. *Cancer Res.* **38:** 1843–1851.

Hightower, L.E. 1980. Cultured animal cells exposed to amino acid analogues or puromycin rapidly synthesise several polypeptides. *J. Cell. Physiol.* **102:** 407–427.

Hightower, L.E. and M.D. Smith. 1978. Effects of canavanine on protein metabolism in Newcastle disease-virus infected chicken embryo cells. In *Negative strand viruses and the host cell* (ed. B.W.J. Mahey and R.D. Barry), pp. 395–405. Academic Press, London.

Hightower, L.E. and F.P. White. 1981. Cellular responses to stress: Comparison of a family of 71-73 kilodalton proteins rapidly synthesised in rat tissue slices and canavanine-treated cells in culture. *J. Cell. Physiol.* **108:** 261.

Ish-Horowicz, D., J.J. Holden, and W.J. Gehring. 1977. Deletions of two heat-activated loci in *Drosophila melanogaster* and their effect on heat-induced protein synthesis. *Cell* **12:** 643–652.

Ish-Horowicz, D., S.M. Pinchin, J. Gausz, H. Gyurkovics, G. Bencze, M. Goldschmidt-Clermont, and J.J. Holden. 1979. Deletion mapping of the two *Drosophila melanogaster* loci that code for the 70,000 dalton heat induced protein. *Cell* **17:** 565–571.

Kase, K. and G.M. Hahn. 1975. Differential heat response of normal and transformed human cells in tissue culture. *Nature* **255:** 228–230.

Kelly, P. and M.J. Schlesinger. 1978. The effect of amino-acid analogues and heat shock on gene expression in chicken embryo fibroblasts. *Cell* **15:** 1277–1286.

Leenders, H.J., H.D. Berendes, P.J. Helmsing, J. Derksen, J.F.J.G. Koninkx. 1974. Nuclear-mitochondrial interactions in the control of mitochondrial respiratory metabolism. *Subcell. Biochem.* **3:** 119–147.

Leigh Brown, A.J. and D. Ish-Horowicz. 1981. Evolution of the 87A and 87C heat shock loci in *Drosophila*. *Nature* **290:** 677–682.

Lemeaux, P.G., S.L. Herendeen, P.L. Bloch, and F.C. Neidhardt. 1978. Transient rates of synthesis of individual polypeptides in *E. coli* following temperature shifts. *Cell* **13:** 427–434.

Lewis, M.J., P. Helmsing, and M. Ashburner. 1975. Parallel changes in puffing activity and patterns of protein synthesis in salivary glands of *Drosophila*. *Proc. Natl. Acad. Sci.* **72:** 3604–3608.

Li, G.C. and G.M. Hahn. 1978. Ethanol-induced tolerance to heat and to adriamycin. *Nature* **274:** 699–701.

Lis, J., L. Prestige, and D.S. Hogness. 1978. A novel arrangement of tandemly repeated genes at a major heat shock site in *Drosophila melanogaster*. *Cell* **14:** 901–919.

Livak, K., R. Freund, M. Schweber, P.C. Wensink, and M. Meselson. 1978. Sequence organization and transcription at two heat shock loci in *Drosophila*. *Proc. Natl. Acad. Sci.* **75:** 5613–5617.

McAlister, L. and D.B. Finkelstein. 1980. Heat shock proteins and thermal tolerance in yeast. *Biochem. Biophys. Res. Commun.* **93:** 819–824.

McAlister, L., S. Strausberg, A. Kulaga, and D.B. Finkelstein. 1979. Altered patterns of synthesis induced by heat shock in yeast. *Curr. Genet.* **1:** 63–74.

McKenzie, S. Lindquist, and M. Meselson. 1977. Translation in vitro of *Drosophila* heat-shock messages. *J. Mol. Biol.* **117:** 279-283.

McKenzie, S. Lindquist, S. Henikoff, and M. Meselson. 1975. Localization of RNA from heat-induced polysomes at puff sites in *Drosophila melanogaster*. *Proc. Natl. Acad. Sci.* **72:** 1117–1121.

Miller, M.J., N-H. Xuong, and E.P. Geiduschek. 1979. A response of protein synthesis to temperature shift in the yeast *Saccharomyces cerevisiae*. *Proc. Natl. Acad. Sci.* **76:** 5222–5225.

Mirault, M.E., M. Goldschmidt-Clermont, L. Moran, A.P. Arrigo, and A. Tissières. 1978. The effect of heat shock on gene expression in *Drosophila melanogaster*. *Cold Spring Harbor Symp. Quant. Biol.* **42:** 819–827.

Mitchell, H.K., G. Moller, N.S. Peterson, and L. Lipps-Sarmiento. 1979. Specific protection from phenocopy induction by heat shock. *Dev. Genet.* **1:** 181–192.

Ritossa, F. 1962. A new puffing pattern induced by temperature shock and DNP in *Drosophila*. *Experientia* **18:** 571–573.

———. 1963. New puffs induced by temperature shock, DNP and salicilate in salivary chromosomes of *D. melanogaster*. *Drosophila Information Service* **37:** 122–123.

———. 1964. Experimental activation of specific loci in polytene chromosomes of *Drosophila*. *Exp. Cell Res.* **35:** 601–607.

Sin, Y.T. 1975. Induction of puffs in *Drosophila* salivary gland cells by mitochondrial factor(s). *Nature* **258:** 159–160.

Spradling, A., M.L. Pardue, and S. Penman. 1977. Messenger RNA in heat shocked *Drosophila* cells. *J. Mol. Biol.* **109:** 559–587.

Spradling, A., S. Penman, and M.L. Pardue. 1975. Analysis of *Drosophila* mRNAs by in situ hybridization: Sequences transcribed in normal and heat shock cultured cells. *Cell* **4:** 395–404.

Suit, H.D. 1977. Hyperthermic effects on animal tissue. *Radiobiology* **123:** 483–487.

Suit, H.D. and M. Shwayder. 1974. Hyperthermia: Potential as an anti-tumor agent. *Cancer* **34:** 122–129.

Tissières, A., H.K. Mitchell, and U. Tracy. 1974. Protein synthesis in salivary glands of *Drosophila melanogaster*. Relation to chromosome puffs. *J. Mol. Biol.* **84:** 389–398.

Walsh, C. 1980. Appearance of heat shock proteins during the induction of multiple flagella in *Naegleria gruberi*. *J. Biol. Chem.* **225:** 2629–2632.

White, F.P. 1980. Differences in protein synthesised in vivo and in vitro by cells associated with the cerebral microvascalature. A protein synthesised in response to trauma? *Neuroscience* **5:** 1793–1799.

Yamamori, T., K. Ito, Y. Nakamura, and T. Yura. 1978. Transient regulation of protein synthesis in *Escherichia coli* upon shift-up of growth temperature. *J. Bacteriol.* **134:** 1133–1140.

Drosophila, Yeast, and E. coli Genes Related to the Drosophila Heat-shock Genes

**Elizabeth Craig, Thomas Ingolia,
Michael Slater, Lynn Manseau,
and James Bardwell**
*Department of Physiological Chemistry
University of Wisconsin-Madison
Madison, Wisconsin 53706*

The major heat-shock polypeptides (hsp) in *Drosophila* are hsp83, 70, 68, 27, 26, 23, and 22. hsp70 is the most abundant and is encoded at two cytological locations, 87A and 87C. The gene-copy number per haploid genome is usually five—two at 87A and three at 87C. Primary sequences of the protein-coding regions of the gene copies from each loci are very similar, about 97% conserved. The hsp68 gene found at cytological locus 95D has been shown to be partially homologous with the hsp70 gene. No intervening sequences have been found in either the hsp70 or hsp68 genes. Genes encoding the four small heat-shock proteins—hsp27, 26, 23, and 22—are found at cytological locus 67B within an 11-kb DNA segment. These four genes are not transcribed in the same direction and appear to contain no intervening sequences.

A heat-shock response has been observed in species of bacteria, plants, and animals. The major induced protein in many species is between 70K and 75K and the induced proteins all have very similar isoelectric points. Kelley and Schlesinger (1982) recently demonstrated that antibodies to the chicken hsp70 cross-react with proteins of *Drosophila, Xenopus,* mouse, human, and yeast. We have analyzed the relationship of *Drosophila* heat-shock genes to (1) other genes of *Drosophila* normally expressed during development and (2) genes of yeast and *E. coli*.

11

SMALL HEAT-SHOCK PROTEINS ARE RELATED TO EACH OTHER AND MAMMALIAN α-CRYSTALLIN

The primary base sequence of the protein-coding region of the four small heat-shock genes was determined. A single open reading frame was found for each gene. The molecular weights of the predicted proteins 27.1K, 26.6K, 23.5K, and 22.7K are in good agreement with experimentally determined values obtained by gel electrophoresis. The amino acid sequences are similar for over half of their length of approximately 200 amino acids. The most extensive region of homology extends from about amino acid 85 to amino acid 195 (Fig. 1). The amino acid sequence of mammalian α-crystallin is surprisingly similar to that of the four heat-shock proteins. The homologies are contained within the same 40% of the amino acid sequence near, but not extending to, the carboxyl termini, where the heat-shock proteins are most similar to each other.

DROSOPHILA GENES RELATED TO THE hsp70 GENE

We have isolated, on the basis of homology with the hsp70 gene, several previously unidentified *Drosophila* genes. The results of in situ hybridization experiments localized the *heat-shock cognates* (hsc) at cytological loci 70C, 87D, and 88E, which are not sites of heat-shock puffs. The primary DNA sequence of portions of the homologous region of each gene was determined. The predicted amino acid sequences of the amino termini of the 3 hsc genes and an hsp70 gene of 87C are shown in Figure 2. It is clear from this analysis that, although the genes

```
                                                                        aa124
hsp27    Glu Leu Thr Val Lys Val Val Asp Asn Thr Val Val  Δ  Val Glu Gly Lys His Glu Glu Arg
hsp26     *   *  Asn  *   *   *   *   *  Ala Ser Ile Leu  Δ   *   *   *   *   *   *   *   *
hsp23     *   *  Val  *  Gly  *  Gln  *   *  Ser  *  Leu  Δ   *   *  Asn  *   *   *   *
hsp22     *   *  Lys  *   *   *  Leu  *  Gly Ser  *   *  Leu  *  Gly  *   *  Ser  *  Gln Gln
α-crys    *   *  Lys  *   *   *  Leu Gly Asp Val Ile Glu  Δ   *  His  *   *   *   *   *   *


                                                                        aa144
hsp27    Glu Asp Gly His Gly Met Ile  Δ  Gln Arg His Phe Val Arg Lys Tyr Thr Leu Pro Lys Gly
hsp26    Gln  *  Asp  *   *  His  *   Δ  Met  *   *   *   *   *  Arg  *  Lys Val  *  Asp  *
hsp23     *   *  Asp  *   *  Phe  *   Δ  Thr  *   *   *   *   *  Arg  *  Ala  *   *  Pro  *
hsp22     *  Ala Glu Gln  *  Gly Tyr Ser Ser  *   *   *  Leu  *  Arg Phe Val  *   *  Glu  *
α-crys   Gln  *  Glu  *   *  Phe  *   Δ  Ser  *  Glu  *  His  *   *   *  Arg Ile  *  Ala Asp
```

Figure 1
Comparison of a portion of the amino acid sequence of the four small heat-shock proteins and α-crystallin. The bovine α-crystallin sequence is that of the B$_2$ chain (Vander Ouderaa et al. 1973). The crystallin sequence comprises amino acids 89−128 (out of 173 total). The hsp sequences are taken from Ingolia and Craig (1982b). (*) The same amino acid as in hsp27 is found.

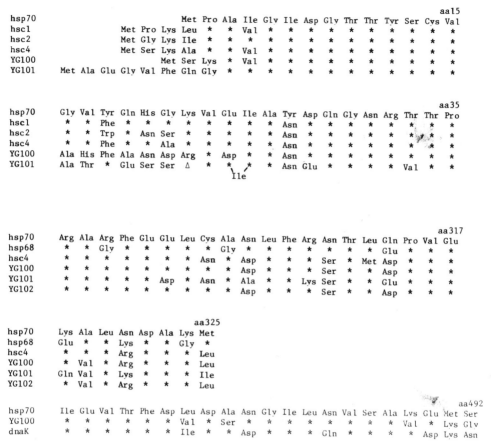

Figure 2

Comparison of the deduced amino acid sequences of hsp70-related sequences from *Drosophila,* yeast and *E. coli.* The hsp70 sequence is from Ingolia et al. (1980) and the hsc1 sequence from Ingolia and Craig (1982a). The amino acids are numbered according to the hsp70 sequence.

are homologous, each could encode a different protein. Among the regions analyzed, the predicted amino acid sequences are about 75% conserved.

An analysis of the transcription of these genes in adult flies was performed because of variation in size of the 5′ noncoding regions. cDNA extension experiments (Ingolia and Craig 1981) have allowed us to distinguish unambiguously transcripts of each of the genes. Primer fragments of each gene from the amino terminal portion of the protein-coding region were isolated, end-labeled in vitro, denatured, and hybridized to RNA from adult flies harvested immediately or after heat shock. The primer was extended using reverse transcriptase and the size and

Table 1
Characteristics of *Drosophila* hsp 70-related Genes

Gene	Locus	Intervening sequence (first 200 amino acids)	Size of 5' noncoding region (bases)	RNA (adult flies)	
				25°C	37°C heat shock
hsp70	87A (2 copies) 87C (3–4 copies)	none	294	+	++++++
hsc1	70C	amino acid 65	379	++	++
hsc2	87D	amino acid 55	185	+	+
hsc4	88E	none	120	+++	+++

the relative amounts of the cDNA synthesized determined by electrophoresis on a denaturing acrylamide gel. The sizes of the 5' noncoding regions of the genes range from 120 to 379 bases (Table 1). Under normal, nonheat-shock conditions, the amount of RNA homologous with the different genes varied at least two orders of magnitude. The transcript from the cognate at 87D was the least abundant and at a concentration similar to that of the transcripts from the hsp70 genes. The 88E transcripts were the most abundant, and the 70C were present at intermediate levels, about 10 times more abundant that those from 87D and 10 times less abundant than those from 88E. After 1 hour of heat shock, the proportional amount of transcripts from the hsc genes had not changed significantly, while the amount of hsp70 transcripts had increased at least 1000-fold.

Comparison of primary sequence data suggests that at least two of the cognates contain intervening sequences. The 70C cognate has an insertion of 1.7 kb in the codon for amino acid 66; the 87D cognate has an insertion of 0.6 kb in the codon for amino acid 55. The 88E cognate has no insertion in the region encoding the first 150 amino acids. The junctions of the insertion sequence with the hsp70-coding regions— 5' GG $_c^T$ GAGT...T $_c^T$ NCAG 3'—are very similar to intron-exon junctions of many eukaryotic genes.

YEAST GENES RELATED TO THE *DROSOPHILA* hsp70 GENES

Yeast genes homologous with the hsp70 gene of *Drosophila* were isolated based on homology with the *Drosophila* gene. Hybridization of portions of the protein-coding regions of the yeast genes to total yeast DNA restriction fragments separated by electrophoresis indicates that yeast contains a multigene family of approximately 10 members homologous with hsp70. Transcription of four isolated family members has been analyzed. Hybridization of RNA from heat-shocked and control cells to the cloned DNAs suggests that transcription of two members (YG100 and YG102) is enhanced after heat shock. However, the amount of RNA sequences homologous with two others (YG101 and YG103) is reduced after heat shock.

The primary DNA of three-quarters of the protein-coding region of YG100 has been determined and compared with the *Drosophila* hsp70 gene. The predicted amino acid sequences are 72% homologous. The sequence of over half of YG101 was also determined; the predicted amino acid sequence was 64% homologous with that of the *Drosophila* gene. The yeast genes are about 65% homologous. The fragmentary sequence data of YG102 accumulated thus far shows 91% homology with YG100, but only 70% homology with YG101. The sequence data, as well as hybridization to total yeast DNA under varying stringencies, indicate that the hsp70-related sequences of yeast differ in the extent of their homology with each other. Yeast genes under similar regulation may well be more homologous with one another than those that respond differently to heat shock.

The predicted hsp70 amino acid sequences encoded at 87A and 87C are 97% homologous. The hsp68 gene is homologous with the hsp70 genes (Holmgren et al. 1980); a comparison of a portion of the amino acid sequence (Fig. 2 and unpublished data) indicates that the hsp68 protein is about 73% homologous (about the same as the hsc proteins) with the hsp70. Therefore, the amount of homology between 87A and 87C hsp70 (97%) is much greater than that between hsp68 and hsp70 (75%).

dnaK GENE OF *E. COLI* IS RELATED TO THE *DROSOPHILA* hsp70 GENE

The *dnaK* gene of *E. coli* has been shown by hybridization studies and DNA sequence analysis to be related to the hsp70 gene of *Drosophila*. DNA sequence analyis of about one-fourth of the *dnaK* gene indicates that the predicted amino acid sequences of the *dnaK* protein is 45–50% homologous to the *Drosophila* and yeast heat-shock genes. Some segments, however, are more highly conserved (see Fig. 2). When the sequence of the entire protein-coding region is completed, an analysis of the distribution of homology among the genes of the three species may provide insight into the functionally important regions of the proteins.

SUMMARY

The major heat-shock proteins of *Drosophila* may be involved in fewer functions than first envisioned when it was thought that seven totally distinct proteins were induced. The fact that the four small heat-shock proteins are partially homologous and all located within an 11-kb DNA segment suggests that they have evolved from a single gene and may be involved in the same or similar functions. The hsp70 and hsp68 genes are related to one another. Therefore, the major heat-shock genes may be grouped into three classes. hsp83 is the sole member of the first class, hsp68 and hsp70 form the second class, and the four small genes form the third.

The small heat-shock proteins are homologous with mammalian α-crystallins, a major component of the vertebrate eye lens. The α-crystallins form large aggregates that perform a major structural role in determining the unique properties of the eye lens. hsp27, 26, and 23 have been shown to exist as aggregates in the nucleus of heat-shocked *Drosophila* cells (Arrigo and Ahmad-Zadeh 1981). A likely reason for the similar amino acid sequences is that this common domain serves to facilitate aggregation. Such complexes could serve a structural role in the nucleus, perhaps stabilizing and protecting DNA against stresses known to induce the heat-shock proteins.

In *Drosophila,* we have found a number of genes related to the hsp70 gene that are normally expressed during development but not induced by heat shock. Perhaps the proteins encoded perform the same or an analogous function under normal growth conditions as the hsp70 does under conditions of stress. We have found no evidence of other genes related to those of the small heat-shock proteins in the *Drosophila* genome. Work from other laboratories, however, has shown expression independent of heat induction of some of the small heat-shock proteins during different developmental stages and after ecdysterone stimulation. However, little evidence exists for expression of the hsp70 and hsp68 genes during normal development. These results are consistent with the idea that the function of heat-shock or related genes is required throughout development and that both normal and heat stress-related functions are performed by the product of the small heat-shock genes, while the different functions required of the hsp70 and related proteins are carried out by the differentially regulated members of the multigene family.

Yeast contains a multigene family of sequences, about 10 in number, related to the hsp70 gene of *Drosophila.* As in *Drosophila,* this family contains both heat-shock-inducible and normally expressed members. Therefore, not only the nucleotide sequence, but also the differential control has been highly conserved in evolution. The maintenance of genes for a heat-shock protein and normally expressed related proteins in evolution points to a probable biological importance of both the heat-shock response and the function of the normally expressed genes.

The remarkable homology between a prokaryotic high-temperature-responsive gene, *dnaK,* and the major heat-shock genes of *Drosophila* and yeast further establishes the conservation of these hsp genes in evolution. The ubiquity of this conserved protein permits the questions of function and regulation to be approached utilizing the tremendous resolving power of yeast and *E. coli* genetics.

REFERENCES

Arrigo, A.-P. and C. Ahmad-Zadeh. 1981. Immunofluorescence localization of a small heat shock protein (hsp 23) in salivary gland nuclei of *Drosophila melanogaster. Mol. Gen. Genet.* **184:** 73–79.

Holmgren, R., K. Livak, R. Morimoto, R. Freund, and M. Meselson. 1980. Studies of cloned sequences from four *Drosophila* heat shock loci. *Proc. Natl. Acad. Sci.* **77:** 5390–5394.

Ingolia, T.D. and E.A. Craig. 1981. Primary sequence of the 5′ flanking regions of the *Drosophila* heat shock genes in chromosome subdivision 67B. *Nucleic Acids Res.* **9:** 1627–1642.

————. 1982a. *Drosophila* gene related to the major heat shock induced gene is transcribed at normal temperatures and not induced by heat shock. *Proc. Natl. Acad. Sci.* **79:** 525–529.

————. 1982b. Four small *Drosophila* heat shock proteins are related to each other and to mammalian α-crystallin. *Proc. Natl. Acad. Sci.* **79:** 2360–2364.

Ingolia, T.D., E.A. Craig, and B.J. McCarthy. 1980. Sequence of three copies of the gene for the major *Drosophila* heat shock induced protein and their flanking regions. *Cell* **21:** 669–679.

Kelley, P.M. and M.J. Schlesinger. 1982. Antibodies to two major chicken heat shock proteins cross-react with similar proteins in widely divergent species. *Mol. Cell. Biol.* **2:** 267–274.

Vander Ouderaa, F.J., W.W. de Jong, and H. Bloemendal. 1973. The amino acid sequence of the αA$_2$ chain of bovine α-crystallin. *Eur. J. Biochem.* **39:** 207–211.

Extensive Regions of Homology Associated with Heat-induced Genes at Loci 87A7 and 87C1 in *Drosophila melanogaster*

István Török*, Philip J. Mason†,
François Karch†, Ibolya Kiss*,
and Andor Udvardy*
**Institute of Biochemistry
Biological Research Center
Hungarian Academy of Sciences
6701 Szeged, Hungary*

*†Department of Molecular Biology
University of Geneva
1211 Geneva 4, Switzerland*

Genes coding for the major 70,000 M_r heat-shock polypeptide (hsp70) are found at two loci, 87A7 and 87C1, in *Drosophila melanogaster*. In most strains there are five copies of the gene encoding hsp70. At 87A7 there are two copies of the gene in diverging orientation separated by a spacer of about 1.5 kb. The arrangement at 87C1 is more complex, with two tandemly repeated distal genes separated from a single proximal copy in diverging orientation by a spacer region of about 40 kb. Within this spacer are found the $\alpha\beta$ sequences that are transcribed in response to heat shock into a number of polyadenylated RNAs, which do not appear to code for any proteins.

Each copy of the hsp70 gene is organized as a basic conserved unit, Z, consisting of a 2.2-kb segment encoding hsp70 mRNA, Zc (Z coding), and a region of about 0.35 kb, Znc (Z noncoding), to the 5′ end of Zc. Though the Z element is highly conserved, there are some characteristic sequence differences between copies from the two loci. Recent studies have shown that the Znc element is part of the element γ, which is interspersed amongst the $\alpha\beta$ sequences at 87C1 in the form of $\alpha\gamma$ units. The association of Znc with two classes of heat-induced genes suggests

19

that this sequence may contain *cis*-acting regulatory elements involved in controlling the expression of these genes (Hackett and Lis 1981).

In *Drosophila simulans* and *Drosophila mauritiana,* two species closely related to *D. melanogaster,* hsp70 genes are also present at the two loci, 87A and 87C, and at both loci they are arranged as two copies in diverging orientation, as at 87A7 in *D. melanogaster.* In these species, $\alpha\beta$ sequences are not present at either locus. The presence of a second distal gene and tandem arrays of $\alpha\beta$ units at 87C1 in *D. melanogaster,* therefore, seem to be the result of recent events at this locus (Leigh-Brown and Ish-Horowicz 1981).

Recently we have isolated a cloned DNA segment containing the proximal hsp70 gene from 87C1 and determined the DNA sequence to the 5' end of this gene. We have also determined the sequence of the spacer between the two divergently orientated genes in a cloned DNA

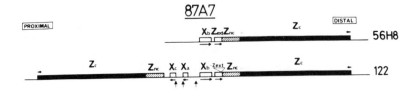

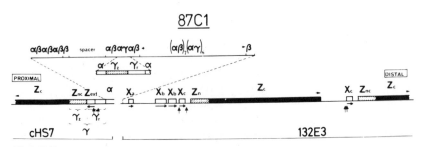

Figure 1

The arrangement of homologous sequences at hsp70 gene loci 87A7 and 87C1 in *D. melanogaster.* This diagram summarizes the sequence data presented in Hackett and Lis (1981), Karch et al. (1981), and Mason et al. (1982). The Zc element is 2.2 kb long and complementary to hsp70 mRNA. The various Znc elements are about 0.35 kb long and approximately 97.5% conserved within each locus and 90% conserved between loci. The other homologous elements (Zext, Xa, Xb, Xc) are at least 85% conserved. The details of the organization of α, β, and γ elements in the region designated $(\alpha\beta)_7$ $(\alpha\gamma)_6$ are not known. The arrows above Zc show the direction of transcription of the hsp70 genes. The arrows below the homologous elements show their relative orientation. The upward pointing arrows show stretches of simple DNA found at this locus and described by Mason et al. (1982). The names of recombinant DNA clones used in sequencing are indicated.

segment isolated from 87A7. In this paper we compare these sequences with those previously reported from 87A7 and 87C1 (Hackett and Lis 1981; Karch et al. 1981). We have determined the 3′ end sequences of the different hsp70 genes. The comparison of these sequences is also presented.

DISCUSSION

Figure 1 shows the arrangement of homologous sequences at 87A7 and 87C1. This complex pattern of homologous blocks of DNA sequence extends far upstream of the hsp70 genes. The two loci are found on the right arm of chromosome 3, separated by about 500 kb of DNA. Since hsp70 genes are found in the corresponding positions on the chromosomes of *Drosophila pseudoobscura* (Pierce and Lucchesi 1980), the original duplication event must be very ancient. *D. pseudoobscura* and

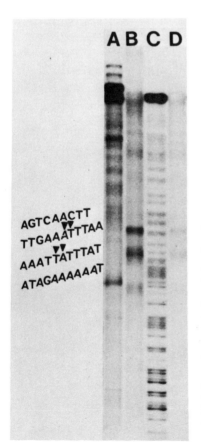

Figure 2
S1 mapping of the 3′ end of hsp70 mRNA. The *Sal* to *Alu* fragment of clone 56H8 (87A7 Di. of Fig. 3; Török and Karch 1980) was labeled by nick translation (slot A), hybridized with hsp70 mRNA, and digested with S1 nuclease for 2 hours (slot B) or 4 hours (slot D). Slot C shows an A>C Maxam-Gilbert reaction of the same fragment labeled at the *Sal* site (Fig. 3). The *Alu* site is some 50 bases downstream from the end of the shown sequence. The sequence presented here is shown between coordinates 172 to 213 of 87A7 Di. in Fig. 3. Since the marker noncomplementary DNA strand was labeled at the *Sal* site, which protrudes four bases over the 3′ end of the S1-resistant DNA strand, we have to localize the real position of the 3′ends 4 bases downstream from the indicated places. In this way the 3′ ends occur at coordinates 194 and 204 of 87A7 Di, in Fig. 3.

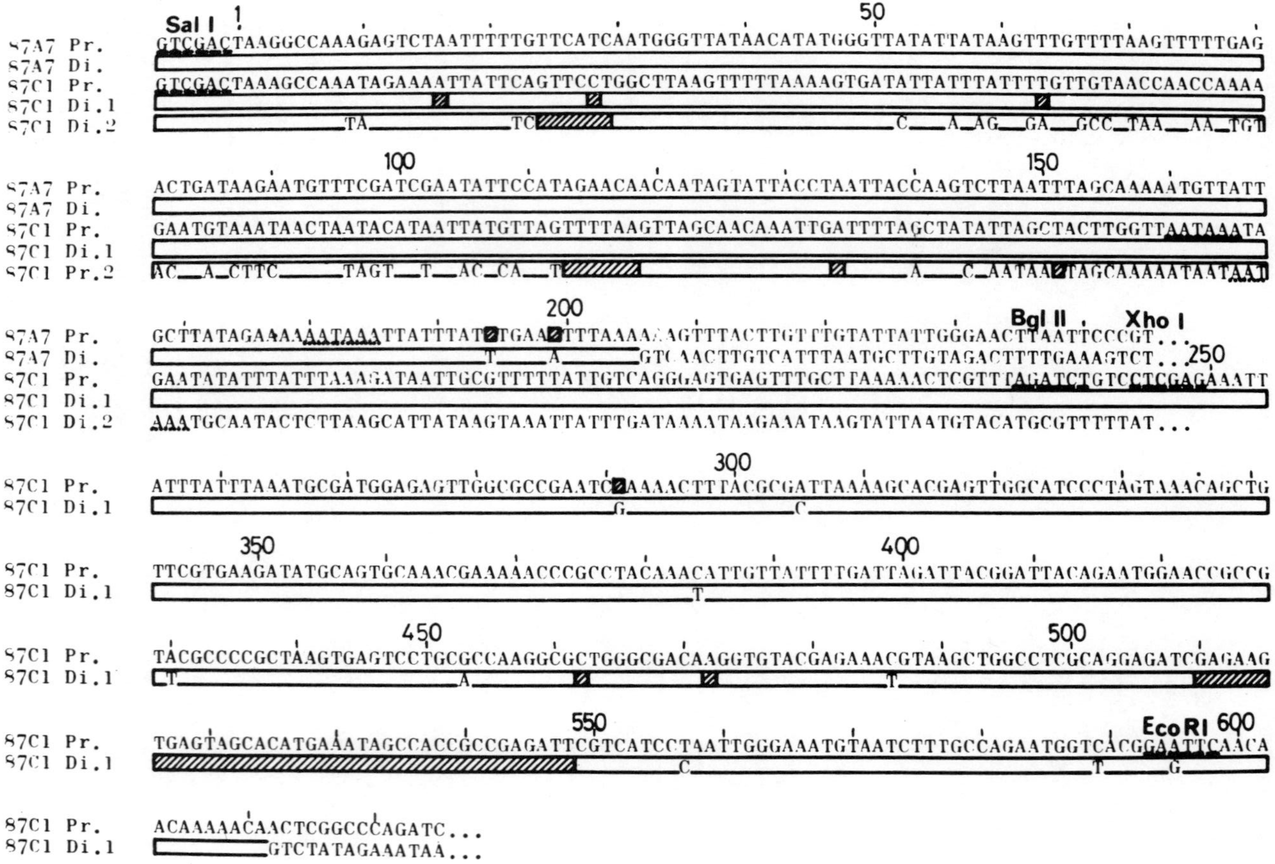

Sal I

50   100   150   200   250   300   350   400   450   500   550   600

Bgl II   Xho I   EcoRI

Sal I
GTCGACTAAGGCCAAAGAGTCTAATTTTTGTTCATCAATGGGTTATATATTATAAGTTGTTTTAAGTTTTTGAG

87A7 Pr.
87A7 Di.
87C1 Pr.
87C1 Di.1
87C1 Di.2

87A7 Pr.
87A7 Di.
87C1 Pr.
87C1 Di.1
87C1 Di.2

87A7 Pr.
87A7 Di.
87C1 Pr.
87C1 Di.1
87C1 Di.2

87C1 Pr.
87C1 Di.1

87C1 Pr.
87C1 Di.1

87C1 Pr.
87C1 Di.1

87C1 Pr.
87C1 Di.1

87C1 Pr.
87C1 Di.1

D. melanogaster diverged some 35 million years ago (Throckmorton 1975). What mechanism, then, has led to the conservation of these sequences?

One possibility is that the homologous sequences have a vital biological function. If they contained signals recognized by regulatory proteins, for example, we may expect them to be conserved. Whilst we cannot, at this time, rule out this hypothesis, two lines of evidence lead us to believe it unlikely. First, there is a different arrangement of external homologous sequences in front of each hsp70 gene, whereas all appear to be transcribed in response to heat shock in a similar manner (Holmgren et al. 1979; Ish-Horowicz and Pinchin 1980). Second, a mutant fly possessing only one hsp70 gene with no flanking X elements appears to be perfectly viable and to respond normally to heat shock (Udvardy et al. 1982).

We favor the hypothesis that these external homologous regions have arisen from recent exchanges of DNA between the two loci. Leigh-Brown and Ish-Horowicz (1981) compared the restriction sites in and around hsp70 genes in several different species of *Drosophila.* They concluded that the various copies of hsp70 genes were not evolving independently and they postulated (1) intralocus gene conversion events and (2) more infrequent interlocus gene conversion events to explain their evolution. The external homologous regions that we find may have arisen by the same interlocus gene conversion events that have led to the conserved evolution of hsp70 genes. On this model, sequence homogeneity of some upstream noncoding regions has arisen as a consequence of the physical association of the two loci promoted by the conserved coding regions.

Figure 1 shows that a complete γ element precedes the proximal hsp70 gene at 87C1 and that this is directly preceded by the element α. The tandem arrays of $\alpha\beta$ and $\alpha\gamma$ sequences at 87C1 in *D. melanogaster* appear to be the result of recent insertion and rearrangement events. Since these sequences are not found associated with hsp70 loci in *D. simulans,* these events must have taken place since the divergence of these two species. Lis et al. (1981) observed that different strains of *D. melanogaster* contain different numbers of $\alpha\beta$ and $\alpha\gamma$ sequences at 87C1. This suggests that unequal crossing over is operating at this

Figure 3
Comparison of the DNA sequences downstream from the *Sal* sites at the 3′ end of each hsp70 gene (Török and Karch 1980; Karch et al. 1981). The shown sequences correspond to the RNA strand. Coordinate 1 is the translation termination triplet. The sequences of the proximal copies of the hsp70 genes are presented at both loci (87A7 Pr., 87C1 Pr.) and the homologies with the distal ones are shown in the empty frames only within one locus. The base substitutions and deletions are indicated in the frames. (□) Deletions; (· · ·) proposed polyadenylation signals; (- - -) main restriction sites.

locus. This could be the mechanism by which the complex array of $\alpha\beta$ and $\alpha\gamma$ elements has rapidly spread through the 87C1 locus—the regulatory sequences (Znc) from the proximal hsp70 gene having been interspersed amongst $\alpha\beta$ sequences, leading to these repeated elements being brought under heat-shock control.

We have recently extended the sequence analysis to the 3′ end of the hsp70 genes at both loci. First, we have determined the extent of the 3′ transcribed but untranslated region by S1 mapping localizing the 3′ end of the hsp70 mRNA in the DNA sequence at locus 87A7. Use of a specific DNA fragment from clone 56H8, which overlaps with the 3′ end of the hsp70 mRNA, yielded two S1-resistant DNA fragments of almost similar size (Fig. 2). One of them was found some 19 nucleotides downstream from the possible polyadenylation signal (coordinate 204 of 87A7 Di. in Fig. 3). The other one was found about 10 bases upstream from this presumed 3′ end and it might be due to the 3′-end heterogeneity of the hsp70 mRNA at this locus (Karch et al. 1981). Since the polyadenylation signals were found among the 3′-end sequences of the hsp70 genes at 87C1, we could estimate the extent of these regions at this locus, too. Second, we compared the DNA sequences at the 3′ end of the hsp70 genes with each other (Fig. 3). Surprisingly, the 3′ transcribed but untranslated region of hsp70 genes derived from 87A7 and 87C1 loci displayed complete heterogenity. Within one locus, however, a high degree of homology could be observed. The only exception is the most distal copy of 87C1, which shows only 57% homology with the other two copies of this locus. In the 3′-flanking region of the hsp70 genes, the longest homology (more than 600 bp) was found between the proximal and the first distal gene of 87C1. The border of this 3′ homologous sequence is just upstream from the Xc element (Mason et al. 1982) found in front of the second distal hsp70 gene in 132E3 (Fig. 1 and coordinate 611 of Fig. 3). Therefore, the middle gene at 87C1 is homologous in the 3′ region with the proximal gene and in the 5′ region with the distal gene. Interestingly, the proximal and distal genes are probably the ancestral genes at this locus since their restriction maps correspond to those found at 87C in related species (Leigh-Brown and Ish-Horowicz 1981).

SUMMARY

We have examined in detail the DNA sequence organization at the hsp70 loci in *D. melanogaster.* Some features of this organization have led us to speculate that gross genomic changes have recently occurred at these loci. We hope to find out more about the nature of these changes by studying hsp70 loci in related species.

REFERENCES

Hackett, R.W. and J.T. Lis. 1981. DNA sequence analysis reveals extensive homologies of regions preceding hsp70 and $\alpha\beta$ heat shock genes in *Drosophila melanogaster. Proc. Natl. Acad. Sci.* **78:** 6196–6200.

Holmgren, R., K. Livak, R. Morimoto, R. Freund, and M. Meselson. 1979. Studies of cloned sequences from four *Drosophila* heat shock loci. *Cell* **18:** 1359–1370.

Ish-Horowicz, D. and S.M. Pinchin. 1980. Genomic organization of the 87A7 and 87C1 heat-induced loci of *Drosophila melanogaster. J. Mol. Biol.* **142:** 231–245.

Karch, F., I. Török, and A. Tissières. 1981. Extensive regions of homology in front of the two hsp70 heat shock variant genes in *Drosophila melanogaster. J. Mol. Biol.* **148:** 219–230.

Leigh-Brown, A.J. and D. Ish-Horowicz. 1981. Evolution of the 87A and 87C heat shock loci in *Drosophila. Nature* **290:** 677–682.

Lis, J.T., D. Ish-Horowicz, and S.M. Pinchin. 1981. Genomic organization and transcription of the $\alpha\beta$ heat shock DNA in *Drosophila melanogaster. Nucleic Acids Res.* **9:** 5297–5310.

Mason, P.J., I. Török, I. Kiss, F. Karch, and A. Udvardy. 1982. Evolutionary implications of a complex pattern of DNA sequence homology extending far upstream of the hsp70 genes at loci 87A7 and 87C1 in *Drosophila melanogaster. J. Mol. Biol.* **156:** 21–35.

Pierce, D.A. and J.C. Lucchesi. 1980. Dosage compensation of X-linked heat shock puffs in *Drosophila pseudoobscura. Chromosoma* **76:** 245–254.

Throckmorton, L.H. 1975. The phylogeny, ecology, and geography of *Drosophila.* In *Handbook of genetics* (ed. R.C. King), vol. 3, pp. 421–470. Plenum Press, New York.

Török, I. and F. Karch. 1980. Nucleotide sequences of heat shock activated genes in *Drosophila melanogaster.* I. Sequences in the regions of the 5′ and 3′ ends of the hsp70 gene in the hybrid plasmid 56H8. *Nucleic Acids Res.* **8:** 3105–3123.

Udvardy, A., J. Sümegi, E. Csordás-Tóth, J. Gausz, H. Gyurkovics, P. Schedl, and D. Ish-Horowicz. 1982. Genomic organization and functional analysis of a deletion variant of the 87A7 heat shock locus of *Drosophila melanogaster. J. Mol. Biol.* **155:** 267–280.

Approximate Localization of Sequences Controlling Transcription of a *Drosophila* Heat-shock Gene

**Victor Corces, Angel Pellicer*,
Richard Axel†, Shang-Yun Mei,
and Matthew Meselson**
*Department of Biochemistry
and Molecular Biology
Harvard University
Cambridge, Massachusetts 02138*

**Department of Pathology
New York University Medical Center
New York, New York 10016.*

*†College of Physicians and Surgeons
Institute of Cancer Research
Columbia University
New York, New York 10032.*

Heat shock and a variety of other stimuli coordinately induce in *Drosophila melanogaster* vigorous transcription of genes for seven characteristic proteins. These features and the availability of cloned genes for each of the heat-shock proteins make this an attractive system for investigating the sequences and interactions controlling eukaryotic gene expression. One approach to identifying the nucleotide sequences involved in the control of transcription by heat shock is to examine different heat-shock genes for possibly significant homologies and common symmetries. In comparing the 5′ regions of a number of different *D. melanogaster* heat-shock genes, one finds several such similarities (Holmgren et al. 1981). Without functional tests, however, their significance is unclear. A more direct approach is to test the effect of sequence alterations on heat-shock gene expression.

We have previously reported that heat shock induces abundant accumulation of transcripts of the *Drosophila* hsp70 gene with correct 5′ and 3′ termini in a line of mouse L cells transformed with plasmid pPW229

27

(Corces et al. 1981). The *Drosophila* insert of pPW229 contains the 2.25-kb hsp70 mRNA sequence and 1.1 kb and 0.2 kb of 5′ and 3′ flanking DNA, respectively. Apparently, the heat-shock transcriptional control signals and the sites with which they interact have been highly conserved, and the regulatory sequences controlling transcription of the hsp70 gene lie within the 3.6-kb *Drosophila* segment of pPW229. The ability to construct such cell lines transformed with various deleted or otherwise altered heat-shock genes makes possible experiments to identify sequences involved in heat-shock gene control. Here, we describe experiments of this type indicating that the 251-bp region extending from 51 bp upstream of the hsp70 mRNA initiation site to 198 bp downstream is sufficient to determine heat-shock transcriptional control. Also, it appears that at least some of the sequences involved in normal control and or promotion lie between −51 and −41.

RESULTS

Figure 1 presents a restriction map of pPW229, containing the cloned hsp70 gene used in the present study (Livak et al. 1978). In order to define the region of pPW229 sufficient to determine the control of heat-shock transcription, we constructed a chimeric gene containing the 1.3-kb *Bgl*II to *Mbo*II fragment of pPW229 joined via a *Bam*HI linker to a *Bam*HI site 59 bp upstream of the first ATG in the protein-coding region of a human growth hormone gene. A map of the resulting plasmid, pH1, is shown in Figure 1. Mouse Ltk⁻ cells were transformed with a mixture of pH1 and a plasmid containing the herpes simplex thymidine kinase (*tk*) gene (Wigler et al. 1979). One of two lines derived from tk⁺ colonies produced transcripts homologous to pPW229. Upon heat shock, it accumulated only the expected 1.0-kb RNA homologous to the cloned growth hormone gene from which pH1 was constructed. Thus, the 1.3 kb of *Drosophila* DNA present in pH1 appears to be sufficient to impart heat-shock transcriptional control.

A similar experiment was conducted using the chimeric plasmid pT1 containing the same 1.3-kb segment of pPW229 joined via a *Bam*HI linker to the *Bgl*II site 57 bp upstream of the first ATG of the protein-coding region of the herpes simplex *tk* gene, as depicted in Figure 1. Mouse Ltk⁻ cells were transformed with a mixture of pT1 and pNeo3 carrying a neomycin-resistance gene (I. Pack, R. Sweet, and B. Wold, pers. comm.). Selection for resistance to the antibiotic G-418 (Schering Corporation) yielded a cotransformed line which, without heat shock, produced a 0.9-kb transcript of the herpes simplex *tk* gene initiated from a secondary promoter located within the thymidine kinase protein (TK) coding region (Roberts and Axel 1982). As may be seen in Figure 2, lane 1, this is the only RNA homologous to the herpes *tk* gene detected

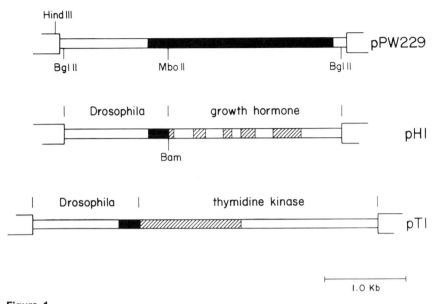

Figure 1
Maps of *D. melanogaster* plasmid pPW229 and the chimeric plasmids pH1 and PT1. The transcribed region of the *D. melanogaster* hsp70 gene is shown filled. (▨) The five exons of the human growth hormone gene in pH1 and the transcribed region of the herpes simplex *tk* gene in pT1. Transcription is from left to right. The distance from the hsp70 mRNA initiation site in pH1 to the end of the transcribed region of the growth hormone gene is 1.0 kb, 0.2 kb greater than the transcribed region of the intact growth hormone gene. Since the latter gives rise to a 0.82-kb mRNA, the size of cytoplasmic RNA derived from the chimeric plasmid is expected to be 1.0-kb (DeNoto et al. 1981; Robins et al. 1982). Similarly, a 1.45-kb cytoplasmic RNA is expected from chimeric plasmid pT1 (McKnight 1980; Roberts and Axel 1982). Only those restriction sites referred to in the text are indicated.

in these cells when cultured at 37°C. After heat shock, however, a second component is found, as shown in the second lane of the figure. Its size is 1.45 kb, as expected for a transcript initiated at the *Drosophila* hsp70 mRNA origin of pT1. Of a total of five tk⁺ lines cotransformed with pT1 DNA, three responded in this way. Two others gave only the 0.9-kb transcript both before and after heat shock.

We conclude from the heat-shock inducibility of transcription in lines transformed with pH1 and pT1, giving rise to RNA molecules of the expected lengths, that the 1.3-kb *Bgl*II to *Mbo*II segment of pPW229 is sufficient to determine heat-shock transcriptional control.

A related approach to identifying control sequences is to investigate the effect of terminal deletions. Figure 3 presents the nucleotide sequence of a portion of the 5′ region of pPW229, showing the probable

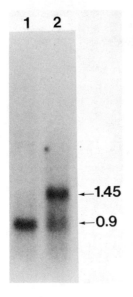

1   2

←1.45

←0.9

Figure 2
Electrophoretic analysis of *Drosophila* hsp70 RNA from mouse L
cells transformed with pT1. Plasmid pT1 was introduced into
mouse Ltk⁻ cells by cotransformation with the plasmid pNeo3
which confers resistance to the antibiotic G-418. Positive transformants were selected and poly (A)-containing RNA was made
from cells grown at 37°C and from cells kept at 43°C for 15
minutes and then at 37°C for 90 minutes. RNA samples were run
on a 0.8% agarose gel containing 2.2 M formaldehyde (Corces et
al. 1981) and transferred to nitrocellulose. The filter was probed
with a 2.7-kb *Bgl*II to *Bam*HI fragment of a plasmid containing
the transcribed region of the herpes simplex *tk* gene (McKnight
1980). Lane 1 contains 3 μg of poly(A)-containing RNA from
cells grown at 37°C. Lane 2 contains an equal amount of poly(A)-
containing RNA from heat-shocked cells. Heat shock induces
the accumulation of the expected 1.45-kb RNA.

mRNA initiation site, the TATAA box, and certain sequence and symmetry features common to a number of *D. melanogaster* heat-shock genes
(Holmgren et al. 1981). The *Mbo*II cleavage site which defines the
downstream end of the *Drosophila* segment of pH1 and pT1 is also
shown. The sequence in the figure was determined for pPW229 itself (V.
Corces, unpubl.) and agrees fully with the independently cloned hsp70
sequences reported by Ingolia et al. (1980).

A series of deletion derivatives of pPW229 was made using *Bal*31
nuclease to remove increasing amounts of DNA starting at the *Hin*dIII
site in pMB9 adjacent to the 5′ junction of the *Drosophila* insert (Fig. 1).
The deleted plasmids were then cut at the *Bgl*II site just beyond the 3′
end of the hsp70 gene, and fragments containing the gene were subcloned into *Hin*dIII to *Bam*HI-cut pBR322. This gave rise to plasmids
designated pL1, pL2, etc., eight of which were chosen for functional
assay. Also tested was 229.4, the 2.6-kb *Xho*I to *Xho*I fragment of
pPW229 (Livak et al. 1978) subcloned at the *Sal*I site of pBR322. The
deletion end-points which lie within 150 bp of the hsp70 mRNA initiation
site are indicated in Figure 3.

Mouse Ltk⁻ cells were cotransformed with the herpes *tk* gene and
each of the deleted plasmids. Lines selected as tk⁺ were tested by dot
blots before and after heat shock for RNA homologous to the 3.4-kb
*Bgl*II fragment of pPW229. The results are summarized in Table 1. Of
the 15 lines transformed with plasmids pL6–pL26 expressing hsp70
RNA, 13 were inducible. That is, in these lines the level of hsp70 RNA,

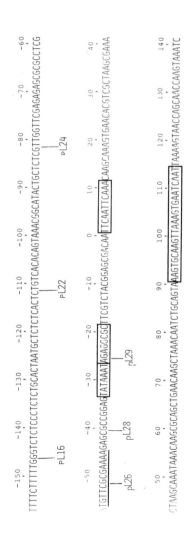

Figure 3

Nucleotide sequence at the 5′ end of the hsp70 gene showing deletion end-points. pPW229 DNA was cut at the *Hin*dIII site of pMB9 and digested with *Bal*31 nuclease for different times under conditions described by Legerski et al. (1978). *Hin*dIII linkers were ligated to the digested ends and the DNA was cut with *Bgl*II. The resulting *Hin*dIII to *Bgl*II fragments containing the transcribed region of the hsp70 gene were then cloned into *Hin*dIII to *Bam*HI-cut pBR322. The resulting plasmids were named pL1, pL2, etc. The exact amount of *Drosophila* DNA remaining was determined by sequencing the appropriate portions of each of the plasmids except for the pL6 and pL9. Boxes enclose homologies found in this and other heat-shock genes. A dyad symmetry centered at −51, −50 is underlined (Holmgren et al. 1981). The probable origin of hsp70 mRNA transcription is at zero.

31

Table 1

Expression of a D. melanogaster hsp70 Gene with 5' Deletions in Transformed Mouse Cells

		Number of	Expression		
Deletions	Last remaining nucleotide	tk⁺ lines tested	none	heat-shock inducible	noninducible
pL6	ca. −850	1		1	
pL9	ca. −650	1		1	
229.4	−195	3		1	2
pL16	−146	2		2	
pL22	−111	4	2	2	
pL24	−81	5	1	4	
pL26	−51	2		2	
pL28	−40	4	1		3
pL29	−25	3		1	2

Mouse Ltk⁻ cells were cotransformed with plasmids pL6, pL9, etc. and a cloned herpes *tk* gene. Lines selected as tk⁺ were tested by dot blots (Kafatos et al. 1979) for accumulation of RNA homologous to the 3.6-kb *Bgl*II to *Bgl*II fragment of pPW229 before and after heat shock. Heat shock was accomplished by keeping cells at 43°C for 15 minutes and then at 37°C for 90 minutes. The deletion end-points, taking the probable hsp70 origin of transcription as zero, are exact except where indicated. Lines are scored as heat-shock inducible if the level of hsp70 RNA was at least doubled by heat shock.

as estimated from dot blots, was increased by a factor of 2–20 by heat shock. The longest deletion represented in this group, that in pL26, leaves only 51 bp of *Drosophila* DNA upstream of the hsp70 mRNA initiation site. In contrast, only one of the six RNA-producing lines transformed with plasmids pL28 and pL29 was inducible. The other five producing lines transformed with pL28 and pL29 all contained substantial levels of hsp70 RNA at 37°C, but no significant increase was found after heat shock.

CONCLUSIONS

On the basis of these observations with deleted plasmids, we conclude that sequences upstream of −51 are not required for heat-shock control. Taken together with the results from chimeric plasmids pH1 and pT1, this indicates that the 249-bp region extending from −51 to the *Mbo*II cut site at +199 is sufficient to determine the control of transcription by heat shock.

The noninducible production of RNA homologous to the hsp70 gene in five of the six producing cell lines transformed with pL28 and pL29 may indicate that sequences involved in control have been deleted while leaving the promoter intact. This provides no explanation, however, for the apparent inducibility (by a factor of 2−4) of one of the pL29 lines. Alternatively, it may be that the deletions in plasmids pL28 and pL29 allow an upstream promoter to govern transcription of hsp70 sequences. In any event, experiments that are now possible will almost certainly make major advances in our understanding of transcriptional control in the heat-shock family of eukaryotic genes.

ACKNOWLEDGMENTS

This work was supported by the National Institutes of Health and by a Fogarty International Research Fellowship awarded to Victor Corces.

REFERENCES

Corces, V., A. Pellicer, R. Axel, and M. Meselson. 1981. Integration, transcription and control of a *Drosophila* heat shock gene in mouse cells. *Proc. Natl. Acad. Sci.* **78:** 7038−7042.

DeNoto, F.M., D.D. Moore, and H.M. Goodman. 1981. Human growth hormone DNA sequence and mRNA structure: Possible alternative splicing. *Nucleic Acids Res.* **9:** 3719−3730.

Holmgren, R., V. Corces, R. Morimoto, R. Blackman, and M. Meselson. 1981. Sequence homologies in the 5′ regions of four *Drosophila* heat shock genes. *Proc. Natl. Acad. Sci.* **78:** 3775−3778.

Ingolia, T.D., E.A. Craig, and B.J. McCarthy. 1980. Sequence of three copies of the gene for the major *Drosophila* heat shock induced protein and their flanking regions. *Cell* **21:** 669−679.

Kafatos, F.C., C.W. Jones, and A. Efstratiadis. 1979. Determination of nucleic acid sequence homologies and relative concentrations by a dot hybridization procedure. *Nucleic Acids Res.* **7:** 1541−1552.

Legerski, R.J., J.L. Hodnett, and H.B. Gray. 1978. Extracellular nucleases of *Pseudomonas* Bal 31 III. Use of the double-strand deoxyribonuclease activity as the basis of a convenient method for the mapping of fragments of DNA produced by cleavage with restriction enzymes. *Nucleic Acids Res.* **5:** 1445−1464.

Livak, K.J., R. Freund, M. Schweber, P.C. Wensink, and M. Meselson. 1978. Sequence organization and transcription at two heat shock loci in *Drosophila*. *Proc. Natl. Acad. Sci.* **75:** 5613−5617.

McKnight, S.L. 1980. The nucleotide sequence and transcript map of the herpes simplex virus thymidine kinase gene. *Nucleic Acids Res.* **8:** 5949−5964.

Roberts, J.M. and R. Axel. 1982. Gene amplification and gene correction in somatic cells. *Cell* **29:** 109−119.

Robins, D., I. Pack, P. Seeburg, and R. Axel. 1982. The regulated expression of human growth hormone genes in mouse cells. *Cell* **29:** 623−631.

Wigler, M., A. Pellicer, S. Silverstein, R. Axel, G. Urlaub, and L. Chasin. 1979. DNA-mediated transfer of the adenine phosphoribosyltransferase locus into mammalian cells. *Proc. Natl. Acad. Sci.* **76:** 1373−1376.

A DNA Sequence Upstream of *Drosophila* hsp70 Genes Is Essential for Their Heat Induction in Monkey Cells

Marc-Edouard Mirault, Eric Delwart, and Richard Southgate
Department of Molecular Biology
University of Geneva
1211 Geneva 4, Switzerland

All heat-shock genes in *Drosophila melanogaster* have been cloned, analyzed in detail, and sequenced. In order to investigate the mechanisms controlling these genes, functional tests are required to assay transcription. Here we show that the *Drosophila* hsp70 genes can be effectively and accurately activated by heat shock after replication in monkey COS cells as part of plasmid vectors containing an SV40 origin of replication. Using this inducible transcription assay and various deletion mutants generated in vitro, we have found a sequence, less than 70 nucleotides long, just upstream of the hsp70 gene that is necessary for heat-shock inducibility. A detailed account of these studies will be published elsewhere (M-E. Mirault, R. Southgate, and E. Delwart, in prep.).

HOST AND VECTORS

The COS cells used as transfection recipients in these experiments are CV-1 monkey cells transformed by an SV40 mutant defective in its origin of replication, but normal for the expression of T antigen (Gluzman 1981). COS cells are consequently permissive for multiplication of plasmids containing an SV40 origin of replication (Lusky and Botchan 1981; Mellon et al. 1982). Our initial vector pSVO-Kan (Fig. 1) is derived from pBR322 and contains for cloning purposes, the kanamycin-resistance marker from Tn5 (*Bam*HI to *Hind*III fragment) plus a small *Hind*III to R1

35

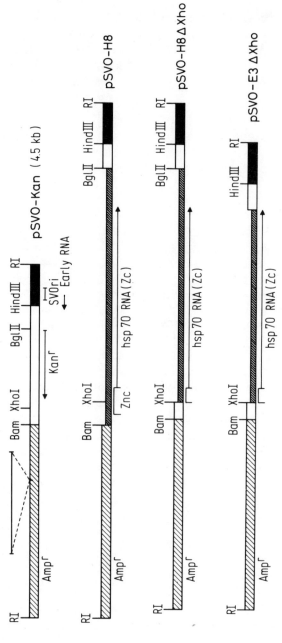

Figure 1
Linear maps of plasmid vectors for COS cells. Plasmid pSVO-Kan derives from pML2-RIIG, itself a derivative from pBR322 (▨) from which a poison sequence has been removed (Lusky and Botchan 1981). The *Bam*HI segment derives from the bacterial transposon Tn5 (□) and the 228-bp *Hind*III to *Eco*RI fragment from pML2-RIIG contains the origin of replication of SV40 (■ double scale). Plasmids pSVO-H8 and pSVO-H8ΔXho contain a *D. melanogaster* hsp70 gene from site 87A, as in 56H8, whilst pSVO-E3ΔXho has a variant copy from site 87C, namely the first gene of 132E3 (see Török et al., this volume).

fragment carrying the SV40 origin of replication. hsp70 genes were cloned into this vector in an opposite polarity to that of the truncated SV40 early promoter. Plasmids pSVO-H8 and pSVO-H8ΔXho contain one hsp70 gene copy from chromosomal site 87A whereas pSVO-E3ΔXho has one variant copy from 87C. Both ΔXho recombinants lack about half of the element Znc which is found adjacent to all *Drosophila* hsp70 genes. None of these clones contains the complete SV40 72-bp repeat unit required for viral transcription.

REPLICATION OF VECTORS

Subconfluent COS-7 cells were transfected in the presence of DEAE-dextran by each of the four plasmids shown in Figure 1. After a 48- to 60-hour incubation, total nucleic acids were extracted from the cells and RNA was separated from DNA. The amount of supercoiled plasmid DNA replicated in the cells was estimated by Southern blot hybridization. Authentic replication of the plasmids in COS cells was ascertained by their complete sensitivity to *Mbo*I restriction (*Mbo*I sites are inactivated by methylation in the *Escherichia coli* strain HB101 used here for plasmid preparation). The four plasmids were found to replicate to the same extent, with transfection efficiency varying from 1% to 10% in different experiments. Supercoiled plasmid DNA usually accounted for 0.01−0.1% of the total DNA which corresponds to approximately 10,000−100,000 copies per cell.

DROSOPHILA hsp70 GENES ARE HEAT-INDUCIBLE IN COS CELLS

COS cells were transfected with plasmid pSVO-H8. After 2.5 days, RNA and DNA were isolated from both heat-shocked cells (1 hour at 43°C, plus 1 hour at 37°C) and cells kept at 37°C. *Drosophila* hsp70 RNA was assayed by both hybridization and protection of complementary DNA from S1 digestion and also by Northern hybridization to nitrocellulose blots. Authentic *D. melanogaster* heat-shock RNA from cultured Kc cells protects a major 2.4-kb DNA fragment in pSVO-H8 and an equivalent fragment was protected by heat-shock RNA from transfected COS cells with pSVO-H8. No protection was observed with RNA from cells transfected but not shocked or from untransfected cells. Equivalent results were obtained in the Northern analysis where hybridization of an hsp70 probe to denatured RNA revealed a strong band at about 2.5 kb detected only in heat-shocked cells (Fig. 2). The same results were obtained with each vector carrying one copy of the hsp70 gene from either 87A or 87C, irrespective of whether Znc sequences upstream from the *Xho*I site were present or not. The slight variations in signal

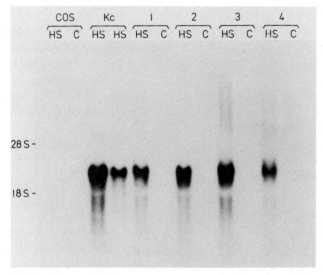

Figure 2
Heat-shock-induced *Drosophila* RNA in COS cells. Formaldehyde-denatured RNA was fractionated in an agarose gel, blotted, and hybridized to a ^{32}P-labeled *D. melanogaster* hsp70 probe. Tracks 1 and 2, RNA from heat-shocked and nonheat-shocked COS cells; 3 and 4, *D. melanogaster* Kc heat-shock RNA; 5 and 6, 7 and 8, 9 and 10, 11 and 12 (pairwise), heat-shock and nonheat-shock RNA from COS cells transfected by pSVO-H8, pSVO-H8ΔXho, pSVO-E3, and pSVO-E3ΔXho, respectively. Autoradiography was for 15 hours with an intensifying screen.

intensity seen in Figure 2 reflect differences in transfection efficiency rather than in transcription. As assessed by dilution tests, the hsp70 RNA pool is at least 100 times greater after heat shock than before. These results indicate first that transcription of the *D. melanogaster* hsp70 gene appears to be strongly induced by heat shock after transfection and autonomous replication in monkey cells. Second, the data suggest that *D. melanogaster* DNA sequences upstream from the *Xho*I site in Znc are not required for heat-shock induction of the gene. Furthermore, we found that 50 μM sodium arsenite added to the culture medium substitutes for a heat shock in inducing hsp70 transcription in both COS and *Drosophila* cells. It is likely that the accumulation of hsp70 RNA observed after induction, either by heat shock or arsenite treatment, represents an increase in the rate of transcription initiation rather than a decrease in turnover rate. Using very sensitive S1 protection assays to detect small RNA fragments, we found no evidence for hsp70 RNA breakdown in COS cells maintained at the normal growth temperature. Induced hsp70 transcription seems to be faithful in COS cells since S1 protection experiments revealed 5′ ends that correspond to those of authentic *D. melanogaster* hsp70 RNA (resolution ±1 nucleotide) and preliminary data with the hsp70 variant from 87A also indicated correct 3′ termini, although with a lower resolution. In order to estimate the transcription efficiency, the amount of *Drosophila* hsp70 RNA made in Kc cells during a 1-hour heat shock at 37°C was compared with that made in transfected COS cells during a 1-hour recovery period at 37°C, following heat shock at 43°C. We observed that much more hsp70 RNA was made during this recovery incubation at 37°C than at 43°C. Both plasmid DNA and hsp70 RNA were titrated by hybridization to DNA and RNA blots. From initial titrations, the number of hsp70 transcripts made in COS cells was calculated on a per gene basis to be about 10% of that made in *Drosophila* cells. This suggests that the fraction of *Drosophila* genes effectively activated in COS cells is at least 10% of the total pool or that each gene is transcribed with a reduced efficiency. The high copy number of these genes in COS cells is clearly an unusual situation and its possible consequence on gene induction needs to be investigated. For example, a positively acting component for the induction of transcription could be titrated out during heat shock. On the other hand, the lack of hsp70 transcription observed at the normal temperature (Fig. 2) would argue against the involvement of a cellular repressor present in limiting amounts.

DNA SEQUENCES REQUIRED FOR INDUCTION

The results presented in the last section suggest that control of induction occurs downstream from the *Xho*I site in Znc. In order to find the sequences critical for induction, DNA was progressively digested by *Bal*31 nuclease from the *Xho*I site of pSVO-H8ΔXho and pSVO-E3ΔXho to yield a series of 5′-deletion clones. A number of these were selected and assayed for hsp70 transcription in COS cells. The deletion point of a critical clone was eventually determined by direct DNA sequence analysis. Deletions up to position −68 (Fig. 3; position +1 is the start of transcription) had repression and transcription inducibility characteristics similar to those presented in Figure 2. In contrast, however, deletions extending to position −53 and beyond were found to completely abolish transcription inducibility. Very little, if any, transcription was detected whether the COS cells transfected with the −53 mutant were shocked or not, whilst endogenous hsp70 synthesis was induced normally following heat shock. Controls indicated that plasmid replication was normal and that a cotransfecting gene from 87A was inducible whereas the −53 deletion mutant from 87C was not (the transcripts from 87A and 87C could be distinguished in S1 protection assays on the basis of sequence divergence in the 5′-nontranslated region). From the results obtained so far, we conclude that heat-shock induction of the *D. melanogaster* hsp70

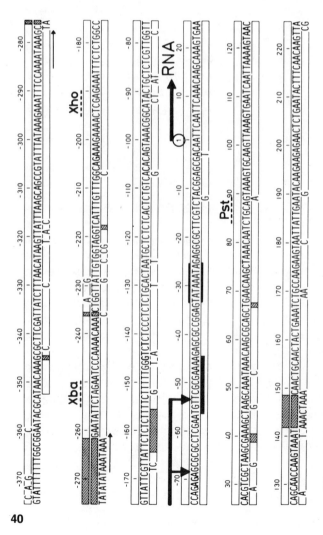

Figure 3

DNA sequences required for heat-shock induction. This figure is derived from Fig. 3 of Karch et al. (1981) which shows a comparison of the DNA sequences flanking the hsp70 genes of plasmids 56H8 and 132E3. The arrows on the left indicate the deletion points of two critical mutants. hsp70 transcription is still inducible with a deletion coming from upstream to position −68. Induction is completely abolished with a deletion up to position −53. A short imperfect dyad is underlined (−44 to −56).

40

gene involves a regulatory sequence element, the upstream boundary of located between position -68 and -53 (Fig. 3); the downstream boundary is presently being determined. A short imperfect inverted repeat present 11 nucleotides upstream from the TATA box at positions -44 to -56, visible in Figure 3, is truncated in the -53 deletion. This dyad, first noted by Ingolia et al. (1980) and Holmgren et al. (1981), is found within a region perfectly conserved in front of all hsp70 gene copies (Karch et al. 1981; Mason et al. 1982). Furthermore, similar palindromes are also detected upstream of the other *D. melanogaster* heat-shock genes (Holmgren et al. 1981; R. Southgate, in prep.). In comparing all these dyads, Pelham and Bienz (this volume) have deduced an apparent consensus sequence which they discuss in relation to control. Mellon et al. (1982) find that constitutive transcription of the human α_1-globin gene in COS cells is drastically reduced by deleting upstream sequences down to position -55. We note that this deletion actually truncates a 14-bp dyad element, 100% G + C, at positions -46 to -60. Are these dyad elements potential binding sites for specific proteins? Interestingly, Wu (this volume) finds a DNase-I hypersensitive region of the chromatin that centers on the TATA box of the hsp70 genes. This region appears to include the control sequences that we have detected and may correspond to an entry site for RNA polymerase II.

CONCLUSION

When *Drosophila* hsp70 genes are replicated in monkey cells by means of SV40 vectors, the transcription of these genes can be activated by raising the normal growth temperature of the cells by a few degrees. Thus, the mechanism of heat induction is, at least in part, conserved in mammalian cells, as was suggested by Corces et al. (1981) who transformed mouse cells with *Drosophila* heat-shock genes. Furthermore, arsenite can substitute for heat-shock in inducing these genes both in COS and *Drosophila* cells. This suggests that different stresses can converge towards a similar mechanism of induction, the activation steps of which are independent of protein synthesis (Arrigo 1980).

Transcription studies of *Drosophila* hsp70 genes in COS cells have revealed that a DNA sequence of less than 70 nucleotides upstream of the genes is required for stress induction of transcription. This sequence includes a small imperfect dyad not far upstream from the TATA box. Whether a positive or negative control of gene activity is implicated, DNA-specific protein interactions or chromatin modifications could possibly lock, or unlock, the polymerase entry site by altering locally the chromatin conformation. A detailed analysis of the chromatin structure in both the induced and noninduced states is therefore required. In this respect, heat-shock genes in SV40-like minichromosomes, as amplified in COS cells, may provide a suitable tool for such experiments.

Finally, it is hoped that similar studies be carried out on homologous cells in which possible long-range structural effects could be observed. This indeed may be necessary in order to understand fully the regulation of the heat-shock genes in their native state. The limiting step so far has been the lack of efficient procedures to transfer genes into *Drosophila*. In this respect, the recent transformation technique developed by Spradling and Rubin (pers. comm.) may be the first step toward this goal.

REFERENCES

Arrigo, A.-P. 1980. Investigation of the function of the heat-shock proteins in *Drosophila melanogaster. Mol. Gen. Genet.* **178:** 517–524.

Corces, V., A. Pellicer, R. Axel, and M. Meselson. 1981. Integration, transcription, and control of a *Drosophila* heat shock gene in mouse cells. *Proc. Natl. Acad. Sci.* **78:** 7038–7042.

Gluzman, Y. 1981. SV40 transformed simian cells support the replication of early SV40 mutants. *Cell* **23:** 175–182.

Holmgren, R., V. Corces, R. Morimoto, R. Blackman, and M. Meselson. 1981. Sequence homologies in the 5′ regions of four *Drosophila* heat shock genes. *Proc. Natl. Acad. Sci.* **78:** 3775–3778.

Ingolia, T.D., E. Craig, and B.J. McCarthy. 1980. Sequence of three copies of the gene for the major heat shock induced protein and their flanking regions. *Cell* **21:** 669–679.

Karch, F., I. Török, and A. Tissières. 1981. Extensive regions of homology in front of the two hsp70 heat shock variant genes in *Drosophila melanogaster. J. Mol. Biol.* **148:** 219–230.

Lusky, M. and M. Botchan. 1981. Inhibition of SV40 replication in simian cells by specific PBR322 DNA sequences. *Nature* **293:** 79–81.

Mason, P., I. Török, I. Kiss, and A. Udvardy. 1982. A complex pattern of DNA sequence homology extending upstream of the hsp70 genes at loci 87A7 and 87C1 in *Drosophila melanogaster:* Evolutionary implications. *J. Mol. Biol.* **156:** 21–35.

Mellon, P., V. Parker, Y. Gluzman, and T. Maniatis. 1982. Identification of DNA sequences required for transcription of the human α-1 globin gene in a new SV40 host-vector system. *Cell* **27:** 279–288.

DNA Sequences Required for Transcriptional Regulation of the *Drosophila* hsp70 Heat-shock Gene in Monkey Cells and *Xenopus* Ooctyes

Hugh Pelham and Mariann Bienz
MRC Laboratory of Molecular Biology
University Medical School
Cambridge CB2 2QH, England

It is well established that in *Drosophila,* as well as in many other organisms, the response to heat shock involves de novo transcription of the heat-shock genes. The mechanism of this induction appears to be similar in insects, mammals, and amphibia since the *Drosophila* hsp70 gene is heat-inducible when introduced into mouse tissue-culture cells (Corces et al. 1981) or *Xenopus* oocytes (Voellmy and Rungger 1982). As a first step towards an understanding of the mechanism of the transcriptional control, we have attempted to define the DNA sequences necessary for the heat-shock response.

The gene we have used is the *Drosophila* hsp70 gene. Mutants of this were constructed in vitro and analyzed both in *Xenopus* oocytes and in the COS cell system. COS cells are SV40-transformed monkey cells that produce T antigen constitutively and will support high-level replication of any plasmid containing an SV40 replication origin (Mellon et al. 1981). Some of this work is published elsewhere (Pelham 1982).

THE hsp70 GENE IS HEAT-INDUCIBLE IN COS CELLS

The basic plasmid used in these experiments contained both the hsp70 gene and, 600 bp downstream, the herpesvirus thymidine kinase (*tk*) gene. Cells were transfected with this DNA at 37°C in the presence of DEAE-dextran, and 30 hours later they were warmed to 42°C for several

43

hours. Transcripts from the two genes were detected by S1 mapping of their 5′ ends (Weaver and Weissman 1979). In a number of experiments, the level of *tk* transcripts was unaltered or reduced by heat shock, whereas the abundance of hsp70 transcripts was increased 20- to 50-fold. S1 mapping confirmed that these transcripts had the expected 5′ and 3′ ends. Thus, the hsp70 gene shows appropriate regulation, despite being present on a small plasmid at thousands of copies per cell and adjacent to a gene that is expressed at low temperature.

CONTROL DOES NOT REQUIRE hsp70 mRNA SEQUENCES

The hsp70 promoter was fused at position −10 (position +1 being the first base of the mRNA) to the corresponding position of the *tk* gene. About 2 kb of 5′ flanking sequence from the hsp70 gene was present. The resulting hybrid gene was fully heat-inducible in COS cells. A further construction in which only bases −10 to −186 of the hsp70 gene were present was also fully inducible. This indicates that control is at the level of transcription initiation. Since *tk* transcripts are not particularly unstable at either low or high temperature, this result rules out regulation of hsp70 mRNA abundance merely by rapid selective degradation of the mRNA at low temperature.

NO MORE THAN 66 BASES OF 5′
FLANKING SEQUENCE ARE REQUIRED

A series of deletion mutants was prepared by *Bal*31 nuclease digestion, starting at the *Xho*I site at −186. *Hin*dIII linkers were ligated to the remaining DNA and the gene was recloned. The deletions were mapped by restriction enzyme digestion, and selected ones checked by DNA sequencing. These were then tested for inducible transcription activity in COS cells.

Figure 1 summarizes the results of S1 mapping of the transcripts derived from some of the deletion mutants. The probe was a 568-bp *Hae*III fragment continuously labeled in the mRNA-complementary strand. Authentic *Drosophila* hsp70 mRNA from heat-shocked flies or tissue culture cells protects about 400 bases of this from S1 nuclease. As shown in the figure, deletion to position −66 did not prevent full induction of the hsp70 gene. A mutant deleted to position −44 was transcribed very poorly at both high and low temperature, but still yielded transcripts with the correct 5′ end. Deletion to position −28 within the TATA box completely prevented the synthesis of such transcripts. We conclude that the TATA-box region is capable of directing accurate initiation of transcription in vivo, but sequences between −66 and −44

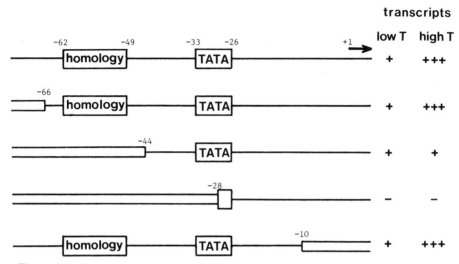

Figure 1
Summary of S1 mapping results with some deletion mutants. Open-ended boxes represent deleted sequences. The 5′ deletions have been tested both in COS cells and in *Xenopus* oocytes; the bottom line represents a fusion to the *tk* gene, which has been tested in both systems. The positions of the TATA box and the homologous sequence element shown in Fig. 2b are indicated.

(at least) are necessary for efficient use of the TATA box at high temperature.

To check that the DNA sequences brought close to the gene do not affect its transcription, the −66 deletion was recloned adjacent to a different sequence. As expected, the recloned mutant retained full inducible activity.

REGULATION IN *XENOPUS* OOCYTES
INVOLVES THE SAME SEQUENCES

We have tested the same deletion mutants by injection into *Xenopus* oocytes. After a 2-hour recovery period at 25°C, the oocytes were heat-shocked at 33°C for 8 hours. As in COS cells, the level of hsp70 transcripts was increased by heat shock, whereas the *tk* transcripts were not. The mRNA was detectable not only by S1 mapping, but also by its translation into *Drosophila* hsp70, either in the injected oocytes or (after phenol extraction) in the reticulocyte cell-free system.

Using 5′ S1 mapping as an assay, we have found that the −66 deletion is still capable of induction in oocytes, but the −44 deletion is not. At low temperature, both of these mutants produce a low level of correct transcripts, while the −28 deletion produces none at either temperature.

Thus, despite the different species and temperatures involved, the basic mechanism of the transcriptional response appears similar in *Xenopus* oocytes and monkey cells.

HOMOLOGIES WITH OTHER HEAT-SHOCK GENES IN THE IMPORTANT REGION

A further mutant was constructed that had a 10-base *Hind*III linker between residues −44 and −50. This mutant retained inducible activity in both assay systems, suggesting that at least one discrete functional unit lies between −50 and −66. In fact, residues −47 and −48 were effectively unchanged, so the important region may extend to these residues. Figure 2a shows that significant homologies exist between this region and similar positions upstream from most of the other *Drosophila* heat-shock genes. Moreover, the homologous sequences are associated with striking inverted repeats, although the actual sequences of these inverted repeats are not strictly conserved. The derived consensus sequence itself forms a separate inverted repeat (Fig. 2b), a feature common to many protein-binding sites on DNA. The hsp23 gene appears to be rather different and does not fit these features convincingly. Despite this, the general correlation between the homology, the inverted repeats, and the hsp70 functional sequence suggests that these features constitute an important component of the transcriptional heat-shock response.

CONCLUSIONS

These results indicate that the hsp70 promoter itself is a site at which regulation occurs in response to heat shock and, in particular, that a region upstream of the TATA box is essential for this response. This region seems to have an analogous function to that of other "upstream elements" of polymerase-II promoters—namely to stimulate use of the TATA-box region—but it is active only in heat-shocked cells. The functional element is associated with particular sequence features, and ultimately it may be possible to identify a protein that binds to this region. It will also be interesting to see whether the upstream promoter element is the site of regulation of other inducible genes.

The hsp70 upstream element might also be involved in the local organization of chromatin structure. A DNase-I hypersensitive site has been mapped to a diffuse region covering about 200 bp upstream from the transcription start site in the hsp70 gene (Wu 1980), and the availability of deletion mutants will allow a test of the hypothesis that this site reflects the presence of a potentially active promoter element.

a

```
                 *******************
HSP 70   AGCGCGCCTCGAATGTTCGCGAAAAGAGCGCCGGAGTATA
HSP 83   GAAGCCTCTAGAAGTTTCTAGAGACTTCCAGTTCGGTCGGGTTTTTCTATA
HSP 22   GAGAGTGCCGGTATTTTCTAGATTATATGGATTTCCTCTCTGTCAAGAGTATA
HSP 26   TCACTTTCCGGACTCTTCTAGAAAAGCTCCAGCGGGTATA
HSP 27   TCCTTGGTTGCCATGCACTAGTGTGTGTGAGCCCAGCGTCAGTATA
HSP 68   GGGAAATCTCGAATTTTCCCCTCCCGGCGACAGAGTATA
consensus          CTGGAAT  TTCTAGA
```

b

```
C T - G A A - - T T C - A G
                             ------- 14-28 bp -------TATA
G A - C T T - - A A G - T C
```

Figure 2

(a) Homology between *Drosophila* heat-shock genes. Asterisks mark the region of the hsp70 gene (residues −47 through −66) that is expected to contain a functional element; inverted repeats are underlined. Sequences are from Karch et al. (1981), Holmgren et al. (1981), and Ingolia and Craig (1981). (b) Idealized heat-shock element. The key homologous features are shown, namely a symmetric conserved sequence overlapping a larger inverted repeat. This structure is most closely matched by the hsp83, 70, 26, and 22 genes.

The results described in this paper do not rule out the possibility that other levels of control of heat-shock gene expression might operate in vivo. Controls involving changes in long-range chromatin structure may simply be undetectable in the assay systems that we have used. Multiple levels of control would not be surprising: after all, the heat-shock response in *Drosophila* is known to affect both transcription and translation. However, on the present evidence it is possible to account for the transcriptional induction of the hsp70 gene without postulating such extra levels of control.

REFERENCES

Corces, V., A. Pellicer, R. Axel, and M. Meselson. 1981. Integration, transcription, and control of a *Drosophila* heat shock gene in mouse cells. *Proc. Natl. Acad. Sci* **78**: 7038–7042.

Holmgren, R., V. Corces, R. Morimoto, R. Blackman, and M. Meselson. 1981. Sequence homologies in the 5′ regions of four *Drosophila* heat-shock genes. *Proc. Natl. Acad. Sci.* **78**: 3775–3778.

Ingolia, T.D. and E.A. Craig. 1981. Primary sequence of the 5′ flanking region of the *Drosophila* heat shock genes in chromosome subdivision 67B. *Nucleic Acids Res.* **9**: 1627–1642.

Karch, F., I. Török, and A. Tissières. 1981. Extensive regions of homology in front of the two hsp 70 heat shock variant genes in *Drosophila melanogaster. J. Mol. Biol.* **148:** 219-230.

Mellon, P., V. Parker, Y. Gluzman, and T. Maniatis. 1981. Identification of DNA sequences required for transcription of the human α1-globin gene in a new SV40 host-vector system. *Cell* **27:** 279–288.

Pelham, H.R.B. 1982. A regulatory upstream promoter element in the *Drosophila* hsp 70 heat-shock gene. *Cell* (in press).

Voellmy, R. and D. Rungger. 1982. Transcription of a *Drosophila* heat shock gene is heat induced in *Xenopus* oocytes. *Proc. Natl. Acad. Sci* **79:** 1776–1780.

Weaver, R.F. and C. Weissman. 1979. Mapping of RNA by a modification of the Berk-Sharp procedure: The 5′ termini of 15S β-globin mRNA precursor and mature 10S β-globin mRNA have identical map coordinates. *Nucleic Acids Res.* **7:** 1175–1193.

Wu, C. 1980. The 5′ ends of *Drosophila* heat shock genes in chromatin are hypersensitive to DNase I. *Nature* **286:** 854–860.

Heat-induced Transcription of *Drosophila* Heat-shock Genes in *Xenopus* Oocytes

Richard Voellmy*†
and Duri Runggeŗ‡
*Department of Molecular Biology
‡Department of Animal Biology
University of Geneva
1211 Geneva 4, Switzerland*

A variety of eukaryotic and even prokaryotic organisms (this volume) respond to heat treatment in a similar fashion: the synthesis of a few specific proteins, the heat-shock proteins, is strongly induced at elevated temperature.

Heat treatments result in the formation of new puffs on *Drosophila* salivary gland chromosomes in regions containing heat-shock genes and in the accumulation of RNA species coding for heat-shock proteins in the cytoplasm of *Drosophila* cells (Ashburner and Bonner 1979). These observations suggest that the expression of heat-shock genes is regulated at the transcriptional level, probably by a mechanism controlling initiation of transcription of these genes. Very little is known about the nature of this mechanism. Amino acid analogs have been shown to induce heat-shock protein synthesis in a variety of different cell types (Kelley and Schlesinger 1978) including *Drosophila* cells (R. Voellmy, unpubl.). This finding has been taken as evidence for the involvement of a labile protein in the control of heat-shock gene expression. The hypothetical protein, however, has not yet been identified. It is, therefore, also not known whether the protein interacts with DNA sequences within or adjacent to heat-shock genes.

†Present address: Department of Biochemistry, University of Miami, School of Medicine, Miami, Florida 33101.

Drosophila heat-shock genes have been isolated and characterized in detail. The development of transcription systems that could be used to examine the heat-regulated expression of isolated heat-shock genes should allow us to define the DNA regions important in regulating the activity of these genes. Segments containing these regulatory sequences could then be employed as specific tools for identifying other components of the heat-shock gene control mechanism.

The structures of certain heat-shock proteins (R. Voellmy et al., in prep.) as well as the basic features of the regulation of heat-shock genes have been conserved remarkably well throughout evolution. These findings suggest that it might be possible to examine the regulated expression of Drosophila heat-shock genes in heterologous transcription systems. We report here on our experiments designed to determine whether Drosophila heat-shock genes introduced into Xenopus oocytes by microinjection are expressed correctly, and in a heat-regulated fashion, in oocyte nuclei (see also Voellmy and Rungger 1982).

TRANSCRIPTION OF VARIOUS *DROSOPHILA* HEAT-SHOCK GENES IN HEAT-TREATED *XENOPUS* OOCYTES

Substantial changes occur in the pattern of protein synthesis in Xenopus cells at temperatures around 35–37°C (Voellmy and Rungger 1982). Heat-shock genes in Drosophila cells are activated at the same temperatures. Most experiments designed to examine the transcription of cloned Drosophila heat-shock genes were therefore performed with oocytes that had been heat-treated at 35°C.

Several Drosophila DNA segments containing different copies of the gene coding for the 70,000 dalton heat-shock protein (hsp70), or genes coding for the small heat-shock proteins (hsp22, 23, 26, and 27), were circularized and microinjected into oocyte nuclei. The segments used included DNA from phage 122 carrying two identical Drosophila hsp70 genes from cytological region 87A arranged in opposite orientations, the isolated 10.5-kb EcoRI insert fragment of 122, the 6.4-kb EcoRI insert fragment of plasmid 56H8 containing a variant hsp70 gene from region 87A, and the 3.3-kb BglII gene fragment from plasmid 132E3 consisting mainly of sequences coding for an hsp70 gene from cytological region 87C (Fig. 1). DNA from phage 179 or a 12-kb SalI fragment from 179 including the hsp26 and hsp22 genes were also employed.

Following heat treatment, total RNA was extracted from oocytes containing Drosophila genes or from uninjected oocytes and was copied by reverse transcription in the presence of $[\alpha\text{-}^{32}\text{P}]\text{dCTP}$ using random pentadeoxynucleotides as primers. The labeled DNA transcripts were then hybridized to Southern blots of DNA coding for Drosophila hsp70 (SalI digest of plasmid Sal 0; see Fig. 1) or hsp26. Reverse transcripts

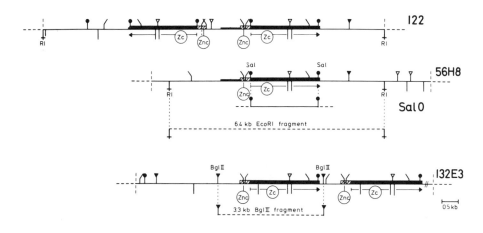

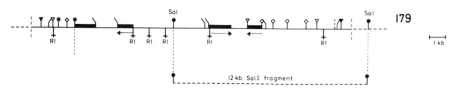

Figure 1
Physical maps of *Drosophila* heat-shock gene segments. 122, 56H8, and 132E3 contain hsp70 genes, 179 the genes coding for hsp22, 23, 26, and 27. Black bars represent RNA coding segments (Zc for hsp70 genes), hatched regions the common nontranscribed sequences at the 5′ ends of the hsp70 genes (Znc). The transcriptional orientations of the genes are indicated by arrows. Symbols: (Y) BamHI; (▼) BglII; (♀) HindIII; (¦) KpnI; (|) PstI; (♥) SalI; (\) XbaI; (ſ) XhoI (Moran et al. 1979; Goldschmidt-Clermont 1980; Voellmy et al. 1981).

of RNA from uninjected oocytes did not hybridize to *Drosophila* heat-shock genes (Fig. 2a–c, lanes 9). All the different *Drosophila* hsp70 gene segments appeared to be transcribed in the oocytes (Fig. 2a–c, lanes 1, 6, 7, and 8). No efficient transcription of the hsp26 gene on phage 179 DNA or the isolated 179 *SalI* fragment could be observed (Fig. 2a, lanes 4 and 5). 122 DNA, which had not been circularized prior to injection, produced only minor amounts of hsp70 gene transcripts (Fig. 2a, lane 2). A fraction of the RNA made in oocytes from 122 DNA appeared to be polyadenylated (Fig. 2a, lane 3).

HEAT-ACTIVATED TRANSCRIPTION OF THE PHAGE 122 hsp70 GENES IN OOCYTES

Oocytes containing phage 122 DNA or the isolated 122 *Eco*RI insert fragment were either heat-treated for 2 hours at 35°C or were incubated

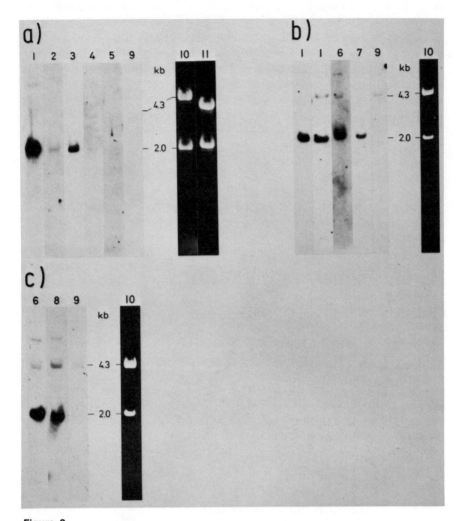

Figure 2
Transcription of *Drosophila* heat-shock genes in heat-treated oocytes. Groups of 10–15 oocytes were injected with 2 ng per oocyte of circularized DNA. Following a 20-hr preincubation, the oocytes were heat-treated for 2 hr at 35°C. RNAs were extracted and reverse-transcribed as described earlier (Voellmy and Rungger 1982). Labeled cDNAs were hybridized to Southern blots of *Sal*I digests of plasmid Sal 0 (lanes 1, 2, 3, 6–9) or of *Bam*HI/*Eco*RI digests of plasmid subclone 179209 (Voellmy et al. 1981) which contains the hsp26 gene (lanes 4 and 5). The results of three independent experiments are shown in *a–c*. DNAs injected: phage 122, lane 1; phage 122 unligated, lane 2; phage 122, RNA was reverse-transcribed using oligo(dT) as primer, lane 3; 12-kb *Sal*I fragment from phage 179 DNA, lane 4; phage 179, lane 5; 6.4-kb *Eco*RI gene fragment from 56H8, lane 6; 3.3-kb *Bgl*II gene fragment from 132E3, lane 7; 10.5-kb *Eco*RI insert fragment from 122, lane 8; no DNA, lane 9. Lane 10 shows *Sal*I digestion patterns of plasmid Sal 0 and 11 the *Bam*HI/*Eco*RI pattern of 179209.

at 20°C for the same period. The synthesis of *Drosophila*-specific RNA was analyzed by the reverse transcription assay as described above. Both the isolated hsp70 genes and the genes flanked by phage λ DNA were transcribed efficiently at 35°C but not at all or only at much lower levels at 20°C (Fig. 3b, lanes 1, 2, 4, and 5). Similar results were obtained from analogous experiments in which the 122 transcripts were labeled in vivo by injection of [α-^{32}P]GTP into the oocytes prior to the heat treatments or control incubations at 20°C (Voellmy and Rungger 1982). Unlike the 122 genes, the hsp70 gene fragments from 56H8 (Fig. 3b, lanes 6 and 7) and 132E3 (data not shown) were found to be transcribed at both temperatures.

DROSOPHILA-SPECIFIC hsp70 RNA OF CORRECT SIZE IS MADE IN HEAT-TREATED OOCYTES ONLY

Total RNA from heat-treated and untreated oocytes containing different *Drosophila* hsp70 gene fragments was run on agarose gels and blotted to nitrocellulose filters (Fig. 3a). Nick-translated plasmid Sal 0 carrying the sequences coding for hsp70 RNA was used to identify *Drosophila*-specific RNA made in oocytes. In heat-treated oocytes, the majority of the *Drosophila* hsp70 gene transcripts were found to be of the correct size (Fig. 3a, lanes 1, 4, and 6). In agreement with our observations using the reverse transcription assay, very little *Drosophila*-specific RNA was made from 122 fragments in untreated oocytes (Fig. 3a, lanes 2 and 5). In contrast, the 56H8 gene fragment was transcribed also in untreated oocytes (Fig. 3a, lanes 7). The 56H8 transcripts made at 20°C however did not have a defined length, and the average molecule was much longer than authentic hsp70 RNA. We do not yet understand why unspecific transcription occurs at a much higher level with the 56H8 fragment than with the 122 gene fragment. Experiments to resolve this problem are being carried out at present.

CONCLUSION

Our experiments strongly suggest that *Drosophila* hsp70 genes are transcribed correctly in heat-treated *Xenopus* oocytes. The genes on the 122 segment are heat-activated in the oocytes. This latter finding indicates that the *Drosophila* sequences important for the heat regulation of hsp70 genes are on the 122 segment and are recognized by components of the *Xenopus* heat-shock gene transcription regulation system. It appears, therefore, that in addition to the structures of the heat-shock proteins the sequences involved in regulating the transcriptional activity of the heat-shock genes also have been conserved during evolution. The

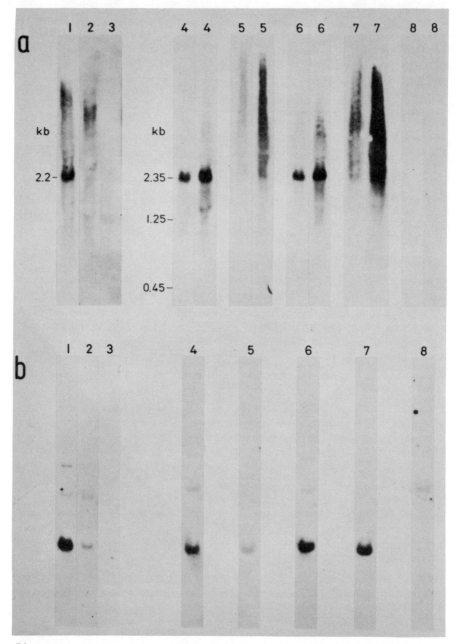

results obtained so far suggest that the *Xenopus* transcription assay may prove a valuable tool in defining the DNA sequences important in the control of *Drosophila* heat-shock gene expression. An alternative in vivo transcription assay based on transformation of mouse cells with *Drosophila* hsp70 genes has recently been developed by Corces et al. (1981).

ACKNOWLEDGMENTS

We wish to thank Dr. P. Bromley for his help with the reverse transcription assay, Dr. M.-E. Mirault for providing 56H8 and 132E3 DNA, and Dr. R. Werner for critical reading of this manuscript. We also thank O. Jenni for the photographs and M. Piedra for preparing the manuscript. This work was supported by grants from the Swiss National Science Foundation to Drs. M. Crippa (3.689.080) and A. Tissières (3.512.79), from the A. and L. Schmidheiny Foundation to A. Tissières, and by a stipend to R. V. from the Emil Barell Foundation of Hoffmann-La Roche.

REFERENCES

Ashburner, M. and J. Bonner. 1979. The induction of gene activity in *Drosophila* by heat shock. *Cell* **17:** 241−254.

Corces, V., A. Pellicer, R. Axel, and M. Meselson. 1981. Integration, transcription, and control of a *Drosophila* heat shock gene in mouse cells. *Proc. Natl. Acad. Sci.* **78:** 7038−7042.

Goldschmidt-Clermont, M. 1980. Two genes for the major heat-shock protein of *Drosophila melanogaster* arranged as an inverted repeat. *Nucleic Acids Res.* **8:** 235−252.

Kelley, P. and M. Schlesinger. 1978. The effect of amino acid analogues and heat shock on gene expression in chicken embryo fibroblasts. *Cell* **15:** 1277−1286.

Figure 3
(a) Size of *Drosophila*-specific RNA made in oocytes. RNA was extracted from oocytes injected with phage 122 DNA (lanes 1 and 2), the 10.5-kb *Eco*RI insert fragment of 122 (lanes 4 and 5), the 6.4-kb *Eco*RI gene fragment from 56H8 (lanes 6 and 7), or from uninjected oocytes (lanes 3 and 8) after 2 hr of heat treatment at 35°C (lanes 1, 3, 4, 6, and 8) or incubation at 20°C (lanes 2, 5, and 7). Equal portions of the RNAs were DNase-treated, glyoxylated, run on a 1.4% agarose gel, and subsequently transferred to a nitrocellulose filter. A *Bam*HI/*Bgl*I double digest of pBR322 served as size marker. Nick-translated Sal 0 was used as hybridization probe. Two different exposures of filters 4−8 are shown. (b) Quantitative analysis of the *Drosophila*-specific transcripts made in oocytes. RNAs from oocytes treated as in *a* were reverse transcribed as in Fig. 2. Equal amounts of the cDNAs were hybridized to Sal 0 Southern blots. Lanes as in *a*.

Moran, L., M.-E. Mirault, A. Tissières, P. Schedl, S. Artavanis-Tsakonas, and W. Gehring. 1979. Physical map of two *D. melanogaster* DNA segments containing sequences coding for the 70,000 dalton heat shock protein. *Cell* **17:** 1−8.

Voellmy, R. and D. Rungger. 1982. Transcription of a *Drosophila* heat shock gene is heat-induced in *Xenopus* ocytes. *Proc. Natl. Acad. Sci.* **79:** 1776−1780.

Voellmy, R., M. Goldschmidt-Clermont, R. Southgate, A. Tissières, R. Levis, and W. Gehring. 1981. A DNA segment isolated from chromosomal site 67B in *D. melanogaster* contains four closely linked heat-shock genes. *Cell* **23:** 261−270.

Transcription and Chromatin Structure of *Drosophila* Heat-shock Genes in Yeast

John Lis, Nancy Costlow,
John de Banzie, Doug Knipple
Deborah O'Connor, and Lesley Sinclair
Section of Biochemistry
Molecular and Cell Biology
Cornell University
Ithaca, New York 14853

Introducing cloned eukaryotic genes into eukaryotic cells by injection or transformation provides a rapid and convenient test of the effects of specific sequence alterations on gene expression. Those sequences which when altered give rise to altered expression are strong candidates for controlling elements. Such techniques have been used to identify sequences involved in regulation of the sea urchin histone (Grosschedl and Birnsteil 1980) and rabbit β-globin genes (Banerji et al. 1981). The ultimate test of the function of *cis*-acting sequences is the homologous system—ideally where the altered gene replaces the wild-type gene at its normal chromosomal location.

When this study began, transformation in *Drosophila* was not available. We therefore decided to look at the expression of *Drosophila* heat-shock genes in yeast. Yeast transformation offers several attractive features for examining the expression of genes of higher eukaryotes: the genetic composition of transformants can be precisely controlled, large numbers of transformants can be handled conveniently, and large cultures can be grown easily and inexpensively. Early results using yeast to express genes of other eukaryotes have been disappointing. Beggs et al. (1980) found that the rabbit β-globin gene is incorrectly initiated and terminated and that the splice sites are ignored in yeast. Moreover, Henikoff et al. (1981) were able to complement only one out of six yeast mutations using a library of *Drosophila* DNA fragments, though this may have been due to the relatively small size of inserts. These results suggest that yeast do not correctly express at least some genes of higher organisms. However, the heat-shock response is highly conserved, occurring in a wide variety of cell types and organisms, including yeast. Therefore it seemed possible that the *Drosophila* heat-shock genes might be correctly expressed in yeast.

We chose to examine the *Drosophila* hsp70 and hsp83 genes because they are well characterized and display similar yet distinct regulatory properties in *Drosophila*. The level of hsp70 transcripts is low to moderate in uninduced cell cultures and increases approximately 50-fold after heat shock (O'Connor and Lis 1981). The hsp83 transcripts are at higher levels in uninduced cell cultures, and the increase is barely detectable in total RNA after heat shock. The level of hsp83 transcripts in uninduced flies is lower than in cell cultures however, and the induction is correspondingly greater. In addition, hsp83 has an intron (Holmgren et al. 1981), unlike the other major heat-shock genes, and it will be of interest to see if this is processed correctly in yeast.

The results presented here address two related questions. Can yeast cells serve as an initial assay system for identification of sequences necessary for regulated transcription of the *Drosophila* genes? Are the regulatory components of the heat-shock response sufficiently conserved such that they are interchangeable between these very distantly related eukaryotes?

INTRODUCTION OF *DROSOPHILA* HEAT-SHOCK GENES INTO YEAST

The hsp70 and hsp83 genes were each joined to the plasmid vector Ylp33 (Botstein et al. 1979) which carries the *leu 2* gene of yeast. This vector is propagated in yeast only if it becomes integrated into the yeast genome. Most integration events occur by homologous recombination between the leu^+ gene of the plasmid and the leu^- gene of the yeast recipient yielding a stable leu^+ transformant (Fig. 1). The characterized transformant containing the hsp70 gene is named Y-70, while the one containing hsp83 is Y-83.

The Y-70 transformant was derived as described in Figure 1a. The *Bgl*II fragment used should contain all *cis*-acting regulatory sequences required for expression in *Drosophila* (Hackett and Lis 1981). The resulting hybrid plasmid (iDmE3.1) was introduced into yeast by transformation (Hinnen et al. 1978). Transformants were obtained at higher than expected frequencies. These transformants produced sectored colonies indicative of unstable transformation. This behavior stems from the presence of an autonomously replicating sequence (ARS) that is associated with hsp70 genes and allows iDmE3.1 to be propagated as a free plasmid (R. Hackett, pers. comm.). Occasionally, smooth fast-growing colonies arose from these

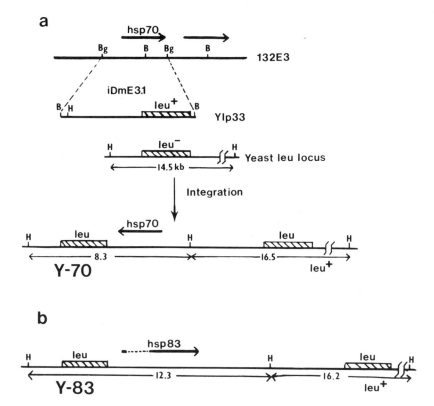

Figure 1

Structure of yeast transformants Y-70 and Y-83. (*a*) The *Bgl*II fragment of *Drosophila* segment 132E3 was ligated to *Bam*HI-cut plasmid YIp33 producing iDmE3.1 (the dashed line connects the restriction sites that were joined). The integration of iDmE3.1 by homologous recombination with the *leu* locus of yeast strain SHY2 produced the *leu*$^+$ transformant Y-70. The integrated iDmE3.1 is bordered by copies of the *leu 2* gene, of which one is mutant and the other is wild type. Bg is *Bgl*II; B is *Bam*HI; and H is *Hin*dIII. (*b*) Using the strategy outlined in *a*, the *Bam*HI to *Bgl*II fragment of the *Drosophila* cloned segment Dm4 (O'Connor and Lis 1981) was joined to *Bam*HI-cut YIp33 and also integrated into the *leu* locus to give strain Y-83.

initial transformants. These were presumed to be stable integrants of the plasmid into the yeast genome. Southern blot analysis of DNA isolated from cultures of sectored and smooth colonies confirmed this.

Using the same strategy, a *Bam* to *Bgl*II fragment carrying the entire hsp83 gene plus flanking sequences was ligated into the *Bam* site of YIp33. In contrast to the hsp70 hybrid, we detected only stable transformants (large smooth colonies) occurring at the expected low frequencies. Restriction-cut DNA from one transformant, Y-83, showed the pattern expected from homologous recombination at the leucine gene.

Table 1

Inducibility of *Drosophila* Heat-shock Genes in Yeast Measured by Dot-blot Hybridization

Gene	Inducibility
hsp70 (Y-70)	1.2
hsp70 (*Drosophila*)	22
hsp83 (Y-83)	6
hsp83 (*Drosophila*)	4

DROSOPHILA GENE TRANSCRIPTS IN YEAST: CONSTITUTIVE hsp70 AND INDUCIBLE hsp83 EXPRESSION

The levels of *Drosophila* gene transcripts in yeast transformants Y-70 and Y-83 were measured after growth under conditions shown to produce strong induction of yeast heat-shock proteins (Kurtz et al., this volume). Typically, cells were grown in minimal acetate medium at 25°C. Half of each culture was grown continuously at 25°C, while the other half was shifted to 40°C for 1 hour before extracting RNA. The level of specific RNAs in the total RNA extract was analyzed both by Northern blot and dot-blot techniques. The hsp70 RNA and hsp83 RNA that were detected were derived from the *Drosophila* genes and not from the corresponding endogenous yeast genes, since no significant hsp70 hybridization signal was detected in Y-83 and no hsp83 signal was detected in Y-70.

The hsp70 RNA of strain Y-70 is abundant and is approximately at the same level under both normal and heat-shock conditions (Table 1). The hsp70 RNA in yeast is at approximately one-tenth the level of induced *Drosophila* cells (calculated on a per gene basis, *Drosophila* possesses five copies of hsp70). The hsp70 RNA of Y-70 is present as a single transcript as detected by Northern blot hybridizations and is equal in size or slightly shorter than the hsp70 RNA produced in *Drosophila* cells. S1 nuclease mapping reveals that the 5' end of the hsp70 transcript made in Y-70 starts several bases downstream of that made in *Drosophila*. If the Y-70 RNA is indeed shorter by the 100−200 bases suggested on some gels, then the bulk of this difference may reside at the 3' end. Since poly(A) tails of yeast message are, in general, over 100 bases shorter than in message of *Drosophila* (McLaughlin et al. 1973), this posttranscriptional modification could account for a slight difference in size.

The hsp83 RNA in strain Y-83 is also abundant, yet in contrast to the hsp70 gene, the level increases severalfold in response to heat shock (Table 1). Here again the hsp83 RNA was detected as a single species by Northern blot hybridization and was of a size equal to that of mature *Drosophila* hsp83 RNA. Although we have not yet mapped the structure of the hsp83 RNA of yeast, its size is consistent with correct initiation, termination, and processing. If S1 mapping experiments verify this result, it is in sharp contrast to that observed for the β-globin expression, in which all three events were incorrect.

The difference in the inducibility of the hsp70 and hsp83 genes in yeast may be due to their opposite orientation at the leucine locus. However, when we cloned hsp70 into yeast on a freely replicating *Trp* plasmid YRp7 (Botstein et al. 1979), the expression of hsp70 also was constitutive. An alternative explanation may lie in the intrinsically different transcriptional properties of hsp70 and hsp83 which have been previously documented in *Drosophila* cells (O'Connor and Lis 1981).

CHROMATIN STRUCTURE OF *DROSOPHILA* hsp70 GENE IN YEAST: 5′ HYPERSENSITIVE SITES ARE CONSERVED

In intact nuclei, the 5′ ends of all *Drosophila* heat-shock genes show a striking sensitivity to the nuclease DNase I (Elgin 1981). This sensitivity is also characteristic of other types of genes that are active or that can be activated in a particular tissue.

We find that the pattern of hypersensitive sites at the 5′ end of the hsp70 gene in the yeast transformant, Y-70, is strikingly similar to sites found for hsp70 in *Drosophila* cells. We have determined by high-resolution gel electrophoresis and Southern blotting that the sites of cleavage in yeast and *Drosophila* cells are the same within the limitations of the technique (10 bp). DNase-I treatment of naked DNA does not preferentially cleave at the hypersensitive sites, suggesting that this hypersensitivity reflects a special feature of chromatin that is conserved between *Drosophila* and yeast.

The precise positions of the hypersensitive regions relative to the hsp70 transcription start, base +1, are worth noting. The major region maps between −120 to −60. This borders a region starting at −65 that is required for induced transcription of *Drosophila* hsp70 in mouse L cells and monkey COS cells (Corces et al.; Mirault et al.; Pelham and Bienz; all this volume). Examination of the hypersensitivity of cloned bits of heat-shock genes that have been introduced into yeast chromosomes should define the DNA sequence that participates in rendering sites hypersensitive in nuclei.

CONCLUSIONS

The strong promoters characteristic of *Drosophila* hsp70 and hsp83 genes are operating efficiently when cloned in yeast cells, as judged from the level of transcripts that accumulate. The level of hsp83 transcript is elevated approximately sixfold by a 1-hour heat shock, while the hsp70 transcript is at the same moderate level in both heat-shock-induced and uninduced cells. The sizes of the RNAs made in yeast appear equivalent to the size of mature RNAs in *Drosophila* cells as determined by Northern blots. Moreover, S1 mapping of the hsp70 transcript demonstrates the transcripts in yeast start within several base pairs of the site used in *Drosophila* cells. Lastly, the hypersensitive sites at the 5' ends of hsp70 gene chromatin are present at the same positions in the yeast transformant, indicating this feature of chromatin structure is maintained on the *Drosophila* gene even when inserted into a yeast chromosome.

Although yeast may prove useful as an initial assay for sequences involved in specifying *Drosophila* hsp70 transcription initiators and chromatin structure, its greater potential lies in the analysis of the hsp83 gene which appears to be regulated by heat shock in yeast.

REFERENCES

Banerji, J., S. Rusconi, and W. Schaffner. 1981. Expression of a β-globin gene is enhanced by remote SV40 DNA sequences. *Cell* **27**: 299–308.

Beggs, J.D., J. van den Berg, A. van Ooyen, and C. Weissman. 1980. Abnormal expression of chromosomal rabbit β-globin gene in *Saccaromyces cerevisiae*. *Nature* **283**: 835–840.

Botstein, D., S.C. Falco, S.E. Stewart, M. Brennan, S. Scherer, D.T. Stinchcomb, K. Struhl, and R.W. Davis. 1979. Sterile host yeast (SHY): A eukaryotic system of biological containment for recombinant DNA experiments. *Gene* **8**: 17–24.

Elgin, S.C.R. 1981. DNAase I-hypersensitive sites of chromatin. *Cell* **27**: 413–415.

Grosschedl, R. and M.L. Birnstiel. 1980. Spacer DNA sequences upstream of the T-A-T-A-A-A-T-A sequence are essential for promotion of H2A histone gene transcription *in vivo*. *Proc. Natl. Acad. Sci.* **77**: 7102–7106.

Hackett, R.W. and J.T. Lis. 1981. DNA sequence analysis reveals extensive homologies of regions preceding hsp70 and $\alpha\beta$ heat shock genes in *Drosophila melanogaster*. *Proc. Natl. Acad. Sci.* **78**: 6196–6200.

Henikoff, S., K. Tatchell, B.D. Hall, and K.A. Nasmyth. 1981. Isolation of a gene from *Drosophila* by complementation in yeast. *Nature* **289**: 33–37.

Hinnen, A., J.B. Hicks, and G.R. Fink. 1978. Transformation of yeast. *Proc. Natl. Acad. Sci.* **75**: 1929–1933.

Holmgren, R., V. Corces, R. Morimoto, R. Blackman, and M. Mesleson. 1981. Sequence homologies in the 5' regions of four *Drosophila* heat-shock genes. *Proc. Natl. Acad. Sci.* **78**: 3775–3778.

McLaughlin, C.S., J.R. Warner, M. Edmonds, H. Nakazato, and M.H. Vaughan. 1973. Polyadenylic acid sequences in yeast messenger ribonucleic acid. *J. Biol. Chem.* **248**: 1466–1471.

O'Connor, D. and J.T. Lis. 1981. Two closely linked transcription units within 63B heat shock puff locus of *D. melanogaster* display strikingly different regulation. *Nucleic Acids Res.* **9**: 5075–5092.

Expression of a Cloned Yeast Heat-shock Gene

**David B. Finkelstein
and Susan Strausberg**
*Division of Molecular Biology
Department of Biochemistry
The University of Texas
Health Science Center at Dallas
Dallas, Texas 75235*

A transient heat-shock response may be elicited in the yeast *Saccharomyces cerevisiae* by rapidly shifting the cultivation temperature of this organism from 23°C to 36°C (both temperatures being within the normal growth range of this yeast). The consequence of this heat shock is a dramatic alteration in the pattern of yeast protein synthesis (McAlister et al. 1979). In addition to an increase in the level of synthesis of a small number of major heat-shock-inducible proteins, one may observe an equally dramatic decrease in the level of synthesis of a large number of heat-shock-repressible proteins. It should be noted, however, that synthesis of some proteins, such as the glycolytic enzyme glyceraldehyde phosphate dehydrogenase (GAPDH) (the single most abundant yeast protein), is unaltered during the heat-shock response.

As a result of the high level of synthesis (as well as the stability) of the major heat-shock-inducible proteins, they become some of the most abundant proteins in the yeast cell. Since the heat-shock response is transient, the heat-shock-inducible proteins are subsequently diluted out by continued cultivation at 36°C. While no role has been assigned to any of the yeast heat-shock proteins, a correlation has been made between the cellular level of these proteins and the degree of thermal tolerance of this organism (McAlister and Finkelstein 1980a). A similar correlation has been observed in other organisms (this volume).

TRANSCRIPTIONAL CONTROL

Previous work has demonstrated that the in vivo changes in protein synthesis that occur during heat shock in yeast may be reproduced in vitro using a rabbit reticulocyte lysate system, suggesting that these alterations in protein synthesis are regulated at the level of transcription (McAlister and Finkelstein 1980b). Examination of poly(A)-containing RNA reveals a major increase in the level of RNA species of 3.5 kb, 2.9 kb, and 2.5 kb within 15 minutes of the imposition of heat shock. In vivo pulse-labeling reveals that these are the major RNA species synthesized during the yeast heat-shock response.

These dramatic transcriptional alterations allow one to use in vivo ^{32}P-pulse-labeled yeast poly(A)-containing RNA as a hybridization probe to isolate heat-shock-responsive genes from recombinant libraries by the technique of differential hybridization. We have isolated a number of heat-shock-responsive genes from a Charon 4-yeast DNA library by this technique and identified the putative gene products of the yeast DNA sequences by in vitro translation of hybrid-selected yeast RNA. One phage (λYhsi1) contains two adjacent EcoRI fragments (3.45 and 2.95 kb) which hybridize to the 2.9-kb heat-shock RNA that encodes yeast hsp90. Hybridization of in vivo pulse-labeled RNA to this phage reveals a 12-fold increase of transcription within 15 minutes of the imposition of heat shock. Transcription levels drop by a similar amount for λYhsr11 which encodes an (as yet unidentified) heat-shock-repressible protein. Finally, it is relevant to mention that transcription of the gene(s) encoding GAPDH is unaltered during heat shock, in agreement with the in vivo translation of this gene product. We conclude from these data that the in vivo alterations in yeast protein synthesis during heat shock are regulated at the level of transcription.

EXPRESSION OF CLONED YEAST hsp90

As our initial characterization of putative heat-shock-responsive genes was by in vitro translation of hybrid-selected RNA, we could not be certain that these DNA segments in fact represented complete, functional heat-shock-responsive genes. We thus used these recombinant phages as hybridization probes to isolate homologous sequences from a library of yeast DNA fragments (obtained by partial digestion of total yeast DNA with Sau3A) inserted in the BamHI site of the plasmid shuttle vector YEp13 (Broach et al. 1979). Figure 1 presents a restriction map of one plasmid, termed pYhsi1, selected by hybridization to λYhsi1, which contains the same 2.95-kb and 3.45-kb EcoRI fragments of yeast DNA as λYhsi1 as part of a 9.2-kb DNA insert in this plasmid.

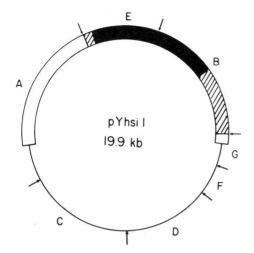

Figure 1
Structure of pYhsi1. The wide line represents the YEp13 portion of this plasmid which contains pBR322 (□), a portion of the yeast 2μ DNA (▨), and a segment of DNA containing the yeast *LEU2* gene (■). The 9.2-kb yeast DNA insert containing hsp90 is represented by a thin line. (→) The sites of digestion of this plasmid by *Eco*RI.

This plasmid was introduced into yeast by transformation of a *leu2* recipient and selection of transformants by plasmid complementation of this chromosomal mutation. When protein synthesis was examined in this transformant, it was observed that prior to heat shock, synthesis of hsp90 was somewhat elevated. Much more dramatic, however, one may observe that within 15 minutes of the onset of heat shock, synthesis of hsp90 accounted for almost one-fifth of total protein synthesis (as compared with 4% of protein synthesis in either an untransformed or YEp13-transformed control) (Fig. 2). Upon growth under nonselective conditions (i.e., in the presence of leucine) plasmid is lost, as is the ability to overproduce hsp90, thus demonstrating that overproduction of hsp90 is a plasmid-borne trait.

It should be noted that no other obvious differences are apparent in the pattern of proteins synthesized by pYhsi1-transformed cells as evaluated by this one-dimensional gel. Thermal tolerance of the pYhsi1-transformed cells is perfectly normal prior to heat shock relative to a YEp13-transformed control, demonstrating that increased levels of hsp90 alone do not cause thermal tolerance. Furthermore, following heat shock the increase in thermal tolerance is identical to that of the YEp13-transformed control, suggesting that whatever heat-shock protein synthesis is responsible for the acquisition of thermal tolerance is not suppressed by introduction of plasmid pYhsi1 into yeast on this multi-copy plasmid.

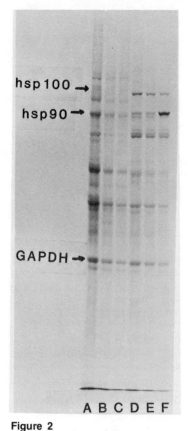

Figure 2
Overproduction of hsp90. *S. cerevisiae* strain DC5 *(MATa, leu2-3, leu2-112, his3, can1-11)* containing either no plasmid (lanes A and D), plasmid YEp13 (lanes B and E), or pYhsi1 (lanes C and F) was grown to mid-log phase at 23°C and aliquots labeled for 15 minutes with [³⁵S]methionine starting either before (lanes A–C) or 15 minutes following a shift in cultivation temperature to 36°C (lanes D–F). SDS-soluble proteins were resolved by electrophoresis on an SDS-10% polyacrylamide slab gel and in vivo translation products visualized by autoradiography. The location of migration of hsp100, hsp90, and GAPDH (36 kD) is indicated.

The apparently otherwise normal heat-shock response of these transformants implies that at this plasmid copy number (~8 per cell) no regulatory elements, either positive or negative, have been titrated out.

DISTINGUISHING PLASMID hsp90 FROM CHROMOSOMAL hsp90

While the above experiments demonstrate that introduction of pYhsi1 results in the overproduction of hsp90, they fail to reveal whether this overproduction is a consequence of transcription of the plasmid-borne

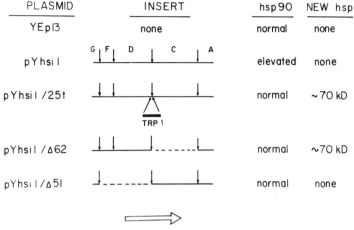

PLASMID	INSERT	hsp90	NEW hsp
YEpl3	none	normal	none
pYhsi l		elevated	none
pYhsil /25t		normal	~70 kD
pYhsil /Δ62		normal	~70 kD
pYhsi l/Δ5l		normal	none

Figure 3
Location and orientation of hsp90 on pYhsi1. The various plasmids listed were introduced into *S. cerevisiae* strain DC5 and protein synthetic patterns of the resulting transformants were examined prior to and during heat shock as in Fig. 2. The letters indicate the *Eco*RI fragments of the yeast insert as in Fig. 1. The open arrow (*bottom*) indicates the direction of transcription of hsp90.

gene. In order to distinguish the chromosomal hsp90 gene product from the plasmid-borne hsp90, a 1453-bp *Eco*RI fragment containing the yeast *TRP1* gene (Tschumper and Carbon 1980) was inserted between fragments C and D of pYhsi1 to generate plasmid pYhsi1/25t. When this plasmid is introduced into yeast by transformation and the heat-shock response of the resulting transformants analyzed, the plasmid-encoded gene product is now observed as a heat-shock-inducible protein of ~70 kD (Fig. 3). It should be noted that this truncated form of hsp90 is the most abundant protein synthesized during heat shock. The level of synthesis of the chromosomal hsp90 gene product (size 90 kD) is now that of an untransformed cell. Thus, the plasmid-borne hsp90 appears to have no effect on the level of expression of its chromosomal counterpart.

DIRECTION OF TRANSCRIPTION

The direction of transcription of hsp90 on pYhsi1 was determined by deletion analysis (Fig. 3). Deletion of the 3.45-kb *Eco*RI fragment C from pYhsi1 results in a plasmid (pYhsi1/Δ62) which directs the synthesis of a heat-shock-inducible protein of similar (but not identical) size to the 70-kD heat-shock protein encoded by pYhsi1/25t. We thus deduce that *Eco*RI fragment C of plasmid pYhsi1 encodes the C terminal portion of hsp90 and that the direction of transcription is as indicated in Figure 3.

As expected for a deletion of the 5′ end of hsp90, plasmid pYhsi1/Δ51 directs increased synthesis of no detectable heat-shock proteins (i.e., pYhsi1/Δ52 transformants behave like untransformed cells). We thus conclude that regulatory sequences responsible for expression of hsp90 are contained on *Eco*RI fragment D (and perhaps F) of pYhsi1.

SUMMARY

The ability to isolate heat-shock-responsive yeast genes has allowed us to demonstrate that the yeast heat-shock response is regulated at the level of transcription.

We have been able to return one of these genes into yeast on a multicopy plasmid shuttle vector. The fact that such a plasmid-encoded gene is expressed in proportion to its gene dosage and results in no other detectable alterations of the yeast heat-shock response leads us to conclude that if any common regulatory elements coordinate the heat-shock response, they are not titrated out by the introduction of extra copies of this single heat-shock gene.

The fact that a C-terminal deletion of hsp90 is a stable protein suggests that fusion to other proteins will be feasible. Furthermore, the availability of a defined, cloned heat-shock gene will allow deletion of the chromosomal gene in order to determine if hsp90 is an essential cell protein.

ACKNOWLEDGMENTS

This work was supported by Grant GM25829 from the National Institutes of Health and Grant I-783 from The Robert A. Welch Foundation.

REFERENCES

Broach, J.R., J.N. Strathern, and J.B. Hicks. 1979. Transformation in yeast: Development of a hybrid cloning vector and isolation of the *CAN1* gene. *Gene* **8:** 121–133.

McAlister, L. and D.B. Finkelstein. 1980a. Heat shock proteins and thermal resistance in yeast. *Biochem. Biophys. Res. Commun.* **93:** 819–824.

———. 1980b. Alterations in translatable ribonucleic acid after heat shock of *Saccharomyces cerevisiae. J. Bacteriol.* **143:** 603–612.

McAlister, L., S. Strausberg, A. Kulaga, and D.B. Finkelstein. 1979. Altered patterns of protein synthesis induced by heat shock of yeast. *Curr. Genet.* **1:** 63–74.

Tschumper, G. and J. Carbon. 1980. Sequence of a yeast DNA fragment containing a chromosomal replicator and the *TRP1* gene. *Gene* **10:** 157–166.

Developmentally Regulated Transcription at the 67B Heat-shock Cluster

Karl Sirotkin
Department of Microbiology
University of Tennessee
Knoxville, Tennessee 37996-0845

The genes that code for the four small heat-shock proteins are all in one cluster at 67B (Corces et al. 1980; Craig and McCarthy 1981; Voellmy et al. 1981). An additional gene (gene 1) is in the middle of the cluster and two more genes (genes 4 and 5) are adjacent to the cluster (Sirotkin and Davidson 1982). The region containing these genes is depicted in Figure 1. The relationship of the three additional genes (1, 4, and 5) to the heat-shock response is unknown. However, the three additional genes and the two heat-shock genes that are bracketed by them (hsp26 and hsp23) are all transcribed during the late larval and early pupal stages (Sirotkin and Davidson 1982). It has not been reported whether the expression from the three additional genes (genes 1, 4, and 5) is increased by heat shock.

In another study of the region containing the genes coding for the four small heat-shock proteins, DNase-I hypersensitive sites in chromatin from embryos were mapped (Keene et al. 1981). Such sites mapped just upstream, transcriptionally, from the beginning of all four of the heat-shock genes. In spite of the fact that no other gene was then known in this region, two additional sites were found. These sites mapped upstream, transcriptionally, from the gene in the center of the cluster (gene 1) that has been shown to be transcribed during the late larval and early pupal stages (Sirotkin and Davidson 1982). The average distance from the sites to the beginning of the coding region was about the same for

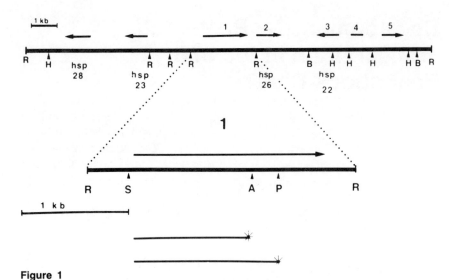

Figure 1
Restriction and transcription map of the cluster of seven genes at 67B. The thick black line denotes the DNA of this region. The capital letters point to restriction endonuclease cleavage sites: R for *Eco*RI, H for *Hin*dIII, B for *Bam*HI, S for *Sal*I, A for *Ava*I, and P for *Pst*I. On the top thick line, only *Eco*RI, *Hin*dIII, and *Eco*RI sites are shown. Additional sites are shown on the bottom thick line for the *Eco*RI fragment that contains gene 1. The thin lines above the thick lines denote transcripts and show the direction of transcription. The numbers above the thick lines identify the genes after Sirotkin and Davidson (1982). The gene names below the thin lines and under the thick lines identify the genes after Corces et al. (1980), who used the same strain of *Drosophila melanogaster*. The thin lines at the bottom of the figure depict the only DNA that would be protected from digestion by transcript 1; top shorter line, after cleavage at the *Ava*I site and 5' labeling; bottom longer line, after cleavage at the *Pst*I site and 5' labeling.

the four heat-shock genes. However, the corresponding distance for gene 1 seemed to be greater than this constant distance.

In this report, I present data about the transcriptional start of the additional gene in the center of the cluster (gene 1), as well as the detection of transcripts from the beginning of this gene in late embryos. No transcription from this gene had previously been reported in embryos, in spite of the fact that RNA from embryos was used in some experiments that could have detected transcription from this gene (Sirotkin and Davidson 1982). The greater distance between the DNase-I hypersensitive sites and the apparent beginning of transcriptional initiation from this gene and the corresponding distance from the heat-shock genes could be explained by the presence of a short exon upstream transcriptionally. I report data that show that no such exon exists.

I also present some preliminary data obtained using RNA from heat-shocked pupae. The conclusions from these data are that none of the three additional genes (genes 1, 4, or 5) is abundantly transcribed as are the heat-shock genes in these pupae, although their transcription may be slightly increased by heat shock.

MAPPING OF 5' ENDS

5' ends were mapped by the technique of Berk and Sharp (1978), as modified by Weaver and Weissmann (1979). Hybridizations of RNA to DNA in high formamide were as reported previously (Sirotkin and Davidson 1982), except reactions were quenched by immersion in liquid nitrogen. Nuclease S1 (Sigma or BRL) digestions were for 10 minutes at 20°C using 300 μl reactions containing 50 μg nucleic acid and 150 units enzyme. Exonuclease-VII (BRL) digestions (300 μl) were at 37°C and were terminated with 100 μl of 3 M ammonium acetate and precipitated with 300 μl isopropanol.

CONFIRMATION OF THE POLARITY AND MAP OF GENE 1

Electron microscope analysis of R loops with marker DNA hybridized to their poly(rA) were used to determine the polarity of genes 1, 2 (hsp26), 3 (hsp23), and 5 by Sirotkin and Davidson (1982). Probably because the structure at gene 1 was usually Y-shaped instead of a loop, very few molecules with markers at the end were interpretable (K. Sirotkin, unpubl.). Furthermore, many fewer markers were seen for this gene, and its polarity was therefore determined with less certainty than the others (Sirotkin and Davidson 1982).

To confirm the polarity of gene 1, the subclone containing the 2.7-kb *Eco*RI fragment was cleaved with *Ava*I or *Pst*I (see Fig. 1). If the previously assigned polarity and map were correct, an end-labeled fragment about 1.2 kb in length would be generated using the *Ava*I-cleaved DNA, as depicted at the bottom of Figure 1. Using *Pst*I-cleaved DNA, a slightly longer fragment would be expected. If the previously assigned polarity were incorrect, then a shorter fragment would be expected using *Pst*I-cleaved DNA. That the previously assigned polarity and map were correct is shown in Figure 2. The *Pst*I-cleaved DNA generates a longer fragment (compare lanes 5 and 7).

In the previous report (Sirotkin and Davidson 1982), the 5' end of transcript 1 was mapped to be just barely to the right of a *Sal*I cleavage site (Fig. 1). To test this conclusion, one aliquot of DNA end-labeled at the *Ava*I site was reassociated and cleaved with *Sal*I. Another aliquot was hybridized to RNA and digested with nuclease S1. The resulting DNA fragments were run on a single gel (not shown). The DNA protected from nuclease S1 digestion ran just ahead of the *Sal*I to *Ava*I fragment. The previous map and polarity determinations were confirmed by these results.

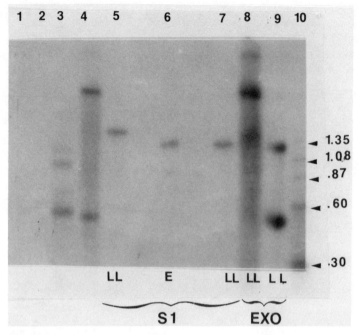

Figure 2

5'-End mapping of gene 1. Cleaved 2.7-kb EcoRI fragment in pKH containing gene 1 (1 ng, about 500–1000 Ci/mMole 5' ends ^{32}P) were hybridized (24 hr, 54°C, 80% formamide) to 40 μg RNA in 11 μl reactions as by Sirotkin and Davidson (1982). The figure is an autoradiogram of a 2% alkaline agarose gel that was neutralized with Tris-HCl (no counter ion) prior to drying. Lane 1 contains DNA cleaved with Ava I and yeast (control) RNA, after nuclease S1 (S1) digestion. Lane 2 contains DNA cleaved with Pst I and yeast (control) RNA, after nuclease S1 digestion. Lane 3 contains DNA cleaved with Pst I and yeast (control) RNA, after exonuclease-VII (EXO) digestion (0.5 hr, 1 unit). Lane 4, as lane 3 (1.5 hr, 2 units). Lane 5 contains DNA cleaved with Pst I, Drosophila late larval RNA (LL), after S1 digestion. Lane 6 contains DNA cleaved with Ava I, Drosophila embryo RNA (E), after S1 digestion. Lane 7 contains DNA cleaved with Ava I, Drosophila late larval RNA, after S1 digestion. Lane 8 contains DNA cleaved with Ava I, Drosophila late larval RNA, after exonuclease-VII digestion (0.5 hr, 1 unit). Lane 9 contains DNA cleaved with Ava I, Drosophila late larval RNA after exonuclease-VII digestion (1.5 hr, 2 units). Lane 10 contains ΦX DNA, cleaved with Hae III and end-labeled.

THERE IS NO NEARBY 5' EXON FOR GENE 1

The previous experiments mapping gene 1 (Sirotkin and Davidson 1982) would not have detected a 5' leader exon. Even if a 5' leader exon were long enough to be stable occasionally under the conditions used for electron microscopy, it could not have been detected, because the DNA used for electron microscopy was cleaved at the SalI site. Furthermore, the S1 nuclease mapping performed previously would also have missed such an exon, because of technical reasons that will not be described here. For these reasons and those stated in the introduction, I undertook experiments that would have detected such an exon, if present.

The rationale for transcript mapping using exonuclease VII is presented by Berk and Sharp (1978). Basically, exonuclease VII digests from both 3' and 5' ends, but it only digests single-stranded DNA. An intron looped out from an RNA:DNA duplex would not be cleaved. These experiments utilize exonuclease VII instead of nuclease S1.

Experiments using exonuclease VII are presented in Figure 2. If a stable 5' exon were present, the length of DNA protected from digestion by exonuclease VII would be greater than the length of DNA protected from digestion by nuclease S1. The increase in length would be due to the extra DNA from both the intervening sequence and the exon. This is not the case; the DNA protected from digestion is the same length whether exonuclease VII or nuclease S1 is used, as is demonstrated by comparing lanes 7 and 9. There is thus no stable leader exon within the subclone used that is far enough away to alter the length of the protected DNA fragment, as observed.

A band is seen in control lanes, without *Drosophila* RNA, even at high exonuclease-VII concentration. This can be contrasted to similar control lanes using nuclease S1 and end-labeled DNA in which I have yet to observe any band, even after very long autoradiographic exposures (not shown). Presumably the band observed after exonuclease-VII digestion is due to a stable hairpin within the DNA itself.

DETECTION OF GENE-1 TRANSCRIPTS IN LATE EMBRYOS

RNA was purified from embryos aged, after deposition, 12–20 hours at 25°C. Transcripts initiating at the same location as in RNA purified from late larvae was detected (Fig. 2, lanes 6 and 7). Gene-1 transcripts had not previously been detected in embryos (Sirotkin and Davidson 1982). The reasons for this difference in detection of gene-1 transcripts is not entirely clear. It is not likely that the sensitivity of the present experiments is greater, since high-specific-activity cDNA was used in the earlier experiments. However, since oligo(dT) was used to prime the previous cDNA reactions and no attempt was made to make full-length transcripts, the cDNA would be more representative of 3' than of 5' ends. If termination of gene-1 transcription is abnormal in embryos, then this skewed cDNA population might be expected to underestimate the level of gene-1 transcription. Moreover, the ages of the embryos may have been different in the two experiments, and the abundance of gene-1 transcripts may depend on the age of the embryo.

TRANSCRIPTION FROM 67B IN HEAT-SHOCKED PUPAE

RNA was purified from early pupae that had been kept at 35.5°C for 1 hour before harvest. Since the additional genes (1, 4, and 5) are expressed at this time in animals that have not been heat-shocked, it is possible that their expression might be further increased by heat shock at that same time. cDNA probes were made as described previously (Sirotkin and Davidson 1982). Preliminary results (Fig. 3) verify the identity of genes 2 and 3 as heat-shock genes (hsp26 and hsp23). Additionally, the other genes (1, 4, and 5) are not induced to the same degree as these heat-shock genes.

SUMMARY AND CONCLUSIONS

Most of the work presented in this report concerns transcription from gene 1, the developmentally regulated gene in the center of the 67B heat-shock cluster. Specifically, the following conclusions can be made: (1) the previous mapping and polarity determinations of this gene are confirmed, (2) transcription from this gene can be detected in late embryos, and (3) there is no nearby 5′ leader exon for this gene.

Preliminary data obtained using RNA from heat-shocked pupae was also mentioned. Transcription from genes 1, 4, and 5 does not appear to be greatly induced by heat shock.

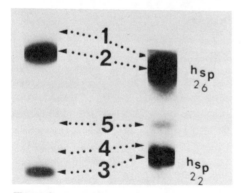

Figure 3
Hybridization of λDmp 67 DNA (Sirotkin and Davidson 1982) to [^{32}P]cDNA from heat-shocked pupae. λDmp 67 (which contains the rightmost two *Eco*RI fragments shown in Fig. 1) was digested with *Eco*RI, *Bam*HI, and *Hin*dIII, electrophoresed, and transferred to nitrocellulose. Positions where DNA containing specific genes would migrate are noted and are named as in Fig. 1. The cDNA reaction produced only short cDNA. Hybridization to *Drosophila* actin gene (not shown) on the same blot produced a signal comparable with and slightly more intense than that seen for gene 5. The figure contains two different exposures from the same experiment.

Several issues remain: (1) RNA abundances at different times during development with and without heat shock, (2) transcription from the same gene in different RNA populations, and (3) protein products from the additional genes (1, 4, and 5).

REFERENCES

Berk, A.J. and P.A. Sharp. 1978. Spliced early mRNA of simian virus 40. *Proc. Natl. Acad. Sci.* **75:** 1274–1278.

Corces, V., R. Holmgren, R. Freund, R. Morimoto, and M. Meselson. 1980. Four heat shock proteins of *Drosophila melanogaster* coded with a 12-kilobase region in chromosome subdivision 67B. *Proc. Natl. Acad. Sci.* **77:** 5390–5393.

Craig, E.A. and B.J. McCarthy. 1981. Four *Drosophila* heat shock genes at 67B: Characterization of recombinant plasmids. *Nucleic Acids Res.* **8:** 4441–4451.

Keene, M.A., V. Corces, K. Lowenhaupt, and S.C.R. Elgin. 1981. DNase I hypersensitive sites in *Drosophila* chromatin occur at the 5' ends of regions of transcription. *Proc. Natl. Acad. Sci.* **78:** 143–146.

Sirotkin, K. and N. Davidson. 1982. Developmentally regulated transcription from *Drosophila melanogaster* chromosomal site 67B. *Dev. Biol.* **89:** 196–210.

Voellmy, R., M. Goldschmidt-Clermont, R. Southgate, A. Tissières, R. Levis, and W. Gehring. 1981. A DNA segment isolated from chromosomal site 67B in *Drosophila melanogaster* contains 4 closely linked heat shock genes. *Cell* **23:** 261–270.

Weaver, R.F. and C. Weissmann. 1979. Mapping of RNA by a modification of the Berk-Sharp procedure: The 5' termini of 15S β-globin mRNA precursor and mature 105 β-globin mRNA have identical map coordinates. *Nucleic Acids Res.* **7:** 1175–1193.

Genetic Analysis of the Region of the 93D Heat-shock Locus

Jym Mohler and Mary Lou Pardue

Department of Biology
Massachusetts Institute of Technology
Cambridge, Massachusetts 02139

Of the heat-shock loci of *Drosophila melanogaster,* 93D is perhaps the most enigmatic. While a puff is induced at 93D by the same agents that induce the other heat-shock loci (such as elevated temperatures or recovery from anoxia), the 93D puff can also be induced independently of the other heat-shock loci by a number of agents: homogenate of heat-shock glands and benzamide (Lakhotia and Mukherjee 1980) and aged media (Bonner and Pardue 1976). Unlike RNA from the other heat-shock loci, the RNA transcribed from 93D is more concentrated in the nucleus than in the cytoplasm. The 93D nuclear RNA contains sequences not found in the cytoplasmic RNA and is both poly(A)$^+$ and poly(A)$^-$, while 93D cytoplasmic RNA is predominately poly(A)$^-$ (Lengyel et al. 1980). None of the heat-shock proteins appear to be encoded at the 93D heat-shock locus. The apparent complexity of this locus makes it an intriguing locus for study. We have adopted a genetic approach to attempt to resolve the function of the 93D heat-shock locus.

DISCUSSION

By screening against two deficiencies for the *ebony* locus and the 93D heat-shock locus (Df[3R]e^{Gp4} and Df[3R]e^{H4}), a number of genetic lesions located in the 93C–D region were isolated. Most were isolated by failure to complement Df(3R)e^{Gp4}, but eight were isolated against

Df(3R)e^{H4}. These mutations were generated with ethylmethanesulfonate, (EMS), diepoxybutane, and gamma irradiation.

These mutations form 13 complementation groups, which were mapped by their complementation behavior against an overlapping series of deficiencies. All the complementation groups, except the previously known *ebony* locus, are capable of mutating to lethality, although not all alleles at all loci are lethal. An additional complementation group, er3, is defined by the inability of Df(3R)e^{GP4} and Df(3R)GC14 to complement fully. This compound deficiency heterozygote survives infrequently to adulthood, and results in a sick, inviable adult (unable to stand, with curved wings).

The 93D heat-shock locus was mapped relative to this overlapping series of deficiencies by three criteria: (1) the ability of the deleted chromosomes to puff at 93D, (2) the ability of the deleted chromosomes to synthesize RNA from the 93D region after a temperature shift, and (3) presence of sequences at 93D complementary to heat-shock RNA as assayed by in situ hybridization. The results are essentially the same by all three criteria. The Df(3R)e^{R1} and Df(3R)e^{H5} chromosomes puffed and incorporated [^{3}H]uridine after a temperature shift; they were labeled at 93D following in situ hybridization of heat-shock RNA from tissue culture cells. All the other deficiencies tested failed to puff and incorporate [^{3}H]uridine following a temperature shift and did not label in this region after in situ hybridization with heat-shock RNA. The 93D heat-shock locus is flanked by three breakpoints: it lies (1) inside of Df(3R)e^{GP4}, (2) inside of Df(3R)GC14, and (3) outside the inverted region of In(3R)GC23. The 93D heat-shock locus is thus located in the overlapping region of Df(3R)e^{GP4} and Df(3R)GC14. This is shown as the shaded region in the deficiency map in Figure 1.

One possible function for the 93D heat-shock locus is regulation of the heat-shock response. The effect of a homozygous deficiency for the 93D heat-shock locus on normal heat-shock protein synthesis was tested in two conditions: in embryos homozygous for Df(3R)e^{GC9} (which extends in both directions from the heat-shock locus) and in adults of the genotype Df(3R)e^{GP4}/Df(3R)GC14 (which appears to delete just the heat shock locus).

Embryos homozygous for Df(3R)e^{GC9} survive to late embryogenesis, and we have chosen to test the heat-shock response in 12-hour-old embryos homozygous for Df(3R)e^{GC9}, which is several hours before the lethal phase. Because the homozygous embryos must be produced by crossing heterozygous flies of the genotype Df(3R)e^{GC9}/TM3, it is necessary to be able to determine the genotype of each embryo. We have accomplished this by marking the Df(3R)e^{GC9} chromosome and the TM3 chromosome with different alleles of a polymorphism of the 22K heat-shock protein, which maps to region 67B on chromosome 3L. Individual 12-hour embryos were heat-shocked at 37°C for 20 minutes and then

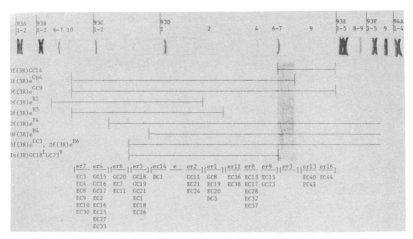

Figure 1

Complementation map of the mutations in the 93D region. The deficiencies are placed relative to the cytological bands that are deleted. Mutations are ordered on the basis of complementation with the deficiencies. The shaded region indicates the limits of the 93D heat-shock locus as defined in this study.

labeled with [^{35}S]methionine (1 mCi/ml) at 37°C for 1 hour. In these experiments, embryos of the expected genotypes—homozygous Df(3R)e^{GC9}, homozygous TM3, and heterozygous Df(3R)e^{GC9}/TM3—were observed in approximately the expected ratios. As seen in Figure 2a, all embryos synthesized the expected set of proteins during the heat shock. Thus a deficiency for the 93D heat-shock locus appears to have no effect on the induction of the major heat-shock proteins in the 12-hour embryo.

The two deficiencies, Df(3R)e^{Gp4} and Df(3R)GC14, appear to overlap only in the region of the 93D heat-shock locus. Thus, flies heterozygous for both these deficiencies are lacking the heat-shock locus and probably very little else. A low percentage of these compound-deficiency heterozygotes survive to adulthood but are inviable as adults. We have tested the ability of these compound-deficiency heterozygotes to induce the heat-shock response and to recover normal protein synthesis after a heat-shock response. Figure 2b shows the proteins synthesized by ovaries from Df(3R)e^{Gp4}/Df(3R)GC14, during and after a heat shock, compared with proteins synthesized by wild-type Canton-S ovaries under the same conditions. Adult ovaries were heat-shocked at 37°C for 0.5 hour, allowed to recover at 25°C for 0, 1, 2, or 4 hours, then labeled for 0.5 hour with [^{35}S]methionine (1 mCi/ml). As in the homozygous Df(3R)e^{GC9} embryos, a deletion of the 93D heat-shock locus does not affect synthesis of the major heat-shock proteins in adult ovaries. The

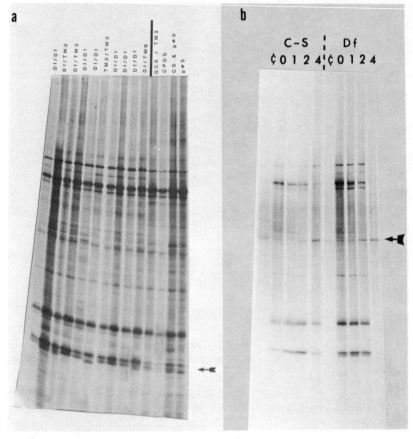

Figure 2
Heat-shock protein synthesis in deficiencies for the 93D heat-shock locus. (a) The first 10 lanes are the heat-shock proteins from individual 12-hour embryos of the Df(3R)e^{GC9}/TM3 stock (20 min at 37°C, then 1 hr label with 1 mCi/ml [^{35}S]methionine at 37°C). The Df(3R)e^{GC9} and TM3 chromosomes were marked with a polymorphism of the 22K heat-shock protein (arrow). The genotype of each embryo, determined by its 22K protein, is indicated above the lane. The rightmost four lanes show the 22K hsp polymorphism in heat-shocked ovaries of the indicated genotype. (b) Protein synthesized during recovery from heat shock in flies heterozygous for two deficiencies for the 93D heat-shock locus. Lanes marked ¢ show proteins synthesized at 25°C in wild-type Canton-S (C-S) and in Df(3R)e^{GP4}/Df(3R)GC14 (Df). Lanes marked 0, 1, 2, and 4 show proteins synthesized by ovaries heat-shocked at 37°C for 0.5 hour, allowed to recover at 25°C for 0, 1, 2, or 4 hours, respectively, and then labeled with [^{35}S]methionine for 0.5 hour. The recovery of 25°C protein synthesis is most clearly reflected by the recovery of synthesis of actin (arrow).

deficiency also does not inhibit recovery of the synthesis of normal protein synthesis. Recovery in the ovaries from the deficiency-bearing flies is at least as rapid as in the wild type, if one compares the recovery of actin synthesis with the loss of synthesis of the heat-shock proteins.

The overlap of the deficiencies Df(3R)e^{Gp4} and Df(3R)GC14 defines one complementation group, er3, for which no point mutations were recovered. Two reservations must be considered before equating this complementation group with the 93D heat-shock locus. First, the overlapping region of these two deficiencies may include more than one undetected locus. This seems unlikely with the level of genetic saturation of the overall region. Second, the 93D heat-shock locus may consist of a multiply repeated gene family (see Lengyel et al. 1980), and these deficiencies may delete most (>90%) but not all of these sequences, since the autoradiographic exposures used to localize the 93D heat-shock locus were of limited duration.

SUMMARY

We have identified 43 new mutations and 3 new deficiencies in the 93C−D region. As determined by the ability to puff, to be transcribed during a heat shock, and to show hybridization of heat-shock RNA, the 93D heat-shock locus lies between the proximal breakpoint of Df(3R)GC14 and the distal breakpoint of Df(3R)e^{Gp4}. None of the point mutations in 93C−D map within this region, which may contain only the heat-shock locus. Both heat-shock protein synthesis and the recovery of 25°C protein synthesis appear to be normal in tissues of flies completely deleted for the 93D heat-shock locus. Thus, the 93D heat-shock locus does not appear to have an effect on protein synthesis during or after heat shock. However, flies heterozygous for small overlapping deficiencies of the heat-shock locus do not survive beyond early adulthood, indicating that the 93D heat-shock locus, or a very closely linked gene, has a distinct effect on the viability of *Drosophila* late in development.

ACKNOWLEDGMENTS

We wish to thank S. Henikoff, Bruce Baker, and F. Ritossa for providing deficiencies. Support was provided by grant NIH-5-R01-GM21874 to M.L.P. and J.M. was supported by an NSF Graduate Fellowship and by an NIH training grant to the Department of Biology.

REFERENCES

Bonner, J.J. and M.L. Pardue. 1976. The effect of heat shock on RNA synthesis in *Drosophila* tissues. *Cell* **8**: 43−50.

Lakhotia, S.C. and T. Mukherjee. 1980. Specific activation of *Drosophila melanogaster* by benzamide and the effect of benzamide treatment on the heat shock induced puffing activity. *Chromosoma* **81:** 125–136.

Lengyel, J.A., L.J. Ransom, M.L. Graham, and M.L. Pardue. 1980. Transcription and metabolism of RNA from the *Drosophila melanogaster* heat shock puff site 93D. *Chromosoma* **80:** 237–252.

Perturbations of Chromatin Structure Associated with Gene Expression

Michael A. Keene
*Department of Biochemistry
and Molecular Biology
Harvard University
Cambridge, Massachusetts 02138*

Sarah C.R. Elgin
*Department of Biology
Washington University
St. Louis, Missouri 63130*

During the last few years, we have been able to learn a considerable amount about the structure of chromatin by using nucleases as probes to delineate patterns of protein-nucleic acid interactions. Evidence of this type contributed to the development of the nucleosome, or "beads-on-a-string" model, of the fundamental 100 Å chromatin fiber (for review, see McGhee and Felsenfeld 1980). This model is now so well established that one can use the generation of a pattern of oligonucleosomal DNA fragments on digestion of a chromatin sample by micrococcal nuclease as a test of normal chromatin structure. More recently, studies using DNase I have identified particular hypersensitive sites along the chromatin fiber. These sites, of 50−200 bp, can be mapped to specific positions in regions of a genome that have been cloned and characterized. It has been of considerable interest to find that one consistently observes such DNase-I hypersensitive sites at or near the 5' end of a gene that is being expressed or can be induced in the cell type under study (for review, see Elgin 1981). In fact, one can suggest that such a chromatin structure is necessary, although not sufficient, for gene activity. Evidence available at present indicates that the DNase-I sites represent a histone-free region (McGhee et al. 1981). Such sites may well be involved in any of several functions requiring access to the DNA in chromatin, such as initiation of DNA replication and DNA rearrangement events, as well as initiation of transcription.

It is a logical extension to enquire as to whether or not perturbations of chromatin structure associated with gene transcription can be detected using nucleases. Such a difference between active and inactive genes was first reported by Weintraub and Groudine (1976), who observed that active globin genes showed a general increase in sensitivity to digestion by DNase I. This has been consistently reported in subsequent studies by others, including an analysis using cDNA probes of genes transcribed at a low level (Garel et al. 1977). In terminally differentiated cells, this nuclease sensitivity is observed for genes such as globin or ovalbumin, even when the gene is not being transcribed (e.g., Palmiter et al. 1978). By using the Southern blot technique to visualize the DNA fragments generated at a specific locus, Wu et al. (1979) observed changes in the pattern resulting from micrococcal nuclease digestion following activation of the major heat-shock locus. The active chromatin was digested to small fragments more rapidly, and the oligonucleosome pattern became less distinct, or "smeared," in fact almost imperceptible. This has also been observed for the active ovalbumin gene in chick oviduct (Bloom and Anderson 1979; for review, see Cartwright et al. 1982).

It is apparent, then, that several alterations in chromatin structure are associated with the process of gene activation. In order to learn more about the process, it is of interest to map these changes in space and time, that is, in relation to the genomic DNA map and in relation to events of differentiation and induction. The heat-shock genes are well suited to this purpose because of the facility with which the activity state can be controlled. Work in our laboratory has been carried out utilizing *Drosophila melanogaster* as the experimental system.

DISCUSSION

In an analysis of the locus 67B using nuclei from embryos maintained at 25°C, we have previously observed the presence of a pair of DNase-I hypersensitive sites at the 5' end of each heat-shock and developmentally regulated gene examined (Keene et al. 1981). A similar analysis using micrococcal nuclease revealed an extensive pattern of specific DNA cleavage, with most prominent cleavage sites falling in the spacers and not in the regions encoding transcripts. The pattern of cleavage sites was very similar for purified DNA and for chromatin, but distinctive differences were observed at or near the 5' ends of the genes (Keene and Elgin 1981). One site, which abuts the 5' end of a developmentally regulated gene R in this cluster, is prominently cleaved in both purified genomic DNA, recombinant plasmid DNA, and nuclei from normal embryos, but appears to be protected in the nuclei of embryos subjected to a 20-minute heat shock (Fig. 1). Under heat-shock conditions, one

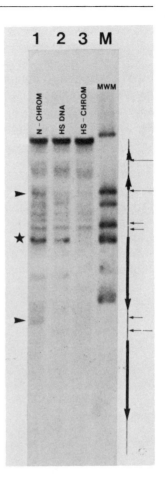

Figure 1
Digestion of locus 67B by micrococcal nuclease. The transcripts for (*top* to *bottom*) hsp28, hsp23, R, and hsp26 are indicated by the large arrows on the map at the right. The small horizontal arrows denote the positions of DNase-I hypersensitive sites of chromatin described previously (Keene et al. 1981). Chromatin-specific sites near the 5′ terminus of hsp26 and hsp23 are indicated by the large arrowheads to the left, and a prominent site that is obscured under heat-shock conditions is marked by a star. Lane 1, DNA from nuclei of normal 6–18-hour embryos digested with 120 U/ml nuclease; lane 2, purified DNA from 6–18-hour embryos heat-shocked at 36°C for 20 min, digested with 6 U/ml nuclease; lane 3, DNA from nuclei of heat-shocked embryos (same as for lane 2), digested with 120 U/ml nuclease; lane 4, molecular weight markers. The DNA from samples 1, 2, and 3 was restricted completely with *Bam*HI. After electrophoresis through a 0.9% agarose gel, a Southern blot was prepared and hybridized with nick-translated 88.3 as probe. 88.3 is a 2.2-kb fragment abutting the *Bam*HI site downstream from hsp26. (Reprinted, with permission, from Keene and Elgin 1981).

anticipates that the R gene will be turned off, while the heat-shock genes will be turned on. Adjacent sites in the chromatin, as well as those at the 5′ ends of several neighboring heat-shock genes, are apparently unaffected. It has now been found that a chromatin-specific, DNase-I-hypersensitive site overlapping this site at the R gene is at least partially protected under those conditions. Preliminary results suggest that related phenomena may be occurring elsewhere in the 67B region. In contrast, investigations of the DNase-I and micrococcal nuclease cleavage patterns in the vicinities of the constitutive ribosomal protein gene, RP49, and the cytoplasmic actin gene at locus 5C failed to reveal similar perturbations. The transcription of actin messages is known to be almost completely repressed during heat shock in the Schneider line-2 *Drosophila* cells used for the latter experiments (Findley and Pederson 1981). The results suggest that within a domain that includes active loci repression might be effected by blocking a specific DNase-I hypersensitive site. However, this does not appear to be a general mechanism of gene inactivation during heat shock.

An altered nucleosomal configuration associated with the fully induced hsp70 heat-shock genes was previously reported using micrococcal nuclease as a probe (Wu et al. 1979). Although the nucleosomal organization of the genome as a whole is not perturbed by heat shock, as revealed by ethidium bromide-staining of the digestion pattern, Southern blots probed with a subclone internal to the hsp70 message showed a virtual obliteration of the crisp nucleosomal ladder seen in the uninduced genes (Fig. 2). It is possible to map the extent of this perturbation by constructing subclones of unique DNA distal to the gene copy at 87A7. The deranged chromatin extends for approximately 2.5 kb beyond the 3′ terminus of the gene (mapped as the poly[A] addition site) and then relatively abruptly returns to a regular nucleosomal array, which is invariant in control and heat-shocked animals. The chromatin structure throughout this region returns to its normal state of organization in tissue culture cells after a 60-minute recovery from heat shock. When nuclei from heat-shocked cells were incubated for 30 minutes with a sufficient concentration of α-amanitin to extinguish heat-shock message produc-

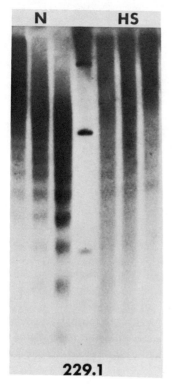

Figure 2
Effect of heat shock on the nucleosome structure within the hsp70 heat-shock gene. Nuclei from normal (N) and heat shocked (HS) *Drosophila* embryos were digested with increasing amounts of micrococcal nuclease, the purified DNA fragments run on a 2% agarose gel, and the resulting Southern blot hybridized with nick-translated pBR 229.1 as a probe. pBR 229.1 contains *Drosophila* DNA sequences that lie entirely within the transcribed region of hsp70. Note that the HS samples (lanes 5−7) are in inverse order relative to the N samples (lanes 1−3); lane 4 contains molecular weight markers.

tion in an in vitro transcription assay, the nucleosome patterns were as smeared as those from control nuclei actively synthesizing RNA, suggesting that the altered conformation had been "frozen" in place.

Analogous studies were performed using subclones of the cytoplasmic actin gene encoded at locus 5C. Actin transcripts are relatively abundant in the *Drosophila* cells (Schneider line 2) used for these experiments, although it can be estimated that the rate of transcription is considerably less than that of the fully induced hsp70 gene (Findley and Pederson 1981). In cells maintained at 25°C, a comparison of the nucleosome pattern 2−3 kb distal to the gene with that within the gene indicates a slight smearing of the nucleosomal array within the transcribed region; the effect is much less pronounced than for the fully induced heat-shock gene. In heat-shocked cells, where transcription is severely repressed, there is no noticeable alteration in the pattern.

CONCLUSION

The sum of the evidence currently available suggests that a set of conditions, each necessary but not sufficient, are required for gene expression in vivo. These include the presence of a DNase-I site at or near the 5′ end of the gene and the presence of an "open" domain, a broad region sensitive to nucleases. The protection of a nuclease-sensitive site proximal to a developmental gene in locus 67B during heat shock suggests a regulatory event associated with the repression of general transcription during heat shock. The finding that a superimposed DNase-I hypersensitive site is similarly protected supports this conjecture. Alternatively, it is possible that whole chromatin domains could be condensed or "protected" in some manner and so be made inaccessible to the cell's transcriptional machinery, thereby eliminating any need for regulating the individual genes within each domain. The cytoplasmic actin and ribosomal protein genes examined could, for example, be in such domains. The 67B gene cluster, on the other hand, must remain open during heat shock; thus, a specific mechanism for repressing nonessential transcription is employed in this case. It will be of great interest to examine other nonheat-shock genes that lie within heat-shock domains. Further work is needed to monitor the general DNase-I sensitivity of all these genes as a function of heat shock and to determine the boundaries of the region of broad DNase-I sensitivity, the "open" domain.

The finding of varying degrees of nucleosomal smearing for the active heat-shock and actin genes corroborates studies of the ovalbumin gene (Bloom and Anderson 1979) that related the effect to the absolute rate of transcription. In the cells used here, the active hsp70 genes are being transcribed at a significantly higher rate than the active actin genes. The fact that inhibition of transcription by either α-amanitin or heat shock fails to restore a regular pattern supports the view that a transient alteration in the local chromatin structure, rather than the passage of the transcriptional machinery per se, is responsible for the effect. While further study is warranted, it seems likely that the extended disorder mapped at 87A indicates a conformational shift propagated to some structural boundary. It is always possible that transcription may continue beyond the 3′ poly(A) addition site, but transcription of this region with sufficient frequency to produce such a dramatic effect on the chromatin structure should have been detected even if the turnover time of that RNA were rapid (Miller and Elgin 1981). It will be of interest in the future to examine more closely the nucleosome structure within regions of transcription both before and after activation.

REFERENCES

Bloom, K.S. and J.N. Anderson. 1979. Conformation of ovalbumin and globin genes in chromatin during differential gene expression. *J. Biol. Chem.* **254:** 10521.

Cartwright, I.L., M.A. Keene, G.C. Howard, S.M. Abmayr, G. Fleischmann, K. Lowenhaupt, and S.C.R. Elgin. 1982. Chromatin structure and gene activity: The role of nonhistone chromosomal proteins. *CRC Crit. Rev. Biochem.* (in press).

Elgin, S.C.R. 1981. DNase I hypersensitive sites of chromatin. *Cell* **27:** 413.

Findley, R.C. and T. Pederson. 1981. Regulated transcription of the genes for actin and heat-shock proteins in cultured *Drosophila* cells. *J. Cell Biol.* **88:** 323.

Garel, A., M. Zolan, and R. Axel. 1977. Genes transcribed at diverse rates have a similar conformation in chromatin. *Proc. Natl. Acad. Sci.* **74:** 4867.

Keene, M.A. and S.C.R. Elgin. 1981. Micrococcal nuclease as a probe of DNA sequence organization and chromatin structure. *Cell* **27:** 57.

Keene, M.A., V. Corces, K. Lowenhaupt and S.C.R. Elgin. 1981. DNase I hypersensitive sites in Drosophila chromatin occur at the 5′ ends of regions of transcription. *Proc. Natl. Acad. Sci.* **78:** 143.

McGhee, J.D. and G. Felsenfeld. 1980. Nucleosome structure. *Annu. Rev. Biochem.* **49:** 1115.

McGhee, J.D., W.I. Wood, M. Dolan, J.D. Engel, and G. Felsenfeld. 1981. A 200 base pair region at the 5′ end of the chicken adult β-globin gene is accessible to nuclease digestion. *Cell* **27:** 45.

Miller, D.W. and S.C.R. Elgin. 1981. Transcription of heat shock loci of *Drosophila* in a nuclear system. *Biochemistry* **20:** 5033.

Palmiter, R.D., E.R. Mulvihill, G.A. McKnight, and A.W. Senear. 1978. Regulation of gene expression in chick oviduct by steroid hormones. *Cold Spring Harbor Symp. Quant. Biol.* **42:** 639.

Weintraub, H. and M. Groudine. 1976. Transcriptionally active and inactive conformations of chromosomal subunits. *Science* **193:** 848.

Wu, C., Y-C. Wong, and S. C. R. Elgin. 1979. The chromatin structure of specific genes. II. Disruption of chromatin structure during gene activity. *Cell* **16:** 807.

Chromatin Structure of *Drosophila* Heat-shock Genes

Carl Wu
Laboratory of Biochemistry
National Cancer Institute
National Institutes of Health
Bethesda, Maryland 20205

The chromatin structure of the heat-shock genes in *Drosophila* has served as a paradigm for studies on how the organization of chromatin might be related to gene function in eukaryotic cells. Analysis of the genes encoding the 70-kD and 83-kD heat-shock proteins (hsp70 and hsp83) originally showed that the orderly compaction of DNA in nucleosomes along the chromatin fiber is punctuated by highly nuclease-sensitive sites located at specific positions (Wu et al. 1979b). Such hypersensitive sites in cellular chromatin are revealed by mild digestion with a nucleolytic enzyme, DNase I, followed by Southern visualization of the partially cleaved DNA fragments using cloned hybridization probes. We further developed this procedure to enable rapid mapping of the cleavage sites in chromatin relative to known restriction endonuclease sites on the DNA (Wu 1980), and showed that the hypersensitivity maps to the 5'-terminal and flanking sequences of the hsp70 and hsp83 genes. Such a surprising correlation with a gene region reported by others in this volume to be necessary for gene expression suggests that the preferentially accessible sites in chromatin could function as entry sites for RNA polymerase and control factors and could be an essential component in the series of regulatory events leading to the initiation of transcription.

HYPERSENSITIVE SITES IN CHROMATIN NEAR THE hsp70 GENES

Figure 1b shows the DNase-I cleavage pattern specific to hsp70 gene chromatin. Nuclei are isolated from *Drosophila* tissue culture cells (Schneider line 2) maintained at 25°C, and digested with increasing concentrations of DNase I until the average double-stranded length of the DNA is on the order of 5 kbp. The purified DNA fragments are electrophoresed on an agarose gel, stained (Fig. 1a), and blotted onto nitrocellulose. The Southern blot is then hybridized with a ^{32}P-labeled, cloned fragment representing approximately the 3′ half of the hsp70-coding sequence (Wu et al. 1979b). We observe a striking difference

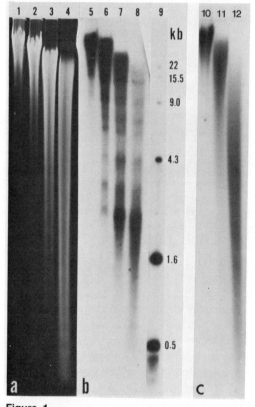

Figure 1

(a)Partial DNase-I digestion pattern of whole chromatin from Schneider cells; increasing cleavage from lanes 1−4. DNA fragments were electrophoresed on an agarose gel and stained with ethidium bromide. (b) Autoradiogram showing the partial DNase-I digestion pattern of hsp70 gene chromatin; increasing cleavage from lanes 5−8. The gel in Fig. 1a was blotted and hybridized with hsp70 structural sequences. (c) Autoradiogram showing the partial DNase-I digestion pattern of hsp70 naked DNA; increasing cleavage from lanes 10−12. DNA fragments were electrophoresed and blotted as in Fig. 1b.

between the general (Fig. 1a) and the specific cleavage pattern (Fig. 1b). The latter displays seven discrete bands ranging from about 2 kb to 14 kb in length; a similar pattern is observed using nuclei isolated from late *Drosophila* embryos. These bands signify that local sites of preferential cleavage must exist in the chromatin on both sides of the DNA sequence homologous with the hybridization probe. The banding pattern is not seen when either naked *Drosophila* DNA (Fig. 1c) or supercoiled plasmid DNA containing the hsp70 gene and flanking sequences (C. Wu, unpubl.) is subjected to a similar Southern analysis. Thus, the specificity of cleavage by DNase I must reflect the properties of the chromatin complex.

MAPPING THE DNase-I-HYPERSENSITIVE SITES

We mapped the positions of DNase-I-hypersensitive sites by a novel, indirect, end-labeling technique (Wu 1980). After mild treatment of chromatin in nuclei with DNase I, the purified DNA is cleaved completely with a rarely cutting restriction enzyme, electrophoresed, and blotted onto nitrocellulose in a typical Southern transfer. The blot is then probed with a short, cloned DNA segment that abuts a chosen restriction cut; the lengths of the labeled subfragments greater than the probe length define the distance between the restriction cut and the partial DNase-I cuts. Nedospasov and Georgiev (1980) have independently used this same procedure to map the location of micrococcal nuclease cuts on the SV40 minichromosome.

Figure 2a shows the partial DNase-I cuts in noninduced Schneider cell chromatin upstream from the internal *Bam* site common to all five hsp70 gene copies; the direction of electrophoresis is from left to right. The *Bam* digest alone, on purified DNA (Fig. 2a, lane 2) locates the different upstream *Bam* sites corresponding with the five hsp70 gene copies. (The fragments obtained in low yield are probably due to polymorphism in the organization of the sequences upstream from the hsp70 coding region.) Increasing extents of partial DNase-I cleavage on chromatin, followed by secondary *Bam* cleavage of purified DNA (Fig. 2a, lanes 3–5) reveal major preferred cuts at two adjacent sites which span about 150 bp and 90 bp, flanking and extending through the 5′ end of the hsp70-coding region. These two preferentially cleaved sites are better resolved on a 50-cm long, 1.4% agarose gel not shown. We have precisely mapped them to within 10 bp by coelectrophoresis on the same gel lane using fragments of known length determined from sequence data. The 5′ upstream flank of the gene is hypersensitive from −38 bp to −215 bp from the 5′ end; the 5′ downstream flank is hypersensitive from −8 to +100 bp. Interestingly, the 30-bp region in between, from −8 to −38 bp, is relatively insensitive. The TATAAAT

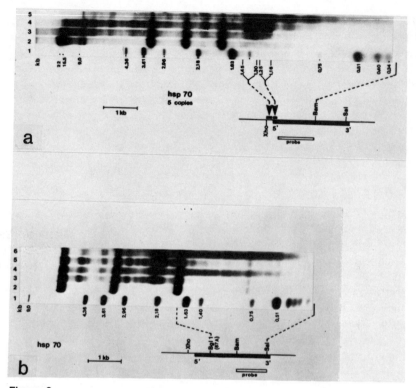

Figure 2

(a)Autoradiogram showing the partial DNase-I cuts in chromatin upstream from the internal
Bam site common to all hsp70 gene copies. The lanes indicate: 1, marker DNA fragments;
2, *Bam* digest on Schneider cell DNA; 3–5, increasing partial DNase-I digests on
chromatin, followed by *Bam* cleavage. The filled bars directly under the arrows indicate the
measured span of hypersensitivity. (b) Autoradiogram showing the state of hsp70 chromatin before and after heat-shock induction. *Sal* and *Bgl*I were used for secondary cleavage.
The lanes indicate: 1, marker DNA fragments; 2, *Sal* plus *Bgl*I digest on Schneider cell
DNA; 3,5, increasing partial DNase-I digests on uninduced Schneider cell chromatin; 4,6,
increasing partial DNase-I digests on chromatin in Schneider cells heat-shocked at 35°C
for 15 min.

sequence at −27 bp lies within this insensitive site. There is also a point
of greatest hypersensitivity at position −93.

The 5′-terminal hypersensitive cuts shown here are averaged over the
five hsp70 gene copies, but further restriction analysis (Wu 1980 and
unpubl.) suggests strongly that the averaged hypersensitivity is exactly
representative of each individual hsp70 gene. We have also shown that
neither the coding sequence nor the 3′ end of the hsp70 gene displays
DNase-I hypersensitivity. Further upstream and downstream from the
hsp70 gene sequence lie other hypersensitive sites, but these are only
specific to some of the five gene copies. Those found upstream from the

87C gene copies map in a region containing repeated sequences responsible for heat-induced, nontranslated RNA of unknown function. Taken together, the mapped positions of the hypersensitive sites near the hsp70 gene sequence can account for the appearance of the discrete bands in Figure 1b, produced by DNase-I cleavage alone on the chromatin.

CHANGES IN CHROMATIN STRUCTURE
UPON HEAT-SHOCK INDUCTION

The 5′-terminal hypersensitive chromatin structure is present in noninduced cells. Figure 2b shows the state of hsp70 chromatin before and after 15 minutes of heat induction at 35°C. Here the DNase-I cuts are mapped upstream from the 3′-terminal *Sal* site common to all five hsp70 gene copies. (The two gene copies at locus 87A have been reduced to a 1.8-kb *Sal* to *Bgl*I fragment; any larger fragments are thus due to those copies at locus 87C.) We find that essentially the whole of the hsp70 transcription unit becomes sensitive to DNase I upon activation, but the level of sensitivity is less than the 5′-terminal hypersensitivity, which is still retained partly (compare Fig. 2b, lanes 3 and 5 with lanes 4 and 6). From preliminary mapping studies, it appears that the distal portion of the hypersensitive region remains essentially unchanged upon induction, whereas the proximal portion becomes less hypersensitive, possibly because of protection by bound polymerase or other factors (C. Wu, in prep.). We have also previously demonstrated that the nucleosome organization of the hsp70 structural sequences is considerably altered upon heat induction (Wu et al. 1979a).

hsp83 GENE CHROMATIN

From a *Sal* restriction site within the single-copy hsp83 coding sequence, we mapped the partial DNase-I cleavages upstream and downstream in a region extending about 40 kbp (Fig. 3a,b). A cluster of preferred cuts lie about 2.3−3.2 kbp upstream from the internal *Sal* site, and two clusters are observed starting from about 1.8 kbp downstream. No preferred cuts lie near the internal *Sal* site. The 5′ terminus of a precursor to the hsp83 mRNA lies about 160 bp downstream from the *Eco*RI site indicated in Figure 3. Thus, a DNase-I-hypersensitive region flanks and extends into the sequences complementary to the 5′ end of this RNA. Another gene, recently discovered by O'Connor and Lis (1981) and by R. Morimoto and J. Jack (pers. comm.), lies just between the two clusters of hypersensitive sites on the downstream side of the

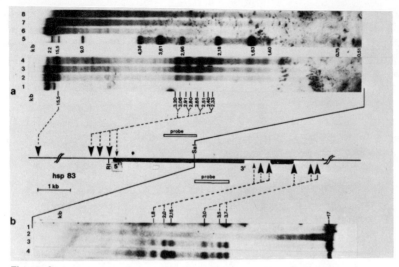

Figure 3

(a)Autoradiogram showing the partial DNase-I cuts in chromatin upstream from the *Sal* site within the hsp83 coding region. The lanes indicate: 1, *Sal*I digest on Schneider cell DNA (a minor polymorphism at the external *Sal* site is observed); 2−4, increasing partial DNase-I digests on chromatin, followed by secondary *Sal* cleavage; 5, marker DNA fragments; 6−8, increasing partial DNase-I digests on naked DNA. (b) Autoradiogram showing the partial DNase-I cuts in chromatin downstream from the *Sal* site in the hsp83 coding region. The lanes indicate: 1, *Sal* digest of Schneider cell DNA; 2, partial DNase-I digest of naked DNA; 3,4, increasing partial DNase-I digests on chromatin, followed by *Sal* cleavage.

hsp83 gene. Both genes are transcribed at a low level in cultured cells under normal conditions. This would account for the mild sensitivity to DNase I shown by sequences homologous to RNA.

CONCLUSION

The studies on hsp70 and hsp83 gene chromatin establish a new, probably nonnucleosomal chromatin structure, often located at the 5′ terminus of eukaryotic genes. Similar studies not shown also demonstrate DNase-I hypersensitivity at the 5′ end of the hsp68 gene. The preferentially accessible structures could function as entry sites for RNA polymerase and control macromolecules. Other hypersensitive sites might signify the 5′ ends of unknown or unmapped genes, or they might be involved with different chromosomal functions.

What could be the basis of this unique sensitivity? A special DNA sequence at these regions could bind other macromolecules such as specific proteins that cause the sensitivity, or the sequence may possess a peculiar base order that renders it sensitive when the DNA in the region is coiled and supercoiled in nucleosomes and higher-order structures. We need to explore further the fine structure of the hypersensitive site in chromatin and determine its relevance for gene function. The accumulating wealth of information and ease in manipulating the heat-shock system continues to make it highly attractive for studying the problem of gene regulation in general.

ACKNOWLEDGMENTS

I thank S. Elgin for supporting the early chromatin work, and P. Bingham, R. Holmgren, K. Livak, R. Morimoto, and M. Meselson for generous gifts of plasmid DNA and for helpful discussions. I am especially indebted to W. Gilbert for providing much stimulus and for supporting the mapping experiments which were carried out in his laboratory during my term in the Harvard Society of Fellows.

REFERENCES

Nedospasov, S.A. and G.P. Georgiev. 1980. Nonrandom cleavage of SV40 DNA in the compact mini chromosome and free in solution by micrococcal nuclease. *Biochem. Biophys. Res. Commun.* **92:** 532.

O'Connor, D. and J.T. Lis. 1981. Two closely linked transcription units within 63B heat shock puff locus of *D. melanogaster* display strikingly different regulation. *Nucleic Acids Res.* **9:** 5075.

Wu, C. 1980. The 5′ ends of *Drosophila* heat shock genes in chromatin. *Nature* **286:** 854.

Wu, C., Y.-C. Wong, and S.C.R. Elgin. 1979a. The chromatin structure of specific genes II. *Cell* **16:** 807.

Wu, C., P.M. Bingham, K.J. Livak, R. Holmgren, and S.C.R. Elgin. 1979b. The chromatin structure of specific genes I. *Cell* **16:** 797.

Chromatin Structure of hsp70 Genes of *Drosophila*

Abraham Levy and Markus Noll
Department of Cell Biology
Biocenter of the University
CH-4056 Basel, Switzerland

An important unsolved question is whether chromatin structure affects the regulation of gene expression. Conceivably, controls exist at the nucleosomal level or at any higher level of structural organization. In an attempt to answer this question, we have started to analyze the structure of transcribed and repressed genes and thus to characterize the structural changes involved in gene activation or repression. In addition, the limits of the structures specific for the active or repressed state of a gene have been determined by comparing the structure of the gene with that of the flanking region. We suggest that these borders of structural transition between the gene and its 5'- and 3'-flanking regions may play a role in gene regulation.

To detect structures characteristic for the transcribed as well as the repressed gene, it may be crucial to select a system in which the repressed gene may be induced to a state of high transcriptional activity. For genes that are only moderately active, transcription-specific structures could remain undetected among the bulk of inactive structures. The heat-shock genes of *Drosophila melanogaster* are turned on to a high degree of activity upon raising the temperature from 25°C to 35–37°C (Lindquist 1980) and hence provide an excellent system for such studies. In particular, we have analyzed the structure of the gene coding for the major protein synthesized at 37°C—hsp70. Six copies of this gene exist in the haploid genome of Kc cells (Mirault et al. 1979).

EXPERIMENTAL APPROACH

Nuclei of a *Drosophila* tissue culture cell line were digested with micro-coccal nuclease before or after heat shock. The DNA fragments obtained were separated according to size by agarose gel electrophoresis, denatured, transferred, and bound covalently to DBM paper (Alwine et al. 1977; Levy et al. 1980). By hybridization with radioactively labeled probes, the DNA sequences of interest were analyzed selectively. Three different probes have been used (Fig. 1) — the *Sal* to *Sal* fragment of 56H8 containing almost the entire coding region yet no flanking regioñ, the *Sal* to *Bgl*II fragment of 56H8 adjacent to the 3' end of the coding region, and the *Bgl*II to *Bgl*II fragment upstream from the coding region in the genomic clone 132E3. The last probe contains the moderately repeated β-sequence element found close to hsp70 genes and in the chromocenter (Lis et al. 1978). It was found to exhibit the same distribution of the nucleosomal repeat pattern (Fig. 2c) as bulk chromatin (revealed by staining the gel with ethidium bromide, Fig. 2a) and hence is designated "bulk." The patterns revealed by these different probes

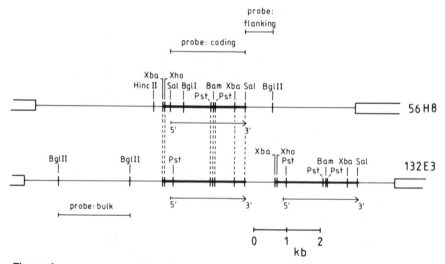

Figure 1
Restriction maps and probes of hsp70 genes. Restriction maps of the *D. melanogaster* sequences in the hybrid plasmids 56H8 and 132E3 have been published (Moran et al. 1979). Sequences complementary to the hsp70 mRNA and the direction of transcription of these genes are indicated by arrows. The protected domain of the repressed hsp70 genes is represented by a thick bar comprising the coding region and most of the Znc region (Artavanis-Tsakonas et al. 1979). The probes for the coding (*Sal* to *Sal*) and the flanking region (*Sal* to *Bgl*II), and the moderately repetitive "bulk" probe (*Bgl*II to *Bgl*II) are shown above and below the corresponding map positions. (Reprinted, with permission, from Levy and Noll 1981.)

have been obtained by hybridizations with the same DBM paper after release of the previous probe by denaturation.

GENE-SPECIFIC STRUCTURES AND ABSENCE OF NUCLEOSOMES IN THE ACTIVE STATE

Hybridization with the probe for the coding region clearly demonstrates that active hsp70 genes are much more sensitive to micrococcal nuclease than repressed hsp70 genes (Fig. 2b), which confirms an earlier report (Wu et al. 1979). Comparing the initial rates of digestion of the active and repressed genes, we have found that the transcribed gene is degraded 30 times more rapidly than the inactive gene (Levy and Noll 1981). However, the most striking observation is that the fragment sizes resulting from digestion of the active genes are not distributed randomly, yet differ from those obtained after micrococcal nuclease digestion of bulk chromatin or of inactive genes. The DNA exhibits some preferential spacings of cleavage sites superimposed on what appears to be a continuous distribution of sizes. The preferential spacings are reflected in bands (hs lanes of B and C, Fig. 2b) at positions between those of the familiar nucleosomal repeat. These bands do not originate from the presence of preferential cleavage sites in the free DNA because digestion of free DNA of a mixture of 56H8 and 132E3 does not produce such bands after hybridization to the same *Sal* to *Sal* probe.

If the digestions after heat shock are relatively extensive (hs lane of D, Fig. 2b), most of the hybridized DNA disappears and only faint bands at the monomer and dimer positions are visible. Most of the DNA has been shown to be degraded to fragments that fail to hybridize under the conditions used that permit detection of hybrids as short as 40 bp (Levy and Noll 1981). Thus, the bulk of the active hsp70 genes is free of nucleosome core particles.

A comparison of the hybridization pattern of the 3′-flanking region (Fig. 2d) with that of the coding region (Fig. 2b) shows that the enhanced sensitivity does not extend much beyond the 3′ end of at least one of the six hsp70 genes. The pattern obtained by hybridization with the 3′-flanking-region probe (Fig. 2d) is very similar to that of the bulk (Fig. 2a,c). Thus, for a first approximation, the region downstream from the right *Sal* site (Fig. 1) is organized similarly to bulk chromatin in heat-shock conditions. However, the patterns revealed by these two probes do differ slightly. Particularly, the heat-shock lanes of B and C in Figure 2d show more hybridization in the regions between monomer and dimer and between dimer and trimer DNA than the same lanes representing bulk chromatin (Fig. 2a,c). This suggests that the *Sal* to *Bgl*II probe overlapped slightly with the chromatin region containing the transcrip-

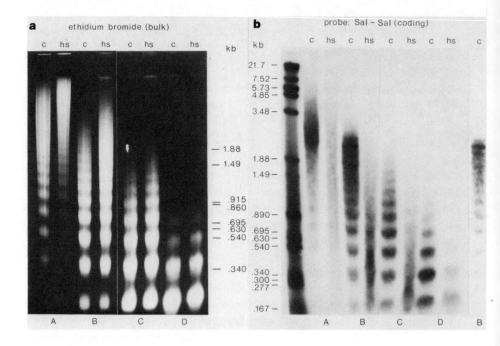

a ethidium bromide (bulk)

b probe: Sal – Sal (coding)

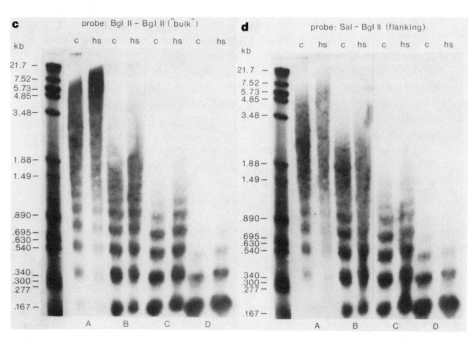

c probe: Bgl II – Bgl II ("bulk")

d probe: Sal – Bgl II (flanking)

tion-specific structures (Fig. 2b). We infer that the sensitive portion of the gene ends close to the right-hand *Sal* site. As the 3' end of the mRNA coding region also maps close to this site (Artavanis-Tsakonas et al. 1979), we consider it likely that the right boundary of the nuclease-sensitive chromatin region is near the termination site of transcription.

PROTECTED DOMAIN OF REPRESSED GENE

It is generally believed that inert genes and bulk chromatin are structurally similar. A comparison of the micrococcal nuclease pattern of the coding region (Fig. 2b) with that of bulk chromatin (Fig. 2a,c) under nonheat-shock conditions is not consistent with such an assumption. The higher ratios of multimers to monomer in the lanes of the repressed gene (Fig. 2b) compared with the corresponding lanes representing bulk chromatin (control lanes, Fig. 2a,c) clearly indicate that at least part of the coding region of the repressed gene is more resistant to micrococcal nuclease than bulk chromatin.

After mild digestions, the DNA of the repressed gene hybridizing with the coding region exhibits a relatively narrow size distribution with an average of about 2.5 kb (control lane of A, Fig. 2b). In a more extensive digestion, three sharp bands at 2.52 kb, 2.34 kb, and 2.16 ± 0.05 kb appear above a background (control lane of B, Fig. 2b). These bands are more clearly visible when the background is reduced by a shorter exposure (lane B at far right, Fig. 2b). Therefore, a region of 2.5 kb, larger than the mRNA-coding region (Artavanis-Tsakonas et al. 1979) but which must contain at least part of the coding region, is more resistant to micrococcal nuclease digestion than its flanking regions when the gene is not expressed.

To determine the right boundary of this protected domain, the *Sal* to *Bgl*II probe specific for the 3'-flanking region of one of the genes (Fig. 1) was used. As evident from Figure 2d, the patterns resemble those of the

Figure 2
Chromatin structure of hsp70 genes: Comparison with structure of flanking region and bulk chromatin. Digests of control (c) and heat-shocked nuclei (hs) were compared pairwise in bulk chromatin by staining the DNA with ethidium bromide (a), in the coding region of the hsp70 gene (b), in a region containing a middle-repetitive sequence (c), and in the 3'-flanking region of the hsp70 gene (d) by hybridization of the DNA transferred from a 1.7% agarose gel to DBM paper with the corresponding labeled probe and subsequent autoradiography. Four levels of digestion are shown in each panel corresponding to about 2.3% (A), 7% (B), 12% (C), and 18% (D) acid-soluble DNA. For calibration, the far-left lanes of each autoradiogram show *Hae*III fragments of PM2 DNA and *Eco*RI fragments of λ DNA labeled at their 3' ends by Klenow DNA polymerase I. The lane on the far right in (b) represents a shorter exposure of the fourth lane from the left in (b). (Reprinted, with permission, from Levy and Noll 1981.)

bulk DNA (Fig. 2a,c) more closely than those of the coding region (Fig. 2b). The three bands between 2 kb and 2.5 kb are barely visible (control lane of B, Fig. 2d), and digestion is slightly inhibited in the repressed gene (ratio of monomer to oligomer DNA in lanes of control, Fig. 2d) compared with the bulk (Fig. 2c). This suggests that the protected domain and the *Sal* to *Bgl*II region overlap only slightly. In other words, the protected region of the repressed gene ends at a site close to the 3′ end of the mRNA-coding region (Fig. 1).

From the size of the protected domain of about 2.5 kb, we predict that its left end reaches beyond the 5′ end of the mRNA coding region (Fig. 1) and maps close to the left boundary of the Znc region (Artavanis-Tsakonas et al. 1979). A more precise localization of the 5′ limit by direct mapping is complicated by the presence of repetitive sequences in this region.

No bands have been observed between 2 kb and 2.5 kb in heat-shock conditions, even when digestion is very mild, so that the DNA containing the coding region exhibits about the same average size as the protected domain of the repressed gene (Levy and Noll 1981).

GENE REGULATION — A MECHANISM
ACTING ON CHROMATIN STRUCTURE?

Altered sensitivity to nuclease of actively transcribed genes may result from changes at various structural levels: (1) changes in the higher structural orders, (2) changes in the structure of the nucleosomes, and (3) the absence of histones from transcribed DNA. Therefore, earlier suggestions made solely on the basis of DNase sensitivity that active genes contain structurally modified nucleosomes are invalid. More detailed information with respect to the structure of active genes may be gained by analyzing the size distribution of the DNA fragments produced by the action of DNase, and this study has revealed transcription-specific structures not previously detected.

The observed changes in chromatin structure depend completely on the level of transcription. After returning heat-shocked cells to the normal temperature (25°C), there is a gradual disappearance of the transcription-specific structures. In some experiments, after micrococcal nuclease cleavage in the mRNA coding region of the active gene, we observed initially a smeared hybridization pattern, which then gave way to a pattern reflecting the nucleosomal repeat. We could clearly correlate this smearing effect with a suboptimal heat-shock response. In even more poorly responding cultures, most of the DNA fragments of the "active" genes obtained by micrococcal nuclease digestion comigrated with the bulk fragments. It thus seems crucial that transcription-specific structures be observed in heavily transcribed genes.

Micrococcal nuclease was found to discriminate between a gene, regardless of its state, and the rest of the chromatin. The repressed genes were protected and the active genes were highly sensitive compared with bulk chromatin or the neighboring nontranscribed region. The levels of sensitivity of the gene are reversible and depend on its transcriptional activity. Hence, models of gene repression need not be based solely on the interaction of regulatory proteins with the 5' end. Similarly, the association of bulk-type nucleosomes with the gene appears, in itself, insufficient to keep the gene in the inactive form. Some mechanism seems to exist which, on repression, alters the chromatin structure of the gene so as to reduce its accessibility (and that of a region of about one nucleosome at its 5' end) to the nuclease considerably below that of the flanking DNA sequence.

The mechanism of protection of the repressed genes could be explained on at least two structural levels. Each linker region joining

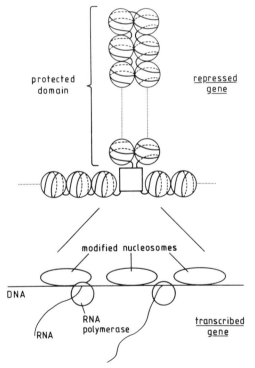

Figure 3

Model of gene regulation. Activation of a repressed gene requires first an unfolding of the protected domain, e.g., by the removal of a protein that ties the two ends of the domain together. Only then is initiation of transcription modifying or even removing the nucleosomes from transcribed sequences possible.

adjacent nucleosomes could be modified in such a way as to reduce its sensitivity to micrococcal nuclease. The possibility that the flanking regions are more susceptible to the nuclease than the gene region because of its higher proportion of AT versus GC base pairs cannot be strictly ruled out. However, it seems unlikely for two reasons. First, brief digestions of free DNA of 56H8 and 132E3 never produce a narrow distribution of DNA sizes (as in the control lane of A, Fig. 2b) after hybridization to the *Sal* to *Sal* probe (not shown). Second, the susceptibility of the flanking region (Fig. 2d) whose AT content is much above the genomic average is the same as that of the bulk DNA (Fig. 2a,c). In a more attractive model (Fig. 3), all the linker regions of the entire domain become collectively resistant to the nuclease by some change in the higher order of chromatin structure. The transition points between the protected domain of the repressed state and the flanking, bulklike structure might then be sequences of regulatory significance. Specific nonhistone protein(s) interacting with these sequences may associate with each other to bring the two ends of this chromatin segment into tight proximity, producing a loop containing 14 nucleosomes. Such a loop may form a supercoiled, more compact, and less accessible form of chromatin. In this speculative model, activation of a gene would require at least two steps. First, the protected domain is unfolded, and only then does initiation of transcription start by the modification or removal of the nucleosomes (Fig. 3).

REFERENCES

Alwine, J.C., D.J. Kemp, and G.R. Stark. 1977. Method for detection of specific RNAs in agarose gels by transfer to diazobenzyloxymethyl-paper and hybridization with DNA probes. *Proc. Natl. Acad. Sci.* **74:** 5350.

Artavanis-Tsakonas, S., P. Schedl, M.-E. Mirault, L. Moran, and J. Lis. 1979. Genes for the 70,000 dalton heat shock protein in two cloned *D. melanogaster* DNA segments. *Cell* **17:** 9–18.

Levy, A. and M. Noll. 1981. Chromatin fine structure of active and repressed genes. *Nature* **289:** 198.

Levy, A., E. Frei, and M. Noll. 1980. Efficient transfer of highly resolved small DNA fragments from polyacrylamide gels to DBM paper. *Gene* **11:** 283.

Lindquist, S. 1980. Varying patterns of protein synthesis in *Drosophila* during heat shock: Implications for regulation. *Dev. Biol.* **77:** 463–479.

Lis, J.T., L. Prestidge, and D.S. Hogness. 1978. A novel arrangement of tandemly repeated genes at a major heat shock site in *D. melanogaster. Cell* **14:** 901–919.

Mirault, M.-E., M. Goldschmidt-Clermont, S. Artavanis-Tsakonas, and P. Schedl. 1979. Organization of the multiple genes for the 70,000-dalton heat-shock protein in *Drosophila melanogaster. Proc. Natl. Acad. Sci.* **76:** 5254–5258.

Moran, L., M.-E. Mirault, A. Tissières, P. Schedl, S. Artavanis-Tsakonas, and W.J. Gehring. 1979. Physical map of two *D. melanogaster* DNA segments containing sequences coding for the 70,000 dalton heat-shock protein. *Cell* **17:** 1–8.

Wu, C., Y.-C. Wong, and S.C.R. Elgin. 1979. The chromatin structure of specific genes. II. Disruption of chromatin structure during gene activity. *Cell* **16:** 807.

Chromatin Structure and Dosage-compensated Heat-shock Genes in *Drosophila pseudoobscura*

Joel C. Eissenberg and
John C. Lucchesi
Genetics Curriculum and
Department of Zoology
University of North Carolina
Chapel Hill, North Carolina 27514

In *Drosophila,* the mechanism used to equalize X-linked gene expression in males and females is dosage compensation (for review, see Stewart and Merriam 1980). Cytological observations suggest a difference in chromatin organization between the male and female X chromosomes, at least in certain tissues. We have tested whether this difference is reflected in the sensitivity of an X-linked gene to DNase-I attack in each sex. In addition, this experiment addresses the more general question of a correlation between levels of transcription and DNase-I sensitivity for a given gene.

EXPERIMENTAL STRATEGY

In *Drosophila pseudoobscura,* four large puffs can be induced by heat shock in the larval salivary gland chromosomes, two on the right arm of the X chromosome and two on chromosome 2. The autosomal puffs contain at least one copy each of hsp70, and one of the X-linked puffs is the site of activity of the hsp83 gene. RNA synthesis at this puff is equivalent in males (with a single X) and in females (with two active X chromosomes) indicating dosage compensation (Pierce and Lucchesi 1980). Furthermore, although its activity is probably enhanced by heat shock, hsp83 seems to be transcribed in nonheat-shocked larvae.

109

Kinetics of Digestion

Nuclei from nonheat-shocked or heat-shocked male and female third instar larvae were prepared and digested to various levels with DNase I. The DNA was then purified, cut with the restriction endonuclease *Eco*RI, and sized on an agarose gel. DNA from this gel was transferred to a nitrocellulose filter and probed simultaneously with labeled cloned sequences homologous to the X-linked and autosomal heat-shock genes (pPW 244.1 and pPW 232.1, respectively). Autoradiography of a filter so probed gave rise to an X-linked fragment of 2.9 kb (Fig. 1A) and five autosomal fragments of 17 kb, 6.1 kb, 5.2 kb, 4.9 kb, and 2.6 kb (Fig. 1, B and C). The 2.9-kb X-linked fragment and the 2.6-kb autosomal fragment—as an internal control—were the subjects of the kinetic analyses.

Computer-assisted densitometric scans were made of these bands, the resulting peaks were integrated, and X/A ratio of autoradiographic densities was obtained at each level of digestion for both sexes. Levels of digestion (compared with an undigested control) were measured by determining weight-average, single-strand sizes for each sample and computing DNase-I-induced, single-strand nicks per thousand bases (Zasloff and Camerini-Otero 1980). The results of six such experiments on heat-shocked nuclei are summarized in Figure 2A. While it is clear that the X-linked fragment is being digested more rapidly than the autosomal fragment, the kinetics of digestion appear to be the same for both sexes. In contrast to the heat-shocked material, the X/A ratio in nonheat-shocked nuclei does not change dramatically with increasing levels of digestion. The results of four such experiments are summarized in Figure 2B. Again, the digestion kinetics for the X-linked fragment were the same for both sexes. Control experiments on naked DNA digested with DNase I gave similar kinetics to nonheat-shock chromatin (data not shown).

The rate of digestion of the X-linked gene hsp83 is significantly different when this gene is transcribing in nonheat-shocked and in heat-shocked nuclei. This is particularly evident in light of the fact that under nonheat-shock conditions the digestion of hsp83 is normalized to that of an apparently inactive autosomal gene, hsp70; digestion of hsp83 in heat-shocked nuclei is compared with that of the active hsp70, which is itself rapidly digested. In spite of this fact, the X/A ratio decreases much more rapidly under heat shock.

Hypersensitive Sites

A constant structural feature of the chromatin of all heat-shocked genes studied is the presence of one or more hypersensitive sites in the proximity of these genes. We observed such a site in association with

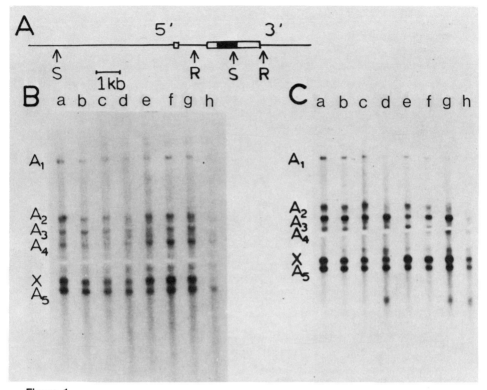

Figure 1
(*A*) Abbreviated restriction map of the hsp83 locus in *D. pseudoobscura*. Boxed regions represent sequences found in processed hsp83 mRNA, with transcriptional polarity indicated above. Gap between boxes indicates the position of the single intervening sequence associated with this locus. Filled area indicates homology of pPW 244.1 probe used in this analysis. S = *Sal*I; R = *Eco*RI. Map information was kindly provided by Mr. R. Blackman, Harvard University. (*B*) Hybridization of ³²P-labeled pPW 232.1 and pPW 244.1 to genomic *D. pseudoobscura* DNA, derived from DNase-I-digested nuclei of heat-shocked males (lanes a–d) and females (lanes e–h). All samples digested with *Eco*RI. (a,e) Control; (b,f) 44 U/ml DNase I; (c,g) 88 U/ml DNase I; (d,h) 220 U/ml DNase I. Fragment sizes are: A₁, 17 kb; A₂, 6.1 kb; A₃, 5.2 kb; A₄, 4.9 kb; A₅, 2.6 kb; X, 2.9 kb. (*C*) Hybridization same as in *B*, except that nuclei were derived from nonheat-shocked larvae. (Lanes a–d) males; (lanes e–h) females; (a,e) control; (b,f) 11 U/ml; (c,g) 22 U/ml; (d,h) 44 U/ml. Recombinant plasmids were made available to us through the kindness of Drs. K.F. Livak, R. Holmgren, and M. Meselson.

the autosomal hsp70 genes during the kinetic analysis of nonheat-shocked chromatin (Fig. 1C). As expected, this site was not observed when naked DNA was digested (data not shown). We wished to test whether the number or position of hypersensitive sites are correlated with the rate of transcription of a gene. To this end, we asked if the hypersensitive sites associated with the X-linked gene differ in males and females, given the differential transcriptional activity of this gene in the two sexes following heat shock. Nuclei were prepared from nonheat-

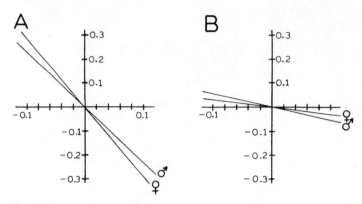

Figure 2
(A) Nuclease digestion kinetics of heat-shock chromatin. For each experiment, mean values for X (DNase-induced nicks/1000 bases) and Y (log [X-fragment density/A_5 fragment density]) were determined. $X - \bar{X}$ and $Y - \bar{Y}$ are calculated and a regression was made on pooled points for males and females. The apparent difference in slopes is not significant ($p > 0.1$). (B) Nuclease digestion kinetics of nonheat-shock chromatin. Data treatment same as A. The apparent difference in slopes is not significant ($p > 0.2$).

shocked male or female larvae and digested with DNase I. Purified DNA was cut with the restriction enzyme *Pst*I, and the resulting fragments were sized on an agarose gel which was probed with the X-specific sequence of clone pPW 244.1. In addition to a single major fragment, the probe revealed four minor bands (Fig. 3) whose pattern was identical in males and females.

CONCLUSIONS

Based on the preliminary detection of hsp83 RNA in nonheat-shocked larvae, our results suggest that the level of sensitivity of this gene to DNase-I digestion is greatly enhanced when transcription changes from the nonheat-shock level to the induced state characteristic of heat shock. A caveat would be that the nonheat-shock RNA is produced in a restricted subset of nuclei whose pattern of digestion does not influence the overall kinetic analysis. It is also possible that this RNA was produced by earlier activity of genes which are now quiescent.

We were unable to demonstrate any difference in sensitivity to DNase I between an active X-linked gene in females and the same gene in males, where its level of activity is twice as great. Furthermore, no difference was found in the number or position of DNase-I hypersensitive sites associated with the dosage-compensated locus in males and females. This suggests that the mechanism of dosage compensation operates at a level that cannot be monitored by these assays.

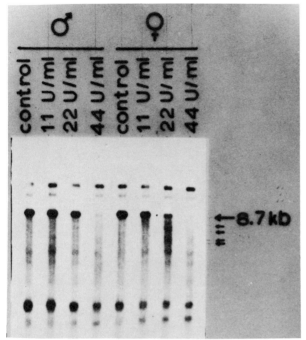

Figure 3
Hypersensitive sites associated with the hsp83 gene in *D. pseudoobscura.* Samples of DNA from the same experiment shown in Fig. 1C were digested with *Pst*I, sized, blotted, and probed with pPW 244.1. A single *Pst*I fragment with homology to the probe appears at 8.7 kb in all samples. Additional bands (small arrows) appear with increasing DNase-I digestion. Other bands are due to spurious homologies between the plasmid probe and λ DNA fragments (λ DNA was included in all digests to monitor completeness of the reaction).

ACKNOWLEDGMENTS

This research was supported by research grant GM-15691 and training grant T32-GM-07121 from the National Institutes of Health.

REFERENCES

Pierce, D.A. and J.C. Lucchesi. 1980. Dosage compensation of X-linked heat shock puffs in *Drosophila pseudoobscura. Chromosoma* **76:** 245–254.

Stewart, B. and J. Merriam. 1980. Dosage compensation. In *The genetics and biology of Drosophila* (ed. M. Ashburner and T.R.F. Wright), vol. 2d, pp. 107–140. Academic Press, New York.

Zasloff, M. and R.D. Camerini-Otero. 1980. Limited DNase I nicking as a probe of gene conformation. *Proc. Natl. Acad. Sci.* **77:** 1907–1911.

Selective Arrangement of Variant Nucleosomes within the *Drosophila melanogaster* Genome and the Heat-shock Response

**Louis Levinger and
Alexander Varshavsky**
*Department of Biology
Massachusetts Institute of Technology
Cambridge, Massachusetts 02139*

Most of the nuclear DNA is organized into nucleosomal and higher-order structures. We sought to determine whether variant nucleosomes are selectively arranged within the *Drosophila melanogaster* genome, depending upon the functional state of the DNA.

We have found that nucleosomes of heat-shock genes in nonshocked cells and of *copia* genes (examples of potentially transcribed or transcribed genes) are more heavily ubiquitinated than the total nucleosome population (Levinger and Varshavsky 1982). In striking contrast, nucleosomes of nontranscribed AT-rich satellite DNA (1.688- and 1.672-density satellites) lack ubiquitin-H2A (uH2A) semihistone but contain close to a stoichiometric amount of a ~50 kD nonhistone protein, D1. Ubiquitination within the transcriptional unit and absence of uH2A from condensed heterochromatin suggest that enzymatic attachment and removal of ubiquitin may be one of the principal effectors of higher-order structural transitions in chromatin. After a brief heat shock, the hsp70 transcriptional units lose most of their nucleosomal organization as probed by staphylococcal nuclease (in agreement with previous results of Wu et al. 1979), while *copia* regions remain nucleosomal and ubiquitinated.

The proteins induced by heat shock of *D. melanogaster* appear in the nucleus (Arrigo et al. 1980; Velasquez et al. 1980). Nuclear heat-shock proteins remain associated with a nuclease- and high-salt-insoluble nuclear fraction (Levinger and Varshavsky 1981), suggesting that they

115

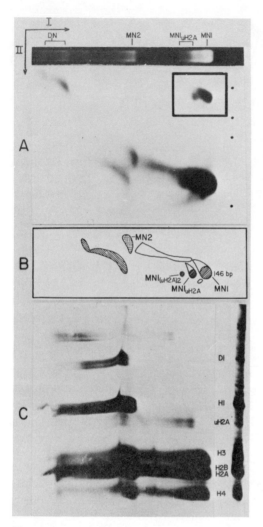

Figure 1

Correlation analysis of *D. melanogaster* nucleosomal DNA length and protein composition. (*A*) *Top strip:* First-dimension nucleoprotein gel. MN1, core mononucleosomes containing 146-bp DNA and two each of the four core histones; MN1$_{uH2A}$, core mononucleosomes containing one or two molecules of ubiquitin-H2A semihistone, a bifurcated protein with C-terminal glycine of ubiquitin, a 76-amino acid protein, covalently linked through the ϵ-amino group of internal lysine 119 of histone H2A (Goldknopf and Busch 1978); MN2, mononucleosomes containing a portion of the linker DNA and a molecule of histone H1 in addition to the components of the core particle; DN, dinucleosomes. *Main panel:* Second-dimension polyacrylamide gel analysis of nucleosomal DNA labeled in vivo with [³H]thymidine and displayed by fluorography. (*B*) Schematic diagram interpreting the DNA and protein spots in (*A*) and (*C*). (*C*) Second-dimension acetic acid-urea gel analysis of nucleosomal proteins labeled in vivo with [³H]lysine and displayed by fluorography. (Reprinted, with permission, from Levinger and Varshavsky 1982.)

protect sensitive genetic elements from damage by heat or metabolic shock through a modification of nuclear structure.

IDENTIFICATION OF NUCLEOSOME VARIANTS FROM *D. MELANOGASTER*

Soluble deoxyribonucleoprotein (DNP) particles are prepared by staphylococcal nuclease digestion of *D. melanogaster* nuclei (isolated from cultured cells) and fractionated on low-ionic-strength polyacrylamide gels (Fig. 1A, top strip). DNA length and protein composition of the nucleoprotein particles are determined by correlation analysis of second-dimension DNA (Fig. 1A) and protein (Fig. 1C) patterns. Several identified spots and regions are highlighted in the schematic diagram (Fig. 1B).

TWO-DIMENSIONAL HYBRIDIZATION MAPPING OF NUCLEOSOMES

The identification of variant nucleosomes by correlation analysis is given functional importance by the results of hybridization analysis (Fig. 2). The patterns of total nucleosomal DNA from nonshocked and heat-shocked cells, respectively, are shown in Figure 2, A and D. Hybridization using hsp70 DNA clones (pPW 229.1 and 232.1, a gift of M. Meselson) in Figure 2B and *copia* (cDM 5002, a gift of G. Rubin; not shown), against mononucleosomal DNA from nonshocked cells, displays a relative intensity in $MN1_{uH2A}$ DNA much greater than that in the bulk pattern (Fig. 2A). In striking contrast, nucleosomes of the nontranscribed, AT-rich (1.688-density DNA) complex satellite (pDM 23, a gift of D. Brutlag) virtually lack uH2A but contain close to a stoichiometric amount of a 50-kD protein, D1 (Fig. 2C). Similar results are obtained with a simple sequence AT-satellite (1.672-density DNA; data not shown).

The bulk pattern of nucleosomal DNA from heat-shocked cells (Fig. 2D) looks similar to that from nonshocked cells (Fig. 2A). However, hybridization with hsp70-specific DNA probe shows that hsp70 genes in heat-shocked cells are digested by staphylococcal nuclease at a rate greatly exceeding that for bulk chromatin (Fig. 2E). Moreover, at intermediate levels of digestion, no nucleosomal organization is detectable in these hsp70 genes with staphylococcal nuclease probe (in agreement with previous results of Wu et al. 1979; data not shown). In contrast, *copia* genes during heat shock remain nucleosomal and more heavily ubiquitinated than bulk chromatin (Fig. 2F).

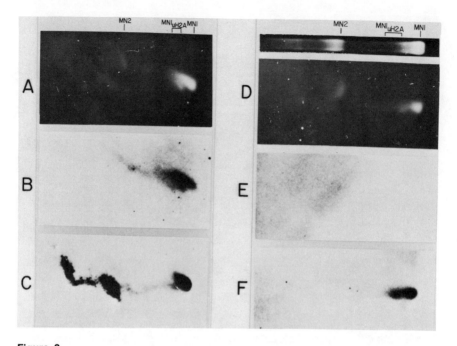

Figure 2
Two-dimensional hybridization analysis of nucleosomal DNA from transcribed and nontranscribed sequences. (*A*) Second-dimension nucleosomal DNA pattern from nuclei of nonshocked cells, displayed by ethidium bromide staining. Designations are the same as in Fig. 1. (*B*) DNA in (*A*) was transferred to DBM paper and hybridized with hsp70-specific DNA probe. (*C*) The same DBM paper was rehybridized with 1.688-density satellite DNA probe. (*D*) A first-dimension strip and second-dimension ethidium bromide-stained DNA pattern of nucleosomes from heat-shocked cells. (*E*) The hybridization pattern obtained from the gel in (*D*) with hsp70-specific DNA probes. (*F*) The same DBM paper was rehybridized with *copia* probe. (Reprinted, with permission, from Levinger and Varshavsky 1982.)

ASSOCIATION OF NUCLEAR HEAT-SHOCK PROTEINS WITH NUCLEASE- AND HIGH-SALT-INSOLUBLE NUCLEAR STRUCTURES

Shocked and nonshocked cells were lysed and resolved into several subfractions. The patterns of proteins labeled with [^{35}S]methionine in vivo were displayed by gel electrophoresis and fluorography (Fig. 3). The heat-shock proteins and histones are prominent in the whole cells (Fig. 3A, lane 1), cytoplasm (Fig. 3A, 2), and nuclei (Fig. 3A, 3). On digestion with staphylococcal nuclease (Fig. 3A, 4) and preparation of soluble chromatin at low ionic strength (Fig. 3A, 5), some histones but very little

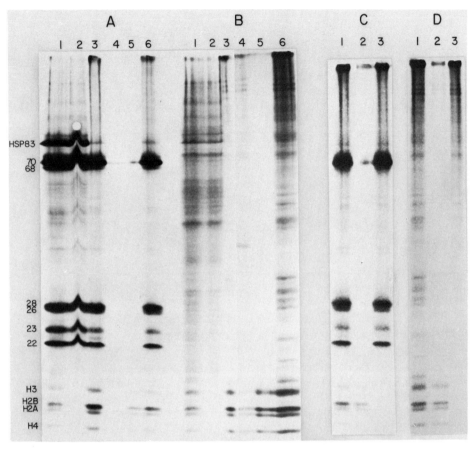

Figure 3
Association of heat-shock proteins with nuclease- and high-salt-insoluble nuclear structures. Designations: hsp83, 70, 68, 28, 26, 23, and 22 refer to nominal molecular weights
of the heat-shock proteins in kilodaltons; H3, H2B, H2A, and H4 are the four core histones.
The images were obtained by [^{35}S]methionine labeling in vivo and gel fluorography. (*A*)
Heat-shocked cells. Lane 1, whole cells; 2, cytoplasm; 3, nuclei; 4, nuclease first supernatant; 5, nuclease second supernatant; 6, nuclease, low-ionic-strength pellet. (*B*) Nonshocked cells. Lanes 1–6 are from the same fractionation scheme as in *A*. (*C*) 2 M NaCl
fractionation of nuclear proteins from shocked cells. Lane 1, the low-ionic-strength-insoluble fraction (same sample as A, lane 6); 2, 2 M NaCl-soluble proteins of C, lane 1; 3, 2 M
NaCl-insoluble proteins of C, lane 1. (*D*) Same as (*C*) but cells were nonshocked.
(Reprinted, with permission, from Levinger and Varshavsky 1981.)

of the heat-shock proteins are released from the nuclei, the bulk of heat-
shock proteins remaining with the low-ionic-strength-insoluble fraction
(Fig. 3A, 6). The only exception is hsp83 protein, which appears to be
almost entirely cytoplasmic (Arrigo et al. 1980). Figure 3B shows the

parallel protein patterns from nonshocked cells. The samples in lane 6, Figure 3, A and B, were further fractionated with 2 M NaCl (Fig. 3C and D). The rest of the histones are released into the 2 M NaCl-soluble fraction (Fig. 3C, 2), while the heat-shock proteins remain insoluble in 2 M NaCl (Fig. 3C, 3). Corresponding images obtained with nonshocked cells are shown in Figure 3D, lanes 1–3 (see Levinger and Varshavsky [1981] for additional details). Thus, relatively well-defined heat-shock proteins can now be used as a probe to study poorly understood relationships between the nuclear lamina, "internal" scaffold, and chromatin.

REFERENCES

Arrigo, A.P., S. Fakan, and A. Tissières. 1980. Localization of the heat shock-induced proteins in *Drosophila melanogaster* tissue culture cells. *Dev. Biol.* **78:** 86–103.

Goldknopf, I.A. and H. Busch. 1978. Modifications of nuclear proteins: The ubiquitin-H2A conjugate. In *The cell nucleus* (ed. H. Busch), part C, pp. 149–180. Academic Press, New York.

Levinger, L. and A. Varshavsky. 1981. Heat shock proteins of *Drosophila* are associated with nuclease-resistant, high salt-resistant nuclear structures. *J. Cell Biol.* **90:** 793–796.

―――――. 1982. Selective arrangement of ubiquitinated and D1 protein-containing nucleosomes within the *Drosophila* genome. *Cell* **28:** 375–385.

Velasquez, J.J., B.J. DiDomenico, and S. Lindquist. 1980. Intracellular localization of heat shock proteins in *Drosophila*. *Cell* **20:** 679–689.

Wu, C., Y-C. Wong, and S.C.R. Elgin. 1979. The chromatin structure of specific genes. II. Disruption of chromatin structure during gene activity. *Cell* **16:** 807–814.

Association of the *Escherichia coli* Nucleoid with Protein Synthesized during Thermal Treatments

Jose R. Pellon*, Reinaldo F. Gomez†, and Anthony J. Sinskey*

*Department of Nutrition and Food Science
Massachusetts Institute of Technology
Cambridge, Massachusetts 02139

†Genentech, Inc.
South San Francisco, California 94080

Mesophilic microorganisms have a maximum growth rate at temperatures around 30–37°C. In the case of *Escherichia coli,* the maximum growth rate occurs at around 37°C (Herendeen et al. 1979). Temperatures above 40°C result in a progressively slower growth rate until growth ceases at 45–48°C, while temperatures of 50–52°C or greater cause cell inactivation or death.

Heat treatment of bacterial cells at lethal temperatures results in a variety of structural and functional alterations, including protein inactivation, RNA degradation, DNA breakage, and changes in membrane permeability (Hurst 1977; Pierson et al. 1978). Repair of cellular damage due to nonlethal thermal injury is well documented, and repaired cells are able to reinitiate growth and division. Nonlethal thermal injury is a temporary condition and may be the initial phase of cell death, since thermally injured cells may die or may repair the damage depending on environmental conditions (Hurst 1977; Pierson et al. 1978).

The role of structural and biochemical changes occurring during thermal injury and repair should be elucidated in order to determine the types of damage that must be repaired before the cell can resume normal growth. In this regard, we have chosen chromosomal DNA and its in vivo-related structures as a system for studying thermal injury and repair. Damaged structures other than DNA could be replaced with newly synthesized ones, provided that the DNA remains functional and provides correct genetic information.

The chromosome of *E. coli* is a circular DNA molecule packaged into a compact structure inside the bacterial cell. Chromosomal particles isolated as compact and nonviscous intact structures have been called folded chromosomes or nucleoids. They are composed of folded and supercoiled DNA organized with RNA and protein. Isolated nucleoids have dimensions, superhelical density, and a number of DNA domains similar to those of the nuclear bodies observed inside the cell (Pettijohn 1976; Sinden and Pettijohn 1981). The nucleoid is the only structure that maintains the DNA supercoiling in the entire bacterial genome after isolation.

Thus, the nucleoid cointegrates relationships between DNA and other macromolecules and, due to the multitarget nature of heat, is a useful system to study thermal injury and repair.

THERMAL DAMAGE TO THE *E. COLI* NUCLEOID

Kinetic studies on *E. coli* AB 2497 cultures heated at 50°C for different periods of time demonstrate that changes in the nucleoid structure occur as a function of heating, as indicated by sedimentation analyses (Fig. 1). The nucleoids isolated from unheated control cells had a sedimentation coefficient of 1800–1900S. Nucleoids isolated from cells heated at 50°C for 5 minutes, had a lower sedimentation coefficient of 500–900S. The decrease in sedimentation velocity in nucleoids isolated from heated cells may reflect one or more of a number of structural alterations including DNA unfolding, DNA breakage that results in the loss of supercoiling, and dissociation of the RNA and/or protein involved in the maintenance of the nucleoid structure from the nucleoid (i.e., a sudden cessation of transcription upon heat shock could result in an unfolding of the DNA in the nucleoids which in turn results in a decrease in their sedimentation coefficient). Heat treatment (50°C, 5 min) also results in the formation of fast-sedimenting, protein-containing structures with a sedimentation coefficient of 4000–5000S. These protein structures contain approximately 25–30% of the protein synthesized during heat treatment. This finding is based upon the total incorporation of [^{3}H]leucine (Fig. 1B). Control sedimentation profiles from unheated cells show less than 3% of the [^{3}H]leucine radioactivity in the 4000–5000S region (Pellon et al. 1980).

Nucleoids isolated from cells treated at 50°C for 15 minutes have an increased sedimentation coefficient and the sedimentation profile broadens (900–4500S) as compared with the sedimentation profile of nucleoids isolated after heating 5 minutes at 50°C. The fast-sedimenting protein structures also contain 25–30% of the protein synthesized during this heat treatment. It is important to point out that around 10% of the total cellular protein synthesized before heating is also integrated in these protein structures (Pellon et al. 1980).

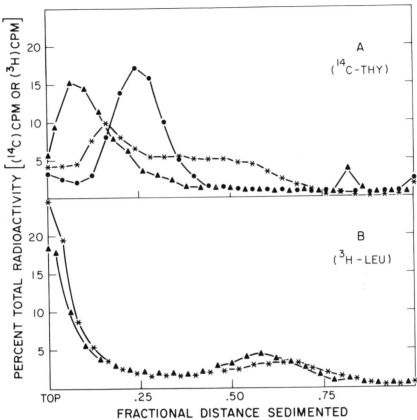

Figure 1

Gradient profiles for nucleoids from heated cells. A culture of *E. coli* AB 2497 was labeled during growth at 37°C in chemically defined medium containing [¹⁴C]thymidine (*A*) (Pellon et al. 1980) and chased with excess thymidine for at least 20 minutes. Then the culture was shifted to 50°C. After holding it for 2 minutes at 50°C, [³H]leucine (*B*) was added. The cultures were maintained for 5 and 15 minutes at 50°C, and then shifted to a 0°C ice bath. Cells were then harvested by centrifugation at 4°C, and nucleoid isolation was carried out as indicated in Pellon et al. (1980). Centrifugation of the cell lysates containing the nucleoids in 10–50% neutral sucrose gradients was performed at 4°C and 3000 rpm for a total $\omega^2 t$ of 5×10^9 rad²/sec. The gradients were fractionated from the top. (●) Control unheated cells; (▲) 50°C, 5 minutes; (∗) 50°C, 15 minutes.

REPAIR OF THERMAL DAMAGE TO THE *E. COLI* NUCLEOID

Incubation of the heated cells at 37°C in liquid medium results in the repair of nonlethal thermal damage. After a thermal treatment at 50°C for 15 minutes, the repair process at the nucleoid level includes at least two steps. The first step is the association of the nucleoids with the fast-sedimenting protein structures (Pellon and Gomez 1981), and the second is their subsequent dissociation resulting in nucleoids with the characteristic sedimentation coefficient of the ones isolated from unheated control cells. The fraction of nucleoids able to regain their control sedimentation coefficient after different time-temperature thermal treatments corresponds in every case to the percentage of cells able to repair thermal damage in liquid medium and form colonies on agar growth medium (Table 1; Pellon and Gomez 1981). This suggests that the disappearance of the fast-sedimenting protein structures and/or the dissociation of the repaired nucleoid from these protein structures is required for the cell to be able to reinitiate growth and division.

Isolation procedures that result in membrane-associated nucleoids (Drlica et al. 1978; Pellon et al. 1980) form fast-sedimenting structures having a similar sedimentation coefficient to the ones observed upon heating. However, the fast-sedimenting protein structures isolated after heating do not contain phospholipids suggesting that they are not truly membrane fragments (Pellon et al. 1980). Furthermore, the nucleoid-protein associations formed during repair increase their sedimentation coefficient accordingly when nucleoid isolation procedures that result in membrane-associated nucleoids are employed. These results indicate that the protein structures to which the nucleoids are associated during repair are not composed of membrane fragments. The fast-sedimenting protein structures may be synthesized de novo as a result of thermal injury and repair. Jones et al. (1982) have observed the development of mesosomelike structures in *Staphylococcus aureus* after heating at 50°C for 45 minutes. These mesosomelike structures could be similar to the fast-sedimenting protein structures that we observe upon heating of *E. coli*. Research to clarify this point is being pursued in collaboration with Dr. S. B. Jones. In addition, preliminary experiments suggest that the proteins being synthesized during and immediately after a 50°C, 15-minute heat treatment are the same proteins synthesized upon heat shock at 43°C.

Table 1
Viability and Nucleoid Integrity after Thermal Injury and Repair

Heat treatment (°C, min)	Cells recovered (%)	Nucleoids repaired (%)
50, 5	90–100	> 95
50, 30	40–50	35–40
50, 120	10–15	5–10
52, 40	< 1	< 1

Pellon and Gomez (1981).

SUMMARY AND CONCLUSION

Heat treatments at 50°C result in measurable changes in the nucleoid structure. They also result in the formation of fast-sedimenting protein structures whose nature is being studied. During the repair process after heat treatment at lethal temperatures, the *E. coli* nucleoid becomes associated with the protein structures. The disappearance of the protein structures and/or the dissociation of the repaired nucleoids from them appears to be required for the cell to reinitiate growth and division. The proteins synthesized during and immediately after heat injury seem to be the same proteins synthesized upon heat shock.

ACKNOWLEDGMENTS

We thank Dr. Carl "BOZO" Batt for valuable discussions. J.R. Pellon was supported by a fellowship from the Instituto Tecnologico para Postgraduados, Madrid, Spain.

REFERENCES

Drlica, K., E. Burgi, and A. Worcel. 1978. Association of the folded chromosome with the cell envelope of *Escherichia coli:* Nature of the membrane-associated DNA. *J. Bacteriol.* **134:** 1108–1116.

Herendeen, S.L., R.A. Van Bogelen, and F.C. Neidhardt. 1979. Levels of major proteins of *Escherichia coli* during growth at different temperatures. *J. Bacteriol.* **139:** 185–194.

Hurst, A. 1977. Bacterial injury: A review. *Can. J. Microbiol.* **23:** 935–944.

Jones, S.B., S.A. Palumbo, and J.L. Smith. 1982. Electron microscopy of heat-injured and repaired *Staphylococcus aureus.* Abs. Ann. Meeting Amer. Soc. Microbiol. J-17,94.

Pellon, J.R. and R.F. Gomez. 1981. Repair of thermal damage to the *Escherichia coli* nucleoid. *J. Bacteriol.* **145:** 1456–1458.

Pellon, J.R., K.M. Ulmer, and R.F. Gomez. 1980. Heat damage to the folded chromosome of *Escherichia coli* K-12. *Appl. Environ. Microbiol.* **40:** 358–364.

Pettijohn, D.E. 1976. Procaryotic DNA in nucleoid structure. *Crit. Rev. Biochem.* **4:** 175–202.

Pierson, M.D., R.F. Gomez, and S.E. Martin. 1978. The involvement of nucleic acids in bacterial injury. *Adv. Appl. Microbiol.* **23:** 263–285.

Sinden, R.R. and D.E. Pettijohn. 1981. Chromosomes in living *Escherichia coli* cells are segregated into domains of supercoiling. *Proc. Natl. Acad. Sci.* **78:** 224–228.

The Heat-shock Phenomenon in Bacteria—A Protection against DNA Relaxation?

Andrew A. Travers
and Hilary A.F. Mace
Laboratory of Molecular Biology
Medical Research Council Centre
Cambridge CB2 2QH, England

The binding of many structural and regulatory proteins to the DNA must be finely balanced. Were this not so, cells would not possess the capacity for a rapid response to a change in environmental conditions. The complex regulatory loops maintaining the cell in homeostasis must ultimately depend on the maintenance of the appropriate DNA-protein interactions. It follows that any environmental insult that has the potential to perturb those interactions could substantially alter the efficiency of any regulatory response. Just as a computer must be protected against a power surge, so the information-processing machinery of a cell might need to be protected against a sudden energy input producing a disruptive destabilization.

Such a sudden energy input occurs during heat shock. In organisms as diverse as *Escherichia coli* and humans, cells respond to rapid temperature rises of 5–15°C by inducing the transient production of a small set of abundant proteins and concomitantly curtailing the synthesis of most other proteins. What triggers this dramatic change in the pattern of gene expression and what is its purpose? We initially considered the formal hypothesis that the direct effect of heat on the structure of DNA might be the signal for the induction of the heat-shock response. The extent of DNA winding is temperature dependent, the average angle, θ, between adjacent base pairs decreasing by $0.012°/C°$. A 15°C heat shock would thus unwind DNA by $0.18°$/base pair or by approximately

127

0.5%. This would be equivalent to melting 1 bp in every 200 or 15,000 bp in the *E. coli* genome as a whole.

Is this degree of unwinding a significant perturbation of DNA structure? Under normal growth conditions, the bacterial chromosome is in a state of torsional stress with, on average, one negatively supercoiled turn for every 200 bp. This structure is essential for normal gene expression and is maintained, in part, by the activity of DNA topoisomerase II, the DNA gyrase. This enzyme thus counteracts the relaxation of the *E. coli* chromosome consequent upon the local melting of DNA required for transcription by RNA polymerase, in which initiation of every RNA molecule requires the melting of 10 bp. At maximum rates of transcription, ~1500 RNA polymerase molecules are actively associated with the bacterial chromosome together unwinding ~15,000 bp, equivalent to reducing the negative supercoiling by 10%. It follows that a 15°C heat shock is equivalent in terms of DNA structure to doubling the maximum observed transcriptional capacity of the bacterial DNA.

A homeostatic response to such a perturbation might be expected, first, to reduce the immediate consequences of the heat-induced relaxation and, second, to produce proteins that would allow the normal DNA structure to be maintained under conditions of higher energy input. The rapid reduction of actively transcribing RNA polymerase molecules to 15% of the normal level following heat shock (Ryals et al. 1982) would counteract the heat-induced unwinding but would probably be inimical to further growth. These considerations thus predict that at least some of the proteins synthesized during heat-shock condition would themselves interact with the chromosomal DNA or associated proteins to allow the restoration of normal transcriptional capacity. Indeed Pellon et al. (this volume) have shown that rapid heating of *E. coli* to 50°C results initially in a lowering of the sedimentation coefficient of the bacterial nucleoid, consistent with either an unfolding or an unwinding of the chromosome, and subsequently, in an association of newly synthesized proteins with the nucleoid.

A further prediction is that agents that disrupt chromosomal structure might themselves induce the synthesis of some or all of the heat-shock proteins. Coumermycin A1, an inhibitor of the B subunit of DNA topoisomerase II, induces the synthesis of a small set of proteins in cells sensitive to the drug but not in cells containing a resistant B subunit (Mirkin et al. 1979). We asked whether the proteins induced by coumermycin were heat-shock proteins. With one exception, a protein of M_r ~ 38 kD, all the coumermycin-induced proteins corresponded to heat-shock proteins (Fig. 1). However, the drug did not induce the full set of heat-shock proteins, suggesting that, in *E. coli* at least, even if the heat-shock proteins are required for the stabilization of chromosome structure, relaxation per se is unlikely to be the signal for the classic heat-shock response.

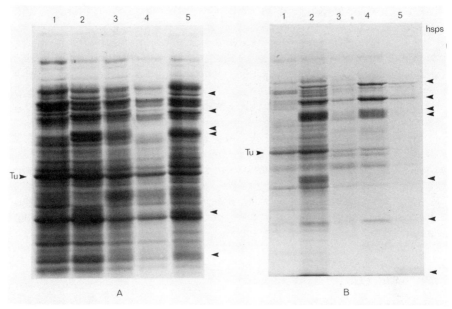

Figure 1

Effect of coumermycin A1 and ethanol on heat-shock protein synthesis. *E. coli* AB1450 was treated with ethanol, coumermycin Al (60 μg/ml) + 3% DMSO, and 3% DMSO alone for 10 minutes at 30°C. Heat shock was for 10 minutes at 43°C. Proteins were then labeled for 15 minutes with [³⁵S]methionine. (*A*) 1, 30°C; 2, 43°C; 3, 30°C + coumermycin A1 + 3% DMSO; 4, 30°C + 3% DMSO; 5, 30% + 4% ethanol. (*B*) 1, 30°C; 2, 43°C; 3, 30°C + 1% ethanol; 4, 30°C + 4% ethanol; 5, 30°C + 10% ethanol. (▶) Heat-shock proteins. The induction of hsp25 is variable (cf. *A*5 and *B*4).

In *Drosophila,* the major heat-shock protein, hsp70, appears to be associated with the interband regions of polytene chromosomes (Linquist et al., this volume) suggesting that in eukaryotes the heat-shock proteins may also play a role in stabilizing chromosome structure. However, the induction of the heat-shock response in eukaryotes by as little as a 2°C rise in temperature again argues that direct heat-induced unwinding of DNA is unlikely to be the primary trigger.

Even if DNA relaxation in bacteria does not in itself induce the heat-shock response, agents which do might be expected to destabilize chromosome structure, possibly by making the maintenance of the supercoiled state energetically less favorable. Under normal intracellular conditions (pH 7.7, 0.2 M KCl), the interactions of many proteins with DNA would be exquisitely sensitive to small changes in ionic strength or pH so that such perturbations of protein-DNA interactions might affect chromosome stability. For example, a raising of either ionic strength by 0.1 unit or intracellular pH by 0.5 unit could decrease by up to an order of magnitude the affinity of at least some DNA binding proteins for DNA

(Strauss et al. 1980). Such a diminution of binding could interfere with the ability of DNA gyrase to supercoil the chromosome.

A clue to the nature of the inducing signal is provided by the observation that ethanol, which is known to induce thermotolerance in CHO cells (Li and Hahn 1978), also induces the heat-shock response in *E. coli* (Fig. 1). Ethanol has the same effect in eukaryotic cells (Li et al., this volume) strongly suggesting that the primary inducing signal might be similar in both prokaryotes and eukaryotes. In eukaryotes it has been suggested (Li and Hahn 1980) that ethanol acts by altering membrane fluidity but this is known not to be the case in *E. coli*. However, ethanol has been shown to affect transmembrane transport in bacteria and might also alter intracellular pH by this means. The effect of ethanol would thus be consistent with the notion that the principal inducer of the heat-shock response is a particular perturbation of the intracellular ionic balance, conceivably an alteration of the intracellular pH.

REFERENCES

Li, G.C. and G.M. Hahn. 1978. Ethanol-induced tolerance to heat and to adriamycin. *Nature* **274:** 699–701.

Mirkin, S.M., E.S. Bogdanova, Zh. M. Gorlenko, A.I. Gragerov, and O.A. Larionov. 1979. DNA supercoiling and transcription in *Escherichia coli:* Influence of RNA polymerase mutations. *Mol. Gen. Genet.* **177:** 169–175.

Ryals, J., R. Little, and H. Bremer. 1982. Control of ribonucleic acid synthesis in *Escherichia coli* after a shift to higher temperature. *J. Bacteriol.* (in press).

Strauss, H.S., R.R. Burgess, and M.T. Record, Jr. 1980. Binding of *Escherichia coli* ribonucleic acid polymerase holoenzyme to a bacteriophage T7 promoter-containing fragment: Evaluation of promoter binding constants as a function of solution conditions. *Biochemistry* **19:** 3504–3515.

Escherichia coli Gene (hin) Controls Transcription of Heat-shock Operons and Cell Growth at High Temperature

**Tetsuo Yamamori, Toshio Osawa,
Toru Tobe, Koreaki Ito,
and Takashi Yura**
Institute for Virus Research
Kyoto University
Kyoto, Japan 606

When a culture of wild-type *E. coli* is transferred from low to high temperature, synthesis of a specific set of polypeptides is markedly and transiently induced while bulk protein synthesis is relatively little affected (Yamamori et al. 1978). Detailed kinetic experiments indicate that heat-shock induction of these polypeptides occurs coordinately at the level of transcription almost immediately after raising temperature to or beyond a certain critical point (Yamamori and Yura 1980). The extent of induction depends both on the preshift temperature and on the size of temperature shift; the maximum induction is obtained within 5 minutes, followed by a gradual decrease in the rates of synthesis. Studies of an amber mutant (Tsn-K165; isolated by Cooper and Ruettinger 1975) have revealed that a product of the gene, designated *hin* (*h*eat-shock *in*duction), is required for induction of these heat-shock polypeptides (hsps) at the level of transcription (Yamamori and Yura 1982). The mutant offers an opportunity to study not only the mechanism of induction but also the role of heat-shock polypeptides in growth and protection against thermal killing in bacteria.

MUTANT DEFECTIVE IN INDUCTION OF THE hsp GENES

The mutant Tsn-K165 grows at 30°C but not at 42°C in complete media; it carries an amber mutation (*hin-165*) which maps at 76 minutes on the *E. coli* chromosome, and a temperature-sensitive suppressor (*supF*-ts) which is barely active at 30°C and virtually inactive at 42°C. When a log-phase culture of Tsn-K165 grown at 30°C was transferred to 42°C, little induction of several major heat-shock polypeptides was observed. In contrast, a marked induction was found in the mutant carrying a temperature-insensitive, highly active supF instead of supF-ts, as well as in the isogenic hin+ strain. The results may be best illustrated by the protein patterns obtained in two-dimensional gel electrophoresis (Fig. 1); the synthesis of polypeptides of 87, 81, 76, 73, and 64 kD was clearly induced in the hin-165 supF strain but hardly induced in the hin-165 supF-ts strain (Tsn-K165). Another polypeptide of 16 kD behaved similarly. Thus, the induction of at least six heat-shock polypeptides is affected by the hin-165 mutation. Similar defective induction of nine different heat-shock polypeptides was reported with the same mutant by Neidhardt and VanBogelen (1981).

THE hin GENE PRODUCT MAY ACTIVATE TRANSCRIPTION OF HEAT-SHOCK OPERONS

One of the major heat-shock polypeptides, the 64K polypeptide, has been identified as a product of the gene groE (= mop, groEL) located at 94 minutes, and can be quantitatively determined by immunoprecipitation using specific antiserum (cf. Yamamori and Yura 1982). mRNA for the groE operon, which presumably contains two genes groEL and groES (Tilly et al. 1981), can be determined by measuring mRNA hybridizable with λgroE transducing phage. In agreement with the results of Figure 1, the synthesis of either groE mRNA or protein was normally induced in the hin+ strain but not in the hin-165 mutant (Tsn-K165). We then lysogenized Tsn-K165 with ϕ80supF or its derivative carrying a gene for the suppressor tRNA^{Tyr} (supF) of different efficiency, and examined the effect of varying levels of hin gene expression on heat-shock induction. The typical results for groE protein are shown in Figure 2A. Evidently, both the rate of induction and the maximum level attained at 42°C are proportional to the level of hin gene expression. The maximal induction is about 10-fold the rate at 30°C for the hin+ strain, and it is 7-fold, 4-fold, or less than 2-fold for the strains with suppressor activities 100%, 50%, or 20% of supF, respectively. This strongly supports the notion that the gene hin (htpR of Neidhardt and VanBogelen 1981) codes for a positive regulatory protein that quantitatively activates transcription of several hsp genes or operons at high temperature (Yamamori and Yura 1982).

THE ROLE OF HEAT-SHOCK POLYPEPTIDES IN CELL GROWTH

The growth rates of the above set of strains were found to be similar at 30°C, but those after transfer to 42°C varied depending on the suppres-

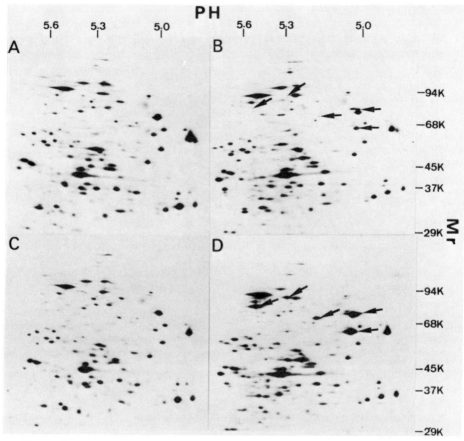

Figure 1
Two-dimensional gel electrophoresis of proteins. Log-phase cultures of Tsn-K165 (*hin-165 supF*-ts) (*A* and *B*) or Tsn-K165 *supF* (*hin-165 supF*) (*C* and *D*) were pulse-labeled with [³H]leucine at 30°C (*A* and *C*) or 5 minutes after shift to 42°C (*B* and *D*). Whole-cell proteins were subjected to isoelectric focusing (first dimension) and to SDS-gel electrophoresis (second dimension). Arrows indicate the heat-shock polypeptides. (Reprinted, with permission, from Yamamori and Yura 1982.)

sor efficiency (Fig. 2B). Taken together with the above results on the synthesis of heat-shock polypeptides, these results suggest that the amount of *hin* gene product determines the degree of heat-shock induction, which in turn affects the growth rate at 42°C. The maximum temperature that permits growth of these strains was also found to correlate well with the suppressor activity and with the level of heat-shock polypeptides. Besides, the amounts of *groE* and other heat-shock polypeptides in the *hin*⁺ strain under steady-state growth increase proportionally with increasing temperature. These results suggest that the

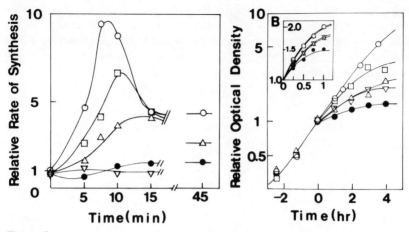

Figure 2
Synthesis of *groE* protein and growth of Tsn-K165 strains carrying *supF* genes of various suppression efficiencies. (*A*) Log-phase cultures grown at 30°C were shifted to 42°C at time 0. Samples were taken at intervals and pulse-labeled with [³H]leucine. Crude extracts were treated with anti-*groE* protein serum, and the precipitates were subjected to SDS-gel electrophoresis. (o) *hin*⁺; Tsn-K165 strains with *supF* (□), a mutated *supF* 50% as active as *supF* (Δ), a mutated *supF* 20% as active as *supF* (∇), and *supF*-ts 3.4% as active as *supF* at 30°c (●), respectively. (*B*) Cultures were grown as in (*A*), optical density was followed by a Klett-Summerson colorimeter and normalized to the zero-time value for each strain (ca. 10⁸ cells/ml). Symbols are as in (*A*). (Reprinted, with permission, from Yamamori and Yura 1982.)

synthesis of heat-shock polypeptides plays a critical role in supporting growth of *E. coli* cells at high temperature. It also appears that the higher the temperature, the higher the amounts of heat-shock polypeptides that are required for normal cell growth.

INVOLVEMENT OF THE *hin* GENE
IN RESISTANCE TO THERMAL KILLING

In further investigation of the physiological significance of heat-shock response in *E. coli*, we have examined whether the synthesis of heat-shock polypeptides is related to the cellular resistance to lethal temperature. It was found that transfer of *hin*⁺ bacteria from 30°C to 42°C transiently conferred on the cell resistance to a subsequent treatment at 55°C. In contrast, the *hin-165* mutant virtually failed to acquire heat resistance under these conditions. However, the increased resistance was observed only for 15 to 60 minutes and gradually disappeared by further growth at 42°C, indicating that factors other than accumulation of heat-shock polypeptides are also involved. In addition, cells of at least some strains of *E. coli* K12 grown at 45°C or above (steady-state growth) in which heat-shock polypeptides are produced at high levels exhibit increased resistance as compared with those grown at 30°C. These results taken together suggest that the *hin* gene, and presumably the synthesis of heat-shock polypeptides, is somehow involved in thermal resistance of *E. coli* cells (Yamamori and Yura 1982).

ENHANCED TRANSCRIPTION OF hsp GENES
UNDER LOW LEVELS OF σ FACTOR

In the course of our studies of amber mutants of RNA polymerase σ subunit, it was revealed that the synthesis of heat-shock polypeptides is markedly enhanced when cellular amounts of σ are specifically reduced by severalfold (Osawa and Yura 1981). These mutants grow normally at 30°C but not at 42°C, because they carry a temperature-sensitive suppressor (*supF*-ts6) and cannot synthesize complete σ polypeptides at 42°C. Upon transfer of a mutant culture from 30°C to 42°C, the amount of σ, but not of core polymerase subunits, decreased to 10% of wild type after about 2 hours. Under these conditions, the synthesis of heat-shock polypeptides was maintained at high levels (5- to 10-fold of wild type) while bulk RNA and protein syntheses were little affected. The enhanced synthesis of heat-shock polypeptides was also observed when such mutants were grown continuously at 34°C or 36°C, in which the σ level was about 20% or 6% of wild type, respectively. Figure 3 summarizes the results of these experiments with respect to the synthesis of *groE* protein. Similar results were obtained with respect to the synthesis of *groE* mRNA as well as to the synthesis of other heat-shock polypeptides (Osawa and Yura 1981 and unpubl.).

The elevated expression of hsp genes requires the *hin* gene product, since the double mutant carrying an amber mutation of σ (*rpoD40*) and *hin-165* as well as *supF*-ts6 produced uninduced levels of heat-shock polypeptides under the conditions used. In addition, an *E. coli* mutant that overproduces σ (Nakamura and Yura 1975) was found to synthesize significantly reduced amounts of heat-shock polypeptides (T. Osawa and T. Yura, unpubl.); the data are also included in Figure 3. It may be surmised that the level of expression of hsp genes is inversely correlated with the cellular amount of σ or RNA polymerase holoenzyme relative to core enzyme. The mechanisms underlying these observations remain to be investigated.

CONCLUSION

We have identified a gene (*hin* = *htpR*) whose product seems to be quantitatively required for heat-shock induction (transcriptional activa-

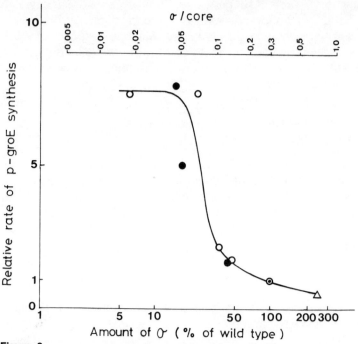

Figure 3

Effects of the cellular amount of σ factor on the synthesis of *groE* protein. The results of three separate experiments (Osawa and Yura 1981 and unpubl.) are summarized. In experiment 1 (●), a log-phase culture of KY1411 (*rpoD40 supF*-ts6) grown at 30°C was transferred to 42°C and samples were taken at intervals (up to 90 min) for determination of σ and *groE* protein. In experiment 2 (○), cultures of KY1411 were grown at 30°C, 32°C, 34°C, or 36°C for several generations in the presence of [³H]leucine. In experiment 3 (△), a mutant that overproduces σ (2.4-fold of wild type) was grown at 30°C. In all the experiments, the amount of σ and the rate of synthesis of *groE* protein were determined as described by Osawa and Yura (1981), and were normalized to the respective value at 30°C for wild type.

tion) of several hsp genes or operons in *E. coli*. The synthesis of heat-shock polypeptides seems to play a critical role in supporting cell growth at high temperature. The involvement of the *hin* gene product, and presumably the synthesis of heat-shock polypeptides in resistance to thermal killing was revealed under certain conditions. On the other hand, the level of hsp gene expression appears to be inversely correlated with the amount of σ factor present in cells of *E. coli*. Structural and functional analyses of the *hin* gene and its product should prove rewarding for further elucidation of the mechanism and physiological roles of heat-shock response in bacteria. Finally, such studies with bacteria would be of interest from the comparative viewpoints, and may provide information that is useful in deciphering analogous problems in various eukaryotic organisms.

REFERENCES

Cooper, S. and T. Ruettinger. 1975. A temperature sensitive nonsense mutation affecting the synthesis of a major protein of *Escherichia coli* K12. *Mol. Gen. Genet.* **139:** 167–176.

Nakamura, Y. and T. Yura. 1975. Hyperproduction of the sigma subunit of RNA polymerase in a mutant of *Escherichia coli. Mol. Gen. Genet.* **141:** 97–111.

Neidhardt, F.C. and R.A. VanBogelen. 1981. Positive regulatory gene for temperature-controlled proteins in *Escherichia coli. Biochem. Biophys. Res. Commun.* **100:** 894–900.

Osawa, T. and T. Yura. 1981. Effects of reduced amount of RNA polymerase sigma factor on gene expression and growth of *Escherichia coli:* Studies of the *rpoD40* (amber) mutation. *Mol. Gen. Genet.* **184:** 166–173.

Tilly, K.M., H. Murialdo, and C. Georgopoulos. 1981. Identification of a second *Escherichia coli groE* gene whose product is necessary for bacteriophage morphogenesis. *Proc. Natl. Acad. Sci.* **78:** 1629–1633.

Yamamori, T. and T. Yura. 1980. Temperature-induced synthesis of specific proteins in *Escherichia coli:* Evidence for transcriptional control. *J. Bacteriol.* **142:** 843–851.

———. 1982. Genetic control of heat-shock protein synthesis and its bearing on growth and thermal resistance in *Escherichia coli. Proc. Natl. Acad. Sci.* **79:** 860–864.

Yamamori, T., K. Ito, Y. Nakamura, and T. Yura. 1978. Transient regulation of protein synthesis in *Escherichia coli* upon shift-up of growth temperature. *J. Bacteriol.* **134:** 1133–1140.

The High-temperature Regulon of *Escherichia coli*

Frederick C. Neidhardt,
Ruth A. VanBogelen, and
Elizabeth T. Lau
Department of Microbiology and Immunology
University of Michigan
Ann Arbor, Michigan 48109

Clear definition of a heat-shock response in bacteria was delayed in part by the adaptibility of bacteria to growth at different temperatures. Cells of *Escherichia coli* grow well over a broad temperature span — 40°C — and adjust the synthesis of a large number of proteins in response to temperature. Within the normal range (20–40°C) for *E. coli*, only a dozen of the 1000 or so proteins made during growth change more than twofold in level, and none change more than fourfold. But steady-state growth above or below these temperatures leads to adjustments that affect the level of all but a few cellular proteins (Herendeen et al. 1979).

Throughout most of the living world, the heat-shock response consists of a transient, nearly exclusive synthesis at high rates of about a dozen proteins. In *E. coli* a shift from any normal growth temperature to, say, 42–46°C elicits quite a different response. Within a minute, and climaxing at 7 minutes, rapid changes occur in the synthesis of most of the 1000 cellular proteins. There is an almost smooth gradient of response extending from some proteins that are hyperinduced transiently a hundredfold to some that for a time virtually cease being made (Lemaux et al. 1978). Furthermore, qualitatively the same protein pattern is obtained for shifts of only a few degrees within the normal growth temperature range. Distinguishing a "heat-shock" set of proteins from amongst this global response has until recently seemed not only difficult, but pointless.

Nevertheless, evidence now indicates that among the many proteins induced by high temperature in *E. coli* is a set of 13 that constitutes a regulon (a group of genes/proteins sharing a common regulator element) that is in many ways analogous to the heat-shock proteins of eukaryotes.

DEFINITION OF THE HTP REGULON

Thirteen proteins are marked on the autoradiogram of a two-dimensional gel of the total cell proteins of *E. coli* strain W3110 (Fig. 1). These proteins constitute the HTP (*H*igh *T*emperature *P*roduction) regulon (Neidhardt and VanBogelen 1981). In Figure 2 is displayed the behavior typical of these 13 proteins upon shifts in the growth temperature: transient hyperinduction on a shift up, repression on a shift down. Measurements concentrated at the early period reveal the response is detectable by 20 seconds after the shift, peaks at 5−7 minutes, and has largely subsided by 20 minutes. Measurements of mRNA synthesis

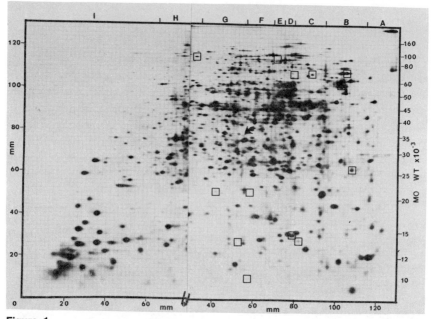

Figure 1
Two-dimensional resolution of total cell proteins of *E. coli* W3110 grown at 37°C. The picture is a composite of autoradiograms of two polyacrylamide gels. On the right, comprising vertical regions A−G, is a standard isoelectric focusing gel; on the left, regions H and I, is a nonequilibrium gel that reveals the basic proteins. The 13 open squares indicate the HTP proteins, or their locations. The arrow points to protein F33.4. Preparation and presentation of these gels are described in Bloch et al. (1980).

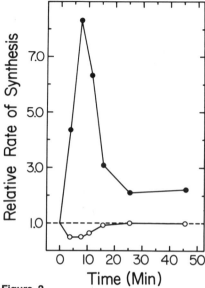

Figure 2
Rates of synthesis of protein B56.5, an HTP protein, after a shift up or down in temperature of cultures of *E. coli* NC3. The ordinate is the differential rate of synthesis of this
protein expressed relative to its rate of synthesis before a shift in temperature. The
abscissa is time measured after a shift. The rates of synthesis were measured by 1-minute
pulses according to the protocol given by Lemaux et al. (1978). (•) Culture shifted from
28°C to 39°C at zero time; (○) culture shifted from 36°C to 28°C at zero time. (Experiment
performed by P. Lemaux.)

confirm early reports (Yamamori and Yura 1980) that these hyperinduc-
tions are transcriptional in origin.

The steady-state levels of these proteins generally rise with growth
temperature, particularly above 37°C. The transient hyperinduction
achieves a rapid attainment of the new level characteristic of steady-
state growth at the higher temperature.

The HTP proteins share another characteristic. They are all "constitutive" proteins. That is, their cellular levels vary less than twofold during
growth at 37°C in media differing widely in their nutritional character;
they are proteins of class IB in the metabolic classes described by
Pedersen et al. (1978).

But the primary criterion for linking these 13 proteins into one regulon
is that their response to temperature is uniquely eliminated by an amber
mutation in a strain (K165) carrying a tRNA that suppresses amber
mutations at low (28°C) but not at high (42°C) temperatures. Work in our
laboratory (Neidhardt and VanBogelen 1981) and that of Yamamori and
Yura (1982) has established that the nonsense mutation in strain K165
is in a gene, called *htpR* (by us) and *hin* (by Yura), that maps near

minute 75 on the *E. coli* chromosome. Reversion analysis and other genetic evidence has confirmed that this mutation eliminates the temperature response of the 13 HTP proteins and leads to the death and lysis of the mutant cells after an hour of growth at 42°C. Table 1 shows the marked difference in the HTP response of an *htpR*[+] and an *htpR* mutant strain.

PROTEINS AND GENES OF THE HTP REGULON

Current interest in constructing a protein index of *E. coli* (cf. Bloch et al. 1980; Phillips et al. 1981) has helped identify some of the 13 HTP polypeptides (Table 2). Currently four have been identified: *groEL* (R.

Table 1
Heat-inducible Proteins

	Rate of synthesis at 42°C[b]	
	Rate of synthesis at 28°C	
Protein[a]	*htpR*[+]	*htpR*
B25.3	8.2	1.5
B56.5	18.3	2.1
B66.0	7.2	1.1
C14.7	8.0	1.7
C15.4	14.5	4.0
C62.5	21.3	2.2
D60.5	5.9	1.4
F10.1	n.d.	n.d.
F21.5	n.d.	n.d.
F84.1	14.7	3.9
G13.5	n.d.	n.d.
G21.0	n.d.	n.d.
G93.0	6.3	1.2
C42.5	2.2	2.6
E24.1	1.8	1.7
F29.0	3.8	3.3

[a]The upper 13 proteins are HTP proteins; the latter 3 are heat-induced but not under *htpR* control.
[b]The values given are the differential rates of synthesis at 42°C divided by those at 28°C. The rates were determined by a 1-minute pulse of [³H]leucine at 28°C and a similar pulse beginning 5 minutes after a shift to 42°C. The results are given for the normal strain SC122 (*htpR*[+]) and for the K165 mutant (*htpR*).

Table 2
Identification of the Proteins and Genes of the HTP Regulon

HTP protein	Identification	Map location (min)
B25.3	unknown	—
B56.5	*groEL* gene product	93.5
B66.0	*dnaK* gene product	0.5
C14.7	unknown	—
C15.4	*groES* gene product	93.5
C62.5	unknown	—
D60.5	*lysU* gene product	93.0
F10.1	unknown	—
F21.5	unknown	—
F84.1	unknown	—
G13.5	unknown	—
G21.0	unknown	—
G93.0	unknown	—

Hendrix and this laboratory, unpubl.), *groES* (K. Tilly, C. Georgopoulos, and this laboratory, unpubl.), *dnaK* (Georgopoulos et al. 1982), and *lysU* (this laboratory, unpubl.).

They are a strange group. The products of the *groEL* and *groES* genes together are necessary for steps in the morphogenesis of most of the phages of *E. coli* (see references in Tilly and Georgopoulos 1982) and seem to be essential for bacterial growth as well. The *dnaK* gene product is thought to be necessary for initiation of phage DNA replication; its function in the uninfected cell is not known. Perhaps the strangest of all is an auxiliary lysyl-tRNA synthetase, a product of the *lysU* gene; the properties of this gene and enzyme have recently been described (Hirshfield et al. 1981).

These four HTP genes are not linked to the putative regulatory gene, *htpR; groEL* and *groES* map at 93.5, *dnaK* at 0.5, and *lysU* at 93.0 minutes. The other nine HTP proteins have not been identified nor have their genes been mapped; the latter seem also unlinked to *htpR,* because their products are not among the translational products of plasmids containing this gene and bordering DNA segments.

THE *htpR* GENE AND ITS PRODUCT

All the phenotypic characteristics of the mutant strain K165 are complemented by plasmid pLC 31-16 of the Clarke-Carbon library of *E. coli.* This plasmid carries a segment of the chromosome around 75 minutes, including the *livJ, livK,* and *ftsE* genes. In minicells and maxicells, this

144 *F.C. Neidhardt, R.A. VanBogelen, and E.T. Lau*

Table 3
Correlation of Protein F33.4 with *htpR* Function

Strain	Genotype	HTPR phenotype[a]	Presence of F33.4
SC122	*supC*$_{ts}$ *htpR*$^+$	+	+
K165	*supC*$_{ts}$ *htpR*	−	−
K165R	K165 revertant	+	+
K165 (pLC 31-16)	K165 with plasmid	+	+

[a]The HTPR$^-$ phenotype consists of (1) inability to grow at 42°C and (2) failure to induce the HTP proteins upon a shift to 42°C.

plasmid makes a protein (termed F33.4 in the nomenclature of Pedersen et al. 1978) of 34,000 mw, along with the identified products of *livJ* and *livK*. The presence of F33.4 correlates with *htpR* function in the wild strain, the K165 mutant, revertants of K165, and K165 containing the pLC31-16 plasmid (Table 3).

CONCLUSIONS

E. coli responds to a shift up in growth temperature by transient hyperinduction of a large number of proteins, thereby raising their cellular levels quickly to those found in steady state growth at the higher temperature. The response of 13 of these proteins, which constitute the HTP regulon, depends on the protein product (F33.4) of a gene (*htpR*) unlinked to the HTP protein structural genes.

Our working hypothesis is that protein F33.4 is a positive regulatory element required for heat induction of the HTP regulon. For the future one must verify that *htpR* makes F33.4, and then learn how the latter controls transcription of HTP genes. Sequencing the promoters of cloned HTP genes, testing transcription properties in vitro with purified F33.4 protein, and analyzing of the requirement of the HTP response for cell survival, should shed much light on prokaryotic adaptation to temperature. Perhaps we can also learn the function of the HTP proteins.

REFERENCES

Bloch, P.L., T.A. Phillips, and F.C. Neidhardt. 1980. Protein identification on O'Farrell two-dimensional gels: Locations of 81 *Escherichia coli* proteins. *J. Bacteriol.* **141:** 1409–1420.

Georgopoulos, C., K. Tilly, D. Drahos, and R. Hendrix. 1982. The B66.0 protein of *Escherichia coli* is the product of the *dnaK*+ gene. *J. Bacteriol.* **149:** 1175– 1177.

Herendeen, S.L.. R.A. VanBogelen, and F.C. Neidhardt. 1979. Levels of major proteins of *Escherichia coli* during growth at different temperatures. *J. Bacteriol.* **139:** 185–194.

Hirshfield, I.N., P.L. Bloch, R.A. VanBogelen, and F.C. Neidhardt. 1981. Multiple forms of lysyl-transfer ribonucleic acid synthetase in *Escherichia coli*. *J. Bacteriol.* **146:** 345–351.

Lemaux, P.G., S.L. Herendeen, P.L. Bloch, and F.C. Neidhardt. 1978. Transient rates of synthesis of individual polypeptides in *E. coli* following temperature shifts. *Cell* **13:** 427–434.

Neidhardt, F.C. and R.A. VanBogelen. 1981. Protein regulatory gene for temperature-controlled proteins in *Escherichia coli*. *Biochem. Biophys. Res. Commun.* **100:** 894–900.

Pedersen, S., P.L. Bloch, S. Reeh, and F.C. Neidhardt. 1978. Pattern of protein synthesis in *E. coli*: A catalog of the amount of 140 individual proteins at different growth rates. *Cell* **14:** 179–190.

Phillips, T.A., P.L. Bloch, and F.C. Neidhardt. 1981. Protein identifications on O'Farrell two-dimensional gels: Locations of 55 additional *Escherichia coli* proteins. *J. Bacteriol.* **144:** 1024–1033.

Tilly, K. and C. Georgopoulos. 1982. Evidence that the two *Escherichia coli groE* morphogenetic gene products interact in vivo. *J. Bacteriol.* **149:** 1082–1088.

Yamamori, T. and T. Yura. 1980. Temperature-induced synthesis of specific proteins in *Escherichia coli:* Evidence for transcriptional control. *J. Bacteriol.* **142:** 843–851.

———. 1982. Genetic control of heat-shock protein synthesis and its bearing on growth and thermal resistance in *Escherichia coli* K-12. *Proc. Natl. Acad. Sci.* **79:** 860–864.

Regulation of the *Drosophila* Heat-shock Response

J. Jose Bonner
Department of Biology
Indiana University
Bloomington, Indiana 47405

In our laboratory, we have addressed the question of control of gene activity by assaying the regulatory molecules directly. Nuclei isolated under conditions that preserve their structure and function may be expected to respond to exogenously added transcriptional regulators in a fashion which mimics in vivo regulation. Polytene nuclei prepared from larval *Drosophila* salivary glands according to published procedures (Compton and Bonner 1978; Bonner 1981) appear to retain many of the necessary components, for they exhibit nearly normal characteristics of induction of puffing at the sites of the heat shock-inducible genes when supplied with appropriate factors. By a number of criteria which can be measured cytologically, induction of heat-shock puffs in isolated nuclei mimics the in vivo situation very closely; it is a reasonable conclusion that the puff-inducing molecules identified in this system probably function in a similar role in vivo.

With regard to the regulatory phenomena exhibited by the heat-shock puffs, two primary observations are of interest. First, the puffs are induced in vivo very rapidly following a heat shock. Thus, there must be one class of regulatory molecules whose function is that of the induction of puffing—presumably the induction of transcription at the puff sites. Secondly, the puffs exhibit the behavior of regression after a short time, even during the continuation of a heat shock (Lewis et al. 1975). Although the puffs regress, and thus RNA synthesis slows considerably, the characteristics of protein synthesis continue to reflect the state of heat shock throughout the period of stress (Mirault et al. 1978). Thus, there must be a class of regulatory molecules that function in the cessation of heat-shock transcription despite the presence of the initial stimulus. The experiments described here address both of these aspects of the regulation of the heat-shock puffs and provide an indication of their interrelationship.

INDUCTION OF PUFFING

Tissue culture cells that have been heat-shocked briefly can be used to prepare a cytoplasmic extract which will induce puffs in isolated polytene nuclei (Compton and Bonner 1978). Extracts from nonshocked cells normally do not induce puffs unless the extracts are aerated (Bonner 1981). For example, in a 90-minute incubation, an unaerated extract failed to induce puffs (ratio of chromosome diameter at 87C:87F of 1.08 ± 0.03, compared with 1.06 ± 0.04 for a control, unpuffed sample). By contrast, the same extract, when aerated, induced puffs to a size of 2.13 ± 0.07, very close to the maximum attainable for this puff. An explanation of this observation may be found in the fact that, in vivo, recovery from anoxia induces the heat-shock puffs without a temperature shift. Tissue culture cells are relatively anoxic due to the sealed culture flasks; it is likely that puff induction by aerated extracts represents "in vitro recovery from anoxia" by the extract.

That this in vitro recovery from anoxia is truly occurring in the cytoplasmic extract rather than in the nuclei is shown by the data in Figure 1A. Here, a cytoplasmic extract from normal tissue culture cells was aerated and, at various times, aliquots were boiled for 2 minutes. The precipitated proteins were removed by centrifugation and the resulting extracts tested for puff-inducing ability (according to Bonner 1981). The results show quite clearly that (1) the puff-inducing activity is heat-stable; (2) this activity is not present in normal cells; and (3) the production of active inducer is a cytoplasmic event.

The heat stability of the inducing activity has made possible the analysis of this activity independent from the recovery-from-anoxia activity. As may be seen from Figure 1A, this latter activity is completely eliminated by the heat treatment, thus allowing the early time points to give the low values that are observed. Characteristics of the heat-shock puffs at cytological locations 63C, 87A, 87C, and 93D include: the inducer(s) is heat-stable, trypsin-sensitive, excluded by Sephadex G25, and not bound to hydroxyapatite. It is reasonable to conclude that this activity is a protein or several proteins. Further purification is in progress and cannot be reported at this time.

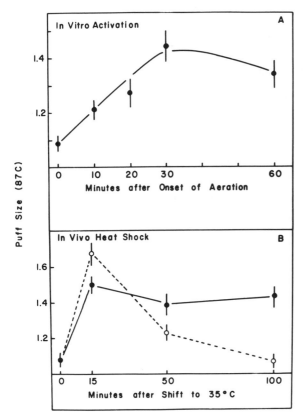

Figure 1
Kinetics of production of the puff inducer. (*A*) A cytoplasmic extract from Kc cells was aerated for the times indicated, boiled 2 minutes, centrifuged to remove precipitated proteins, and tested for the ability to induce puffs in polytene nuclei, according to Bonner (1981). (*B*) Kc cells were heat-shocked at 35°C for the times indicated, and cytoplasmic extracts were prepared and tested for puff-inducing ability either directly (o) or after boiling as in *A* (•). The puff at 87C was measured relative to the reference band, 87F.

INHIBITION OF PUFFING

Figure 1B presents an analysis of puff-inducing activity during the course of a heat shock of tissue culture cells. At various times after shift of the cells to 35°C, cytoplasmic extracts were prepared. A portion of the extracts were boiled for 2 minutes, and the boiled and unboiled samples were tested for puff-inducing activity. The boiled samples most likely provide an assay for the heat-stable protein(s) described above; they are produced rapidly on temperature shift and remain at a fairly constant level. The unboiled extracts, however, show the progressive inability to induce puffs in the isolated nuclei despite the presence of the heat-stable inducer(s). This is true of all of the major heat-shock puffs, although data for 87C alone are presented. This result provides evidence for an inhibitor of puffing, which is certainly present in cells after prolonged heat shock.

One simple interpretation of the results depicted in Figure 1B would be that heat shock results in the production of a molecule, perhaps a heat-shock protein, which represses the activity of the heat-shock-inducible genes. Certainly, many of the heat-shock proteins appear to have a nuclear localization in heat-shocked cells (Arrigo et al. 1980). However, this simple concept is ruled out by the observation that the development of the inhibition activity is insensitive to cycloheximide (not shown). More surprising is the result of the experiment depicted in Figure 2. We had observed that two flasks of "normal" tissue culture cells, presumably identical, were not: extracts from one exhibited the ability to induce puffs in isolated nuclei during "in vitro recovery from anoxia," while extracts from the other did not. The former could be made to resemble the latter by a prolonged heat shock (8 hr at 35°C). As shown in Figure 2, both the extract from the noninducing cells and the extract from the severely heat-shocked cells inhibited puff induction by extracts from the normal cells. Over the past several years, we have observed that the majority of flasks of tissue culture cells are of the noninducing, inhibitory type. Such cells were used for both of the experiments shown in Figure 1.

These data compel us to conclude that there is a heat-labile inhibitor of in vitro puffing that may be produced upon prolonged heat shock and that a similar inhibitor may be produced under ill-defined conditions of tissue culture. This inhibitor does not prevent the "activation" of the puff-inducing molecule(s) (Fig. 1), but rather its action.

The molecular identity of this inhibitor remains to be proven. Its production during heat shock of cells which do not show this activity initially (Fig. 2A) suggests that it may be related to the heat-shock system in a more-than-coincidental way. If it were a heat-shock protein, it would be necessary that it be present in significant quantities in non-shocked but inhibitory cells (such as those of Fig. 2B). The only heat-shock protein that displays such characteristics is hsp84, the 84-kD heat-shock protein which appears to be localized exclusively in the cytoplasm of heat-shocked cells (Arrigo et al. 1980). In an effort to rule out the possibility that hsp84 could be the inhibitor identified in the experiments described above, we added a partially purified preparation of hsp84 to the in vitro puffing system. Figure 3 shows that the preparation inhibited the induction of three of the puffs which were measured, but not of the fourth. The uninhibited puff, 67B, has been shown to be developmentally regulated (Ashburner 1972; Sirotkin and Davidson

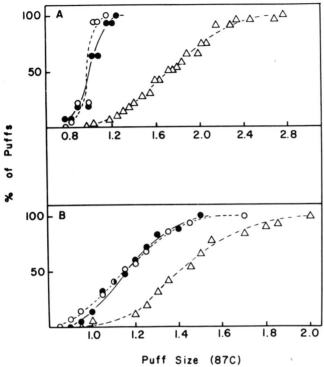

Figure 2

Inhibiton of puff induction. (*A*) Cytoplasmic extracts were prepared from normal cells and from normal cells which had been incubated at 35°C for 8 hours. Salivary glands were homogenized in the extract from normal cells, and the homogenate was mixed 1:1 with more normal extract (o, Δ) or extract from heat-shocked cells (•). The extracts were aerated, incubated, and analyzed cytologically. (o) Puff sizes after no incubation; (Δ, •) puff sizes after 2 hours incubation. (*B*) Similar to *A*, except the inhibitory extract (•) was obtained from nonheat-shocked cells.

1982); conceivably, the continued puffing of 67B in this experiment could reflect this developmental regulation.

It is unlikely that the results of Figure 3 reflect a general effect on puffing ability, for then 67B should have been inhibited as efficiently as the other puffs. Nonetheless, these experiments by no means verify that hsp84 is the inhibitor that we have observed. Indeed, they cannot, for they utilize the inhibition of a system which is extremely sensitive to changes in the buffer composition (Bonner 1981). Verification (or falsification) of this identity for the inhibitor will require the preparation of a high-titer antiserum to this heat-shock protein and/or the genetic analysis of the function of this protein in vivo.

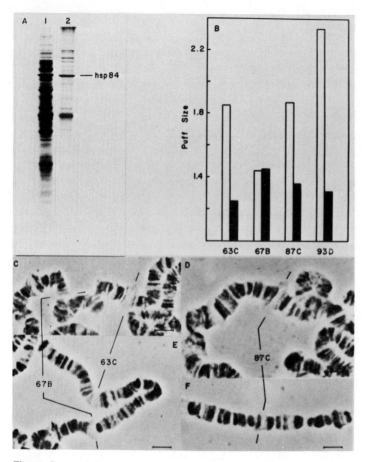

Figure 3
Inhibition of puff induction. hsp84 was partially purified from cells shocked for 8 hours at 35°C. (A) The initial cytoplasmic material run on a 12% SDS-polyacrylamide gel (lane 1) and the partially purified material (lane 2). (B – F) Addition of the hsp84 preparation to the in vitro puffing system results in inhibition of induction of most of the puffs. (B) Puff size measurements for 63C:63A, 67B:66E, 87C:87F, 93D:94A. (□) Without hsp84 preparation; (■) with hsp84 added. (C, D) Sample chromosomes from control incubation without hsp84. (E, F) Sample chromosomes from incubation with hsp84. Bar, 5 μm.

CONCLUSION

From the data presented here, it is reasonable to propose the following model for the induction of puffing by heat shock. In response to metabolic stress—whether it be a temperature shift or an anoxic shock—a purely cytoplasmic system causes the modification of a preexisting protein (or more than one). This protein must exist prior to the shock, for inhibitors of protein synthesis do not prevent induction of heat-shock puffs (not shown) and because the phenomenon occurs in cytoplasmic extracts in which protein synthesis must surely be inefficient. Once activated, this protein is capable of interacting with nuclei in a manner that results in the induction of puffing—with concomitant alterations in transcription and RNA polymerase localizations (Bonner 1981; Bonner and Kerby 1982). The nature of the alterations in the preexisting protein that convert it into the heat-stable inducer identified here remain to be determined.

The regression of puffs seen in vivo (Lewis et al. 1975) is, at least formally, similar to the decline in inducing ability seen in native extracts during the course of heat shock. The heat-labile inhibitor responsible for this effect may act by complexing with the heat-stable inducer or by complexing with the nuclei upon which the inducer must act; these two possibilities cannot be distinguished at this time. On the basis of the characteristics of the inhibitor, it is possible to make a tentative assignment of hsp84 to that role. If on further experimentation this assignment proves to be correct, then it indicates an interrelationship between the inducer and a heat-shock protein, representing a feedback regulatory system.

ACKNOWLEDGMENTS

This work was supported by NIH Grant No. GM26693.

REFERENCES

Arrigo, A.P., S. Fakan, and A. Tissières. 1980. Localization of the heat-shock-induced proteins in *Drosophila melanogaster* tissue culture cells. *Dev. Biol.* **78:** 86–103.

Ashburner, M. 1972. Puffing patterns in *Drosophila melanogaster* and related species. *Results Probl. Cell Differ.* **4:** 101.

Bonner, J.J. 1981. Induction of *Drosophila* heat shock puffs in isolated polytene nuclei. *Dev. Biol.* **86:** 409–418.

Bonner, J.J. and R.L. Kerby. 1982. RNA polymerase II transcribes all of the heat shock induced genes of *Drosophila melanogaster*. *Chromosoma* **85:** 93–108.

Compton, J.L. and J.J. Bonner. 1978. An in vitro assay for the specific induction and regression of puffs in isolated polytene nuclei of *Drosophila melanogaster*. *Cold Spring Harbor Symp. Quant. Biol.* **42:** 835–838.

Lewis, M., P.J. Helmsing, and M. Ashburner. 1975. Parallel changes in puffing activity and patterns of protein synthesis in salivary glands of *Drosophila*. *Proc. Natl. Acad. Sci.* **72:** 3604–3608.

Mirault, M.-E., M. Goldschmidt-Clermont, L. Moran, A.P. Arrigo, and A. Tissères. 1978. The effect of heat shock on gene expression in *Drosophila melanogaster*. *Cold Spring Harbor Symp. Quant. Biol.* **42:** 819–828.

Sirotkin, K. and N. Davidson. 1982. Developmentally regulated transcription from *Drosophila melanogaster* chromosomal site 67B. *Dev. Biol.* **89:** 196–210.

A *Drosophila* DNA-binding Protein Showing Specificity for Sequences Close to Heat-shock Genes

Robert S. Jack and Walter J. Gehring
Biozentrum
University of Basel
CH-4056 Basel, Switzerland

Exposure of *Drosophila* cells or tissues to a brief heat shock results in a rapid and fully reversible change in the pattern of transcriptional activity. Almost all genes active prior to heat-shock induction are switched off and a small set of so called heat-shock genes is switched on. The heat-shock genes code for a set of seven proteins with molecular weights of 84, 70, 68, 27, 26, 23, and 22 kD. The 70-kD protein is produced in amounts much greater than any of the others and is known as the major heat-shock protein (mhsp). It is encoded for by multiple genes which are clustered at two positions on the right arm of chromosome 3 at cytogenetic loci 87A7 and 87C1. The genes coding for the four small heat-shock proteins are located at cytogenetic locus 67B. The clear-cut differential expression of these genes makes the heat-shock system a simple and easily manipulated model for transcriptional regulation. The mechanisms by which such regulation is achieved in eukaryotes are unclear but by analogy with the situation in prokaryotes one might expect sequence-specific, DNA-binding proteins to play a major role. We have used cloned DNA segments originating from 87A7, 87C1, and 67B as probes in a filter-binding assay to search for proteins with specificity for sequences associated with the heat-shock genes.

THE 87A7 LOCUS

The 87A7 locus typically contains two closely spaced genes in opposite orientation, both of which have been recovered on the recombinant clone 122 (Goldschmidt-Clermont 1980). A variant form of the locus exists in which a large segment of DNA has been inserted between the two genes. The distal gene from this variant has been recovered in the recombinant 56H8. We have recently described the partial purification of a protein from *Drosophila* third instar larval nuclei, which in a filter-binding assay showed sequence-specific binding to a short region of DNA about 800 bp upstream from the transcription start of the 87A7-derived mhsp gene on 56H8 (Jack et al. 1981b). A possibility for examining the significance of this protein-binding site for the heat-shock response arose with the demonstration that a third variant form of the locus exists in which most of the DNA between the two genes plus the 5' half of the proximal gene has been deleted. The deletion does not, however, impair the ability of the distal gene to be expressed under heat-shock conditions (Udvardy et al. 1982). This variant has been recovered on two independently isolated clones called 123 and pHS1 (see Fig. 1). From the position of the protein-binding site present on 56H8, we would expect that a binding site would also be present on clone 122 that would be deleted in clone 123. We have used a filter-binding assay to test for the presence of binding sites on these clones. As expected, the region between the two genes on clone 122 contains at least one binding site and this site is deleted in clone 123. The unexpected finding was that both clones contain a second binding site which lies 3' to the proximal gene (Fig. 1). The fact that the binding site identified on 56H8 is absent on clone 123 rules out the idea that this site is uniquely involved in controlling the expression of the distal gene. However, since the binding sites are redundant, the possibility is raised that the protein-binding sites might be involved in marking the starting or stopping points of a controlled change in chromatin structure necessary for the expression of these genes. In this model, the two genes on 122 would be separately controlled. In the case of the 123-type locus, expression of the distal gene would be permitted by projecting the structural change from the binding site on the 3' side of the proximal gene across the distal transcription unit.

THE 87C1 LOCUS

The 87C1 locus contains multiple copies of the gene coding for the major heat-shock protein, two of which have been recovered on the recombinant 132E3. We have previously demonstrated the presence of two binding sites for the partially purified, DNA-binding protein on this

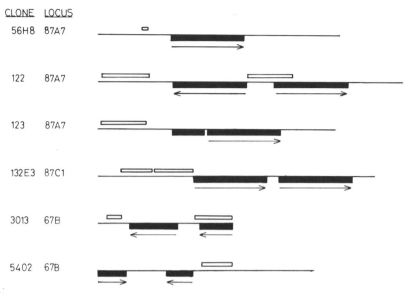

CLONE LOCUS

56H8 87A7

122 87A7

123 87A7

132E3 87C1

3013 67B

5402 67B

Figure 1

Cloned DNA segments on which binding sites for the partially purified DNA-binding protein have been identified. Clones 56H8, 122, and 123 derive from the 87A7 locus and contain genes coding for the major heat-shock protein as does clone 132E3, which derives from the 87C1 locus. Clone 3013 derives from the 67B locus and contains the genes coding for the 28-kD (*left*) and 23-kD (*right*) heat-shock proteins. Clone 5402 also derives from the 67B locus and contains the genes coding for the 26-kD (*left*) and 23-kD (*right*) heat-shock proteins. Transcription units are shown as solid boxes below the line with the direction of transcription indicated by the arrows. Unfilled boxes above the line indicate the positions of the restriction fragments that contain protein binding sites.

recombinant (Jack et al. 1981a). The positions of the restriction fragments which contain these binding sites are shown in Figure 1. In addition to the major heat-shock protein genes, the 87C1 locus also contains the coding sequences for the heat-inducible α β-sequences. We have looked for binding sites associated with these sequences by carrying out filter-binding experiments using the clone pDm704, which contains one β, one Y, and three α elements (Lis et al. 1978). No specific binding to any of these sequences was observed. We have sequenced the fragment from 56H8, which contains the protein-binding site, and compared this with the published sequences of the region of 132E3, which contains the binding site closest to the distal gene (Karch et al. 1981). The only striking similarity we find is a block of 12 nucleotides identical at both loci followed by 5 bp (56H8) or 3 bp (132E3) of unrelated sequence, followed in turn by a 7-bp block of identical sequence. Interestingly, this block of 7 conserved nucleotides is the precise inverse complement of the first 7 bases of the block of 12 and hence there is a palindrome consisting of the two 7-bp sequences separated by approximately one turn of the DNA helix. Whether or not this sequence is indeed the protein binding site will become clear once footprinting experiments have been carried out.

THE 67B LOCUS

The 67B locus contains the four genes that code for the small heat-shock proteins. We have used two genomic RI clones, 3013 and 5402 (R. Levis and W.J. Gehring, unpubl.), to look for specific binding of the protein in this region. Clone 3013 contains the genes coding for the 28-kD and 23-kD heat-shock proteins and clone 5402 contains the genes for the 26-kD and 22-kD proteins. Specific binding sites are found on both of these recombinants. The positions of the restriction fragments within which these binding sites lie are shown in the figure.

SUMMARY AND CONCLUSIONS

The positions of the protein-binding sites we have detected on clones containing heat-shock genes are summarized in Figure 1. We have failed to detect such binding sites on cloned *Drosophila* sequences that contain the histone genes, the alcohol dehydrogenase gene, or on a set of randomly selected clones which together constitute 50 kb of DNA. This does argue weakly for a concentration of binding sites around the heat-shock genes but of course the fraction of all nonheat-inducible sequences that has been tested is minuscule. It is apparent from Figure 1 that the binding sites are not placed at a fixed distance from the coding sequences nor are they invariably on the 5′ side of the gene. Indeed, the only rule to emerge so far is that they do not appear to be present within the coding regions. From this distribution of binding sites, it is clear that the protein does not play a direct role in the processes involved in positioning the polymerase at the precise start of transcription. It is possible, however, that the protein does play a part in controlling a change in chromatin structure necessary for heat-inducible transcription. Whether or not this is the case will emerge once the type of experiment described by Corces et al. (1981) has been extended to recombinants from which all binding sites have been deleted. Should these binding sites turn out not to be involved in controlling stress-inducible expression, then the possibility arises that sequences present close to the heat-shock transcription units are involved in processes unrelated to the heat-shock system. Our approach to the elucidation of the function of this protein will be to recover mutants in its structural gene.

ACKNOWLEDGMENTS

We thank Michel Goldschmidt-Clermont for clones 122 and 123, Mike Goldberg for the cloned *Drosophila* histone genes, David Goldberg for the cloned *Drosophila* ADH gene, and John Lis for pDm704.

REFERENCES

Corces, V., A. Pellicer, R. Axel, and M. Meselson. 1981. Integration transcription, and control of a *Drosophila* heat shock gene in mouse cells. *Proc. Natl. Acad. Sci.* **78:** 7038–7042.

Goldschmidt-Clermont, M. 1980. Two genes for the major heat shock protein of *Drosophila melanogaster* arranged as an inverted repeat. *Nucleic Acids Res.* **8:** 235–252.

Jack, R.S., C. Brack, and W.J. Gehring. 1981a. A sequence specific DNA binding protein from *Drosophila melanogaster* which recognises sequences associated with two structural genes for the major heat shock protein. *Fortschr. Zool.* **26:** 271–285.

Jack, R.S., W.J. Gehring, and C. Brack. 1981b. Protein component from *Drosophila* larval nuclei showing sequence specificity for a sort region near a major heat shock protein gene. *Cell* **24:** 321–331.

Karch, F., I. Török, and A. Tissières. 1981. Extensive regions of homology in front of the two hsp70 heat shock variant genes in *Drosophila melanogaster*. *J. Mol. Biol.* **148:** 219–230.

Lis, J., L. Prestidge, and D.S. Hogness. 1978. A novel arrangement of tandemly repeated genes at a major heat shock site in *D. melanogaster*. *Cell* **14:** 902–919.

Udvardy, A., J. Sümegi, E.C. Toth, J. Gausz, H. Gyurkovics, P. Schedl, and D. Ish-Horowicz. 1982. Genomic organisation and functional analysis of a deletion variant of the 87A7 heat shock locus of *Drosophila melanogaster*. *J. Mol. Biol.* **155:** 267–280.

Transcription In Vitro of *Drosophila* Heat-shock Genes in HeLa Cell-free Extracts

Richard Morimoto, Jean Schaffer, and Matthew Meselson

*Department of Biochemistry and
Molecular Biology
Harvard University
Cambridge, Massachusetts 02138*

Studies with *Drosophila melanogaster* have shown that heat shock almost immediately induces the transcription of several characteristic genes and, at the same time, greatly reduces the transcription of most other genes (for review see Ashburner and Bonner 1979; Findly and Pederson 1981). Thus, it appears that transcription of most genes is subject to heat-shock control, either stimulatory or inhibitory. In order to investigate the factors and interactions involved in these controls, we have begun to study the transcription of cloned heat-shock and nonheat-shock genes in vitro. Here we describe the transcription in vitro of heat-shock and nonheat-shock genes cloned from *D. melanogaster* in the HeLa cell-free transcription system of Manley et al. (1980). Reason to believe that this heterologous combination is suitable for such investigations is provided by the apparent accuracy of transcription and heat-shock control of a cloned *Drosophila* heat-shock gene introduced by transformation into mammalian cells (Corces et al. 1981).

We tested the ability of the HeLa cell-free system to transcribe truncated genes for the *D. melanogaster* heat-shock proteins hsp83, hsp70, and hsp28 and for the *D. melanogaster* nonheat-shock protein actin and a truncated transcription unit containing the major late promoter of

Table 1
DNA Templates and In Vitro Transcription Products

Gene	Plasmid	Reference	Restriction digest	Predicted run-off product (kb)	Observed α-amanitin-sensitive run-off product (kb)		Length of Drosophila or Ad2 DNA upstream from RNA initiation site (kb)
					(major)	(other)	
hsp83	λ6.1 = 3.6-kb HindIII–SalI fragment of λ6 in pBR322	Holmgren et al. (1981)	PstI	1.5	1.4	—	1.5
			SalI	2.1	2.2	2.8	1.5
hsp70	229.2 = 3.4-kb BglII fragment of pPW229 in pBR322	Livak et al. (1978); R. Morimoto (unpubl.)	BamHI	1.25	1.3	—	1.0
			SalI	2.05	2.1	—	1.0
hsp70	229.3 = 1.4-kb XhoI–BamHI fragment of pPW229 in pBR322	R. Morimoto (unpubl.)	BamHI	1.25	1.3	—	0.18

hsp28	88.5 = 4.7-kb EcoRI fragment of λ88 in pBR322	Corces et al. (1980)	SalI	1.4	1.4	0.75, 4.3	2.1
			HindIII	2.0	2.1	3.2	2.1
Actin	DMA4 = 4.2-kb BamHI–EcoRI fragment of λDMA4 in pBR322	Fyrberg et al. (1980)	PstI	0.8	0.8	4.2, 2.5, 2.4	1.2
			BglII	1.0	1.0	—	1.2
Adeno virus late promoter	2.2-kb BalI-E fragment of Ad2 in pBR322	Manley et al. (1980)	BamHI	1.75	1.8	—	0.4
			EcoRI	2.1	2.1	—	0.4

adenovirus 2. Reactions were conducted as described by Handa et al. (1981) for 60 minutes at 30°C in the presence of [^{32}P]UTP. The RNA products were glyoxylated and analyzed by electrophoresis in agarose gels. Each 20-μl reaction mixture contained 10 mM creatine phosphate, 50 μM ATP, CTP, and GTP, 5–10 μCi [^{32}P]UTP, 1.0–1.5 μg total added DNA, and usually 8 μl HeLa cell extract (approximately 2 × 10^6 cell equivalents). Sensitivity of transcription to α-amanitin was tested at 1 μg/ml of the inhibitor. Separate reactions were conducted with purified but unfractionated DNAs from two different restriction digests of each plasmid, chosen to provide truncated transcription templates of different lengths. For all templates, the efficiency of transcription from the promoter of interest was optimal at approximately 50 μg/ml of total added DNA. The efficiency of specific transcription as a function of HeLa extract concentration, measured at 4, 8, and 12 μl extract per 20-μl reaction mixture, was found to depend on the template. For the Ad2-Bal E and DMA 4 templates, the optimum was at 8 μl, while for 229.2 transcription was least at 12 μl and greatest at 4 μl.

As summarized in Table 1, the major α-amanitin-sensitive products produced in all cases were transcripts of sizes corresponding to the known distance from the RNA initiation site to the end of the template. Thus, both heat-shock and nonheat-shock *Drosophila* promoters function in the HeLa cell-free system. It may be noted that such function was observed even for the hsp70 plasmid 229.3, which contains only 183 bp of *Drosophila* DNA upstream from the initiation site.

We next sought to determine whether heat-shock and nonheat-shock templates are distinguishable in their transcriptional response to a heat shock given to the Hela cells just before extraction. Heat shock was accomplished by keeping the culture at 42°C for 15 minutes. Heat-shock and nonheat-shock extracts were prepared in parallel.

A suggestion that extracts of heat-shocked cells and nonshocked cells possess different transcriptional specificities is shown in Figure 1. In these experiments, a mixture of Ad2-Bal E and hsp70-229.2 templates is transcribed with extract made from heat-shocked cells, with extract from nonshocked cells and with different mixtures of the two extracts. It may be seen that transcription of Ad2-Bal E decreases and transcription of hsp70-229.2 increases with increasing percentage of heat-shock extract. Further experiments are needed to confirm that this progression is specifically due to the induction of the heat-shock response in the cells and not to some other difference in the extract and to determine whether the difference in response between hsp70-229.2 and Ad2-Bal E transcription is characteristic of heat-shock and nonheat-shock genes in general. If these conditions are met, the HeLa whole cell extract system will offer a promising means to identify the components and interactions involved in heat-shock transcriptional control.

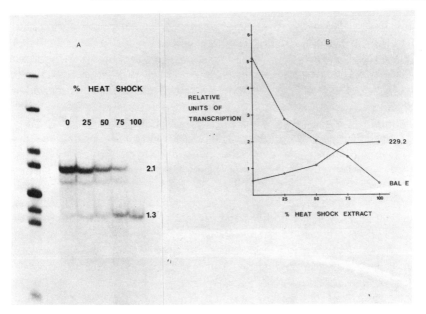

Figure 1
Transcription in mixed extracts. (A) Gel electrophoretic analysis of transcription products. A 1:1 mixture of templates Ad2-*Bal* E–*Eco*RI and 229.2-*Bam*HI was transcribed in control extract, heat-shock extract, and in mixtures of the two extracts. All reactions were performed with a total DNA concentration of 75 μg/ml and with a total of 8 μl of extract in a final volume of 20 μl. The number above each well is the percentage of heat-shock extract in the 8 μl. Ad2-*Bal* E–*Eco*RI gives a transcript of 2.1 kb and 229.2–*Bam*HI gives a transcript of 1.3 kb. Markers, shown on the left, are 7.6, 4.3, 2.5, 2.1, 1.5, 1.3, 1.1, and 0.52 kb. (B) Areas under the 2.1-kb and 1.3-kb bands of the autoradiograph shown in (A) were measured by densitometry under conditions giving a linear response to the amount of radioactivity in each band.

ACKNOWLEDGMENTS

We are grateful to the National Institutes of Health for support. R. Morimoto is supported by a Fellowship from the King Trust.

REFERENCES

Ashburner, M. and J. Bonner. 1979. The induction of gene activity in *Drosophila* by heat shock. *Cell* **17**: 241–254.

Corces, V., R. Holmgren, R. Freund, R. Morimoto, and M. Meselson. 1980. Four heat shock proteins of *Drosophila melanogaster* coded within a 12-kilobase region in chromosome subdivision 67B. *Proc. Natl. Acad. Sci.* **77**: 5390–5393.

Corces, V., A. Pellicer, R. Axel, and M. Meselson. 1981. Integration, transcription and control of a *Drosophila* heat shock gene in mouse cells. *Proc. Natl. Acad. Sci.* **78:** 7038–7042.

Findly, R. and T. Pederson. 1981. Regulated transcription of the genes for actin and heat-shock proteins in cultured *Drosophila* cells. *J. Cell Biol.* **88:** 323–328.

Fyrberg, E., K. Kindle, and N. Davidson. 1980. The actin genes of *Drosophila:* A dispersed multigene family. *Cell* **19:** 365–378.

Handa, H., R. Kaufman, J. Manley, M. Gefter, and P. Sharp. 1981. Transcription of simian virus 40 DNA in Hela whole cell extract. *J. Biol. Chem.* **256:** 478–482.

Holmgren, R., V. Corces, R. Morimoto, R. Blackman, and M. Meselson. 1981. Sequence homologies in the 5' region of four *Drosophila* heat-shock genes. *Proc. Natl. Acad. Sci.* **78:** 3775–3778.

Livak, K., R. Freund, M. Schweber, P. Wensink, and M. Meselson. 1978. Sequence organization and transcription at two heat shock loci in *Drosophila. Proc. Natl. Acad. Sci.* **75:** 5613–5617.

Manley, J., A. Fire, A. Cano, P. Sharp, and M. Gefter. 1980. DNA-dependent transcription of adenovirus genes in a soluble whole-cell extract. *Proc. Natl. Acad. Sci.* **77:** 3855–3859.

Regulation of the Heat-shock Response in *Drosophila* and Yeast

Susan Lindquist, Beth DiDomenico,
Gabrielle Bugaisky, Steven Kurtz,
Lawrence Petko, and Sandra Sonoda
The University of Chicago
Department of Biology
Chicago, Illinois 60637

REGULATION IN *DROSOPHILA*

The induction of heat-shock proteins (hsp) in *Drosophila* cells is accompanied by repression of preexisting transcription and translation to ensure that heat-shock proteins will be produced as rapidly as possible. This commitment to heat-shock synthesis is fully reversible. When cells are returned to 25°C, normal patterns of transcription and translation are restored and heat-shock synthesis is repressed. Careful examination of the recovery process has provided several insights on the regulation of the response.

Characteristics of Recovery

We subjected cells to many different heat shocks of varying severity and found that recovery obeyed certain rules:

1. When cells are returned to normal temperatures, heat-shock synthesis continues for a time which is proportional to the severity of the preceding treatment.
2. When a return to normal synthesis is initiated, it proceeds much more gradually than the initial shift to heat-shock synthesis.
3. Preexisting mRNAs act as a cohort, with messages for different proteins returning to translation at approximately the same rates.

4. Heat-shock mRNAs never act as a cohort. Their repression is invariably asynchronous, with hsp70 always the first and hsp82 always the last to be repressed.
5. Repression of hsp70 is coordinate with reactivation of normal synthesis.
6. Although induction patterns vary with the severity of heat shock, they are highly reproducible for any particular treatment, both in terms of the quantity of heat-shock proteins produced and the kinetics of their repression.

Autogenous Regulation

The extremely reproducible character of these inductions and recoveries suggests that the intensity of the response is tightly regulated. This might be accomplished in two ways: (1) An independent regulator might measure the stress and induce an appropriate but invariant program of synthesis, or, alternatively, (2) the response might be self-regulated, with heat-shock proteins acting to repress their own synthesis when they have accumulated at high enough levels to serve the requirements of stress.

As an initial test for autoregulation, we blocked the function of heat-shock proteins by incorporating amino acid analogs into them during heat shock. Recovery was then monitored at 25°C. Not only did the synthesis of normal proteins remain repressed, but the synthesis of heat-shock proteins continued to increase. This increase was found to be due to the combined effects of continued transcription and an increase in the stability of transcripts. Since heat-shock proteins are induced slowly by amino acid analogs at normal temperatures, extensive controls were performed to ensure that the analogs were not keeping cells locked into heat-shock synthesis simply by providing an additional form of stress. We found that if cells were allowed to synthesize functional heat-shock proteins before the addition of the analogs, heat-shock synthesis was repressed and normal synthesis was restored with standard kinetics. Thus, the response appeared to be autoregulated, with functional heat-shock proteins repressing their own transcription and destabilizing their own messages (B. DiDomenico, S. Lindquist, and G. Bugaisky, in prep.). In the limited space available here, we present data from later experiments which specifically analyzed the effects of autoregulation on these two processes.

Regulation at the level of transcription was investigated by blocking the synthesis of heat-shock proteins with cycloheximide and checking for overinduction of RNAs. Under normal circumstances, when cells are heat-shocked for 1 hour and then returned to 25°C, synthesis of heat-shock mRNAs is repressed and normal transcription is restored (Fig. 1A). In marked contrast, when cycloheximide was added immediately before heat shock, the heat-shock pattern of transcription was maintained indefinitely at

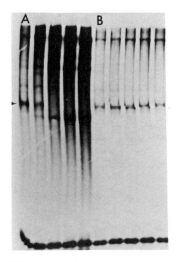

Figure 1
Cells were heat-shocked for 60 minutes at 36.5°C, returned to 25°C, and pulse-labeled with [³H]uridine in 30-minute intervals. RNAs were extracted, electrophoresed on methyl mercury gels, and visualized by fluorography. (A) Standard heat shock; (B) cycloheximide added immediately before heat-shock. The arrow indicates the position of hsp70 mRNA.

25°C (Fig. 1B). In control experiments, cells were allowed to synthesize heat-shock proteins for 1 hour before the addition of cycloheximide; synthesis of heat-shock mRNAs was repressed with standard kinetics.

RNA concentrations were monitored by Northern analysis. In the absence of cycloheximide, hsp70 mRNA was degraded during recovery, at a rate that exactly paralleled declining synthesis of the protein (DiDomenico et al. 1982). In cells treated with cycloheximide prior to heat shock, hsp70 mRNA accumulated in massive quantities. [³H]uridine pulse-chase experiments showed that heat-shock mRNAs were far more stable in the absence of heat-shock protein (B. DiDomenico et al., in prep.).

Posttranscriptional regulation of message stability was confirmed in a different type of experiment. Since heat-shock proteins are made on newly synthesized RNAs, the dose of heat-shock mRNA, and thereby the rate of heat-shock protein synthesis, can be regulated by adding actinomycin at various times during heat shock (Lindquist 1980). When actinomycin was added 60 minutes after temperature elevation (just before the return to 25°C), normal synthesis was restored within 3 hours (Fig. 2A). Synthesis of hsp70 had then declined to less than 25% of its initial value. When actinomycin was added 20 minutes after temperature elevation, production of heat-shock proteins proceeded at a slower rate, in proportion to the lower concentration of heat-shock mRNA. Note that synthesis continued at this level for at least 5 hours. After 7 hours, production of hsp70 had only

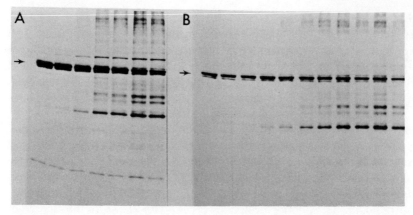

Figure 2
Schneider's cells were heat-shocked at 36.5°C with actinomycin added 60 minutes (A) or 20 minutes (B) after temperature elevation. After 60 minutes, all cells were returned to 25°C and aliquots were pulse-labeled with [³H]leucine in 30-minute intervals. The arrow indicates the position of hsp70.

declined by 50%. If heat-shock mRNAs had fixed rates of decay, heat-shock synthesis should have declined more rapidly in the second set of cells since the initial message concentrations were lower. Instead, the slower accumulation of heat-shock protein led to a slower rate of message decay. Recovery of normal protein synthesis was also delayed in proportion.

Conclusions

Our data suggest that heat-shock proteins are involved in regulating their own synthesis at both transcriptional and posttranscriptional levels. Their accumulation inhibits further transcription of heat-shock genes and destabilizes preexisting heat-shock transcripts in the cytoplasm. We do not know whether the proteins act individually or as a complex to effect this repression. Nor do we know whether they interact directly with their own genes and RNAs, or indirectly through some other aspect of their action in the cell.

The data further suggest that *Drosophila* cells use different translational mechanisms to change their patterns of protein synthesis during heat shock and recovery. At high temperatures, preexisting mRNAs are sequestered from translation and protected from degradation; with return to 25°C, heat-shock messages are degraded as they are repressed (DiDomenico et al. 1982).

It also appears that heat-shock proteins are required to release the block in normal transcription and translation (see Arrigo [1980] for additional data on transcription). Thus, heat-shock proteins appear to play a variety of roles in the cell, protecting them from the toxic effects of heat, acting to inhibit their own synthesis, and helping to reactivate preexisting synthesis. The specific molecular mechanisms by which they accomplish these functions are not known. They have, however, been shown to bind to chromatin and to RNA (Arrigo 1980; Storti et al. 1980; Velasquez et al. 1980). Interestingly, autogenous regulation is emerging as a general properly of nucleic acid-binding proteins in eukaryotic viruses and in prokaryotes.

Finally, it is worth noting that the self-limiting synthesis of heat-shock proteins provides a possible explanation for the fact that a bewildering and diverse array of agents are known to induce the heat-shock response. These agents might well act at separate control points, either by activating inducers, by blocking self-inhibition of heat-shock proteins at the transcriptional or posttranscriptional level, or by creating new sites for the binding of heat-shock proteins on chromatin and RNA, thus releasing a feedback loop.

THE HEAT-SHOCK RESPONSE IN *SACCHAROMYCES CEREVISIAE*

When yeast cells growing at 25°C are shifted to 39°C, heat-shock proteins are rapidly induced and normal protein synthesis is repressed. To date our investigations of the yeast response have focused on three issues.

RNA Metabolism

RNA metabolism in yeast cells differs from that of higher eukaryotes in that mRNAs have much shorter half lives. We wondered whether the mechanisms involved in regulating heat-shock induction would differ in a complementary way. Briefly, we found that yeast cells do not possess the special translational control seen in *Drosophila* cells that specifically discriminates against preexisting messages and sequesters them for later use. Instead, the rapid shift to heat-shock synthesis is accomplished primarily by a shift in transcription, and is facilitated by rapid degradation of many preexisting messages (Lindquist 1981).

Respiratory Metabolism

Because many inhibitors of respiration induce heat-shock RNAs and proteins, the assumption has grown that the response is a homeostatic reaction to respiratory stress. *Saccharomyces* should be an ideal organism to investigate this relationship since it is a facultative anaerobe with genetically and biochemically well-characterized respiratory mutations. Another advantage is that energy metabolism can be controlled simply by growing

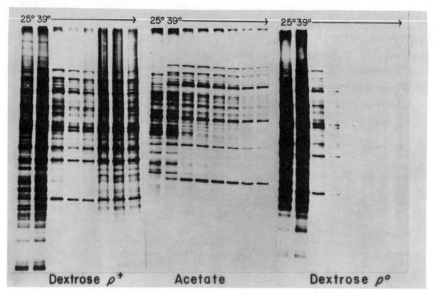

Figure 3
Aliquots of log-phase diploid cells (strain LM-1) were pulse-labeled for 20 minutes at 25°C, and after a shift to 39°C. Cells were grown on minimal medium with dextrose or acetate as a carbon source. The ρ^0 strain was created from LM-1 by growth in ethidium bromide 10 μg/ml. It was ρ^- by several metabolic criteria and ρ^0 by Dapi fluorescence.

cells on different carbon sources. When supplied with high concentrations of dextrose they grow by fermentation; when supplied with nonfermentable carbon sources such as acetate or ethanol, they grow by respiration.

As may be seen in Figure 3, the patterns of induction are distinctly different in these two states. As previously reported (Miller et al. 1979; McAlister et al. 1979), the response of dextrose-grown cells is transient. With continued incubation at high temperatures, heat-shock proteins are repressed and normal synthesis is restored. Cells growing by respiration in acetate display a lower level of constitutive heat-shock synthesis at 25°C than cells growing in dextrose. When shifted to 39°C, however, heat-shock proteins are strongly induced. Furthermore, once the heat-shock pattern is established, it is maintained indefinitely. Total protein synthesis gradually fades away, but heat-shock synthesis is not repressed unless cells are returned to 25°C. In this respect the response of respiring yeast cells is analogous to that of *Drosophila,* while the response of fermenting cells is like that of *Zea maize.*

Yet another pattern is seen in ρ^0 cells which lack mitochondrial DNA and are therefore completely deficient in respiration. The initial induction is identical to that of the isogenic ρ^+ strain, in terms of the number of proteins produced and the kinetics of their induction. With continued incubation at

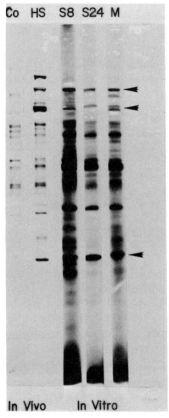

Figure 4
Diploid cells (strain LM-1) were grown to early stationary phase on acetate medium and then transferred to sporulation medium. RNAs were isolated from sporulating cultures after 8 hours (S8) and 24 hours (S24) and translated as previously described (Lindquist 1981). Translation products from a mixed sample of heat-shock and 25°C mRNAs (M) were run on the same gel. In vivo-labeled 25°C heat-shock and proteins from acetate grown cells were run on a separate, but comparable gel. The position of the 70-kD protein doublet and other major heat-induced proteins are marked with an arrow.

39°C, however, heat-shock synthesis is very rapidly reduced, and normal synthesis is not restored. The cells continue a very low but relatively constant level of background synthesis that only fades away with very extended incubations as cells begin to die.

Thermotolerance

We wondered whether heat-shock proteins were associated with the acquisition of thermotolerance which occurs in yeast as a natural consequence of certain physiological processes such as sporulation. Our investigations of this phenomenon are very preliminary, but intriguing. Since sporulating cells do not incorporate label efficiently, we assayed heat-shock gene expression by isolating RNAs from sporulating cultures and translating them in vitro. As may be seen in Figure 4, RNAs from sporulating cells direct the synthesis of heat-shock proteins in reticulocyte cell-free lysates. Note that the two 70-kD protein bands, which are expressed contemporaneously during heat shock, are expressed in a specific temporal order during sporulation.

Conclusions

These experiments raise more questions than they answer. It remains to be seen whether the striking differences in induction patterns observed in different respiratory states reflect an intimate connection between energy metabolism and heat-shock synthesis or have a more trivial basis. It might be argued that ρ^0 cells are unable to recover normal synthesis at 39°C because one of the enzymes of glycolysis is temperature sensitive; ρ^+ cells might be unable to grow in acetate at 39°C because of a temperature-sensitive respiration factor. Since ρ^+ cells do grow in dextrose at 39°C, however, these explanations by themselves are not adequate. Our results place certain restrictions on the role that mitochondria might play in the heat-shock response. Hopefully, yeast will prove more amenable to the analysis of this question than have other organisms. These different induction patterns have already proven of some practical benefit. The low level of constitutive heat-shock expression in acetate-grown cells (compared with dextrose grown cells) is being used, together with the strong induction at 39°C, to screen for yeast heat-shock genes by differential plaque hybridization with cDNAs.

The industrial fermentation literature contains numerous references to the fact that cells growing by dextrose fermentation are more resistant to heat treatment than cells growing on nonfermentable carbon sources. It is tempting to speculate that this phenomenon is correlated with their different constitutive levels of heat-shock synthesis. In this regard it is particularly intriguing that a massive induction of heat-shock genes occurs in cells that have never been exposed to heat, as part of the developmental process of sporulation. This finding greatly strengthens the argument that heat-shock proteins play a natural biological role in the acquisition of thermotolerance. We are currently determining whether the induction is specific to sporulating cells or is also found in haploids and in a/a or α/α diploids under other conditions that induce thermotolerance, such as nitrogen starvation.

REFERENCES

Arrigo, A.P. 1980. Investigation of the function of the heat-shock protein in *Drosophila melanogaster* tissue culture cells. *Mol. Gen. Genet.* **178:** 517–524.

Arrigo, A.P., S. Fakan, and A. Tissières. 1980. Localization of the heat-shock induced proteins in *Drosophila melanogaster* tissue culture cells. *Dev. Biol.* **78:** 86–103.

DiDomenico, B.J., G. Bugaisky, and S.L. Lindquist. 1982. *Proc. Natl. Acad. Sci.* (in press).

Lindquist, S.L. 1980. Varying patterns of protein synthesis in *Drosophila* during heat shock: Implications for regulation. *Dev. Biol.* **77:** 463–479.

———. 1981. Regulation of protein synthesis during heat shock. *Nature* **294:** 311–314.

McAlister, L. and D.B. Finkelstein. 1980. Alterations in translatable ribonucleic acid after heat shock of *Saccharomyces cerevisiae. J. Bacteriol.* **143:** 603–619.

McAlister, L.V., S. Strausberg, A. Kalaga, and D.B. Finkelstein. 1979. Altered patterns of synthesis induced by heat shock in yeast. *Curr. Genet.* **1:** 63–74.

McKenzie, S.L., S. Henikoff, and M. Meselson. 1975. Localization of RNA from heat-induced polysomes at puff sites in *Drosophila melanogaster. Proc. Natl. Acad. Sci.* **72:** 1117–1121.

Miller, M.J., N.H. Xuong, and E.P. Geiduschek. 1979. A response of protein synthesis to temperature shift in the yeast *Saccharomyces cerevisiae. Proc. Natl. Acad. Sci.* **76:** 5222–5225.

Spradling, A., S. Penman, and M.L. Pardue. 1975. Analysis of *Drosophila* mRNAs by in situ hybridization: Sequences transcribed in normal and heat shock cultured cells. *Cell* **4:** 395–404.

Storti, R., M. Scott, A. Rich, and M. Pardue. 1980. Translational control of protein synthesis in response to heat shock in *D. melanogaster* cells. *Cell* **22:** 825–834.

Tissières, A., H.K. Mitchell, and U.M. Tracy. 1974. Protein synthesis in salivary glands of *Drosophila melanogaster.* Relation to chromosome puffs. *J. Mol. Biol.* **84:** 389–398.

Velazquez, J., B. DiDomenico, and S. Lindquist. 1980. Intracellular localization of heat shock proteins in *Drosophila. Cell* **20:** 679–689.

The Heat-Shock Response in *Xenopus* Oocytes and Somatic Cells: Differences in Phenomena and Control

Mariann Bienz
MRC Laboratory of Molecular Biology
University Medical School
Hills Road, Cambridge CB2 2QH, England

Xenopus laevis cells show a typical heat-shock response on exposure to high temperature. Not only somatic cells, but also oocytes, are capable of synthesizing heat-shock proteins (hsp) at temperatures higher than 31°C. However, the heat-shock response of oocytes has features distinct from that of somatic cells. In particular, the underlying control mechanism acts at a different level, reflecting the biological state of the respective cell type.

THE HEAT-SHOCK RESPONSE OF SOMATIC CELLS IS CONTROLLED AT THE TRANSCRIPTIONAL LEVEL

When *Xenopus* tissue-cultured fibroblasts are heat-shocked for 20 minutes at temperatures between 31°C and 37°C, and the proteins are then labeled for an hour with [^{35}S]methionine and subsequently analyzed by SDS-gel electrophoresis, the induction of two new proteins can be observed (molecular weights approximately 70 and 30 kD). Although the proteins are labeled at the high temperature, there is no indication of a cessation of normal (25°C) mRNA translation: the total amount of [^{35}S]methionine incorporation remains the same throughout the different temperatures.

The control mechanism of this heat-shock response is probably transcriptional: *Xenopus* somatic cells accumulate hsp70 and hsp30 mRNA under heat-shock conditions. This was established by heat-shocking cells for 1 or 4 hours, extracting the RNA, and translating it in a reticulocyte cell-

177

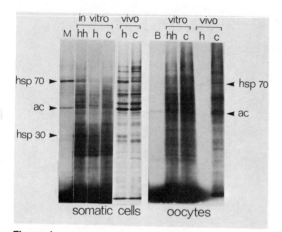

Figure 1
Synthesis of heat-shock mRNAs and proteins in *Xenopus* somatic cells and oocytes. Autoradiogram of a 12% SDS-polyacrylamide gel. Somatic cells or oocytes were heat-shocked at 37°C (h = 1-hr, hh = 4-hr heat-shock) or left at room temperature (c). The proteins were then labeled with [^{35}S]methionine for 1 hour at the same temperature and analyzed on a gel (in vivo) or the RNA was extracted, translated in a rabbit reticulocyte cell-free system, and the translational products were analyzed on a gel (in vitro). (M) Marker containing hsp70 (oocyte proteins, labeled in vivo at 37°C); (B) blank in vitro translation (i.e., no mRNA added). Arrows point to the heat-shock proteins (hsp70, hsp30) or to actin (ac).

free system. A comparison of the translational products to the ones programmed by nonheat-shocked RNA demonstrates very clearly that the hsp70 and hsp30 mRNAs have become the most abundant mRNA species during the 4-hour heat shock (Fig. 1). The concentration of the normal cellular mRNAs is not changed during heat shock. This does not imply that the transcription of normal mRNAs is actually continuing during heat shock, since the half-life of these mRNAs may be several hours. The most likely conclusion is that the heat-shock response of somatic cells is controlled at the transcriptional level as in *Drosophila*. Instead, theoretically, the control could involve a processing event or an increase of mRNA stability during heat shock. In contrast to *Drosophila*, *Xenopus* somatic ribosomes apparently do not undergo a heat-induced alteration which results in selective translation of heat-shock mRNAs (Scott and Pardue 1981).

THE HEAT-SHOCK RESPONSE OF OOCYTES IS CONTROLLED AT THE TRANSLATIONAL LEVEL

Xenopus oocytes subjected to the same heat-shock treatment as somatic cells show a different heat-shock response. In this case, the synthesis of hsp70 is switched on, whereas hsp30 synthesis is not detectable at all. Furthermore, the rate of normal mRNA translation is gradually decreased with increasing temperature. This is not due to mRNA degradation, since normal mRNA translation recovers almost completely within the first 1−2 hours after a shift back to normal temperature.

Considering the fact that *Xenopus* oocytes have an exceptional ratio of cytoplasmic volume to number of genes (at least 10^5 times higher than an average-sized somatic cell), it seems impossible that the synthesis of hsp70 could be dependent on de novo synthesis of the corresponding mRNA as is the case in somatic cells. Indeed, I find that enucleated or α-amanitin-injected oocytes are capable of synthesizing normal amounts of hsp70. This result suggests that the hsp70 mRNA is preformed in every oocyte. From the amount of $[^{35}S]$methionine incorporation into hsp70, it can be calculated that an oocyte must contain about $10^7 - 10^8$ copies or $10 - 100$ pg of hsp70 mRNA. This would require between 10 and 100 days to be synthesized, if the hsp70 genes are not highly reiterated. As expected, if oocytes are heat-shocked for 4 hours and the RNA extracted and translated in vitro, the translational products are identical to those obtained from nonheat-shocked RNA. Only traces of hsp70 can be identified, but the amount is not increased in the case of heat-shocked RNA (results not shown; see Bienz and Gurdon 1982).

In conclusion, the control mechanism of the *Xenopus* oocyte heat-shock response acts entirely at the translational level. It appears that hsp70 mRNA is synthesized during oogenesis and accumulates in an inactive, masked state. Heat shock induces its translation, whereas a shift back to normal temperature leads to slow remasking after a period of 8 hours. In addition, there is a second independent mechanism of translational control that leads to the cessation of normal mRNA translation under heat-shock conditions. This mechanism may be analogous to the one found in *Drosophila* (Scott and Pardue 1981). It is unlikely that the induction of hsp70 translation is solely a consequence of this second translational switch mechanism, because the recovered translational apparatus is capable of translating both normal and hsp70 mRNAs. Details about these two translational switches are published elsewhere (Bienz and Gurdon 1982).

DISCUSSION AND PROSPECTS

A comparison of the heat-shock response of *Xenopus* somatic cells and oocytes demonstrates striking differences. Not only is the phenomenology in each case different, but also the underlying control acts at a different level. This raises the questions: How can the two systems coexisting in one organism be explained in a compatible way and how is the transition from one to the other system achieved? At present, the two reaction mechanisms are described on a crude level and many detailed questions will only be solved when the genes are cloned. Yet I would like to formulate some ideas of what events may be involved during heat-shock gene expression.

The hypothesis is that early in oogenesis the hsp70 genes, but not the hsp30 genes, are transcriptionally activated. A heat shock is unlikely to be the natural trigger for this activation, but alternative inducers such as hormones or different ionic conditions (see other chapters of this volume) may account for it. It is conceivable that the establishment of the unusual lampbrush configuration of the chromosomes has a causal relation to the hsp70 gene activation. During oogenesis, hsp70 mRNA accumulates and is stored in an inactive state as a masked message. This process is thought to take place for a number of genes whose products may be required after fertilization in early development (reviewed by Davidson 1976). In the case of the hsp70 mRNA, the masking mechanism appears to be sensitive to a heat shock allowing hsp70 synthesis during heat shock, thereby fulfilling its function. Although the principal response of oocytes to heat shock does not involve transcription, the transcriptional reaction system is clearly present. Injected *Drosophila* hsp70 genes are transcriptionally induced and controlled in much the same way as they are in monkey cells (Pelham and Bienz, this volume). It is not proven that oocytes induce the transcription of their own heat-shock genes on heat shock: the produced mRNA is likely to be undetectable, since there is only one diploid set of genes per oocyte.

For the same reason, it is difficult to establish when, during embryonic development, the transition to the transcriptional regulation of heat-shock response occurs. Preliminary evidence indicates that neither hsp70 nor hsp30 synthesis is detectable from the fertilized egg to the early gastrula stage, although this latter stage contains at least 10^4 cells or diploid sets of genes. This could mean that the hsp70 mRNA is degraded on fertilization and the heat-shock genes or their inducing mechanism become incompetent to react on heat shock. It is interesting that the stages in question are extremely heat sensitive.

Concerning translational regulation, it is clear that the oocyte translational machinery shows selectivity for heat-shock mRNA translation, whereas the somatic one continues to translate normal cellular mRNAs at the same rate under heat-shock conditions. This implies that at least a part of the translational machinery is altered during development. It is generally observed that the rate of protein synthesis increases substantially after fertilization (Davidson 1976), and in sea urchins this has been attributed partly to the activity of the ribosomes (Danilchik and Hille 1981).

In summary, it seems that the study of the heat-shock response in *Xenopus* may provide insight into mechanisms that operate during oogenesis and early development. Using cloned heat-shock genes, I hope to discover when these genes are transcriptionally activated and how they are regulated. Analysis of the masking mechanism of hsp70 mRNA may serve as a model for understanding the molecular nature of the masking. Following the fate of hsp70 mRNA after fertilization and the transition of the activity of the translational machinery could give clues to which reactions are triggered by the maturation and fertilization of the oocyte.

ACKNOWLEDGMENTS

I thank John Gurdon, Hugh Pelham, and Paul Farrell for helpful discussion during this work and EMBO for providing a long-term fellowship to myself.

REFERENCES

Bienz, M. and J.B. Gurdon. 1982. The heat-shock response in *Xenopus* oocytes is controlled at the translational level. *Cell* (in press).

Danilchik, M.V. and M.B. Hille. 1981. Sea urchin egg and embryo ribosomes: Difference in translational activity in a cell-free system. *Dev. Biol.* **84:** 291–298.

Davidson, E.H. 1976. *Gene activity in early development,* 2nd edition. Academic Press, New York.

Scott, M.P. and M.L. Pardue. 1981. Translational control in lysates of *Drosophila melanogaster* cells. *Proc. Natl. Acad. Sci.* **78:** 3353–3357.

The Subcellular Compartmentalization of mRNAs in Heat-shocked *Drosophila* Cells

Dennis G. Ballinger
and Mary Lou Pardue
Department of Biology
Massachusetts Institute of Technology
Cambridge, Massachusetts 02139

The response of *Drosophila* to growth at elevated temperatures involves an alteration of gene expression at several levels. There is a transcriptional regulation in which most RNA synthesis and processing is interrupted, while the synthesis of a few RNAs is induced or enhanced. Gene expression is also regulated at the translational level.

At elevated temperatures, *Drosophila* tissue culture cells curtail the synthesis of the diverse set of proteins made at 25°C. Instead, the cells make a set of six to nine heat-shock-specific proteins. This alteration in the pattern of protein synthesis is accompanied by a dramatic change in the distribution of ribosomes on polysomes (McKenzie et al. 1975). In spite of these translational changes, however, the messages encoding the proteins normally synthesized only at lower temperatures (25°C mRNAs) are still found in the cytoplasm of heat-shocked cells. For example, mRNA extracted from cells grown at 36–37°C and translated in vitro encodes both the heat-shock proteins and the proteins synthesized in vivo at 25°C (Mirault et al. 1978; Storti et al. 1980). Thus, the alteration in the proteins synthesized during heat shock represents a specific recognition and selective translation of the mRNAs induced by heat shock (36°C mRNAs) from a pool which includes the preexisting cytoplasmic messages. At intermediate temperatures (particularly 33°C; see below), the 36°C mRNAs are efficiently produced but the cells appear to be unable to discriminate between messages and synthesize

183

proteins made at both 25°C and 36°C. Taken with other findings (Pardue et al. 1981), these results indicate that there is some mechanism by which *Drosophila* cells are able to discriminate between 25°C and 36°C mRNAs and that this mechanism is not an inherent property of the mRNAs, but is itself induced by growth at 36°C.

In order to characterize further the mechanisms of message discrimination during heat shock, we are studying the subcellular compartmentalization of mRNAs in heat-shocked and control cells. We have shown that at least part of the translational repression of 25°C mRNAs in heat-shocked cells is due to what appears to be a specific block of elongation on these mRNAs.

CHANGES IN PROTEIN SYNTHESIS IN THE CELL

The proteins labeled with [^{35}S]methionine during a 20-minute pulse after a 1-hour preincubation at 25°C, 33°C, and 36°C are shown in Figure 1B. The patterns of proteins synthesized are clearly different at each of these temperatures. The proteins synthesized at 25°C (in Fig. 1 exemplified by those abundant proteins migrating between actin and the large heat-shock proteins) are also made in a similar amount in cells grown at 33°C, but their synthesis is sharply curtailed at 36°C. In longer exposures of this and other gels it is possible to detect the 25°C proteins among the translation products of cells grown at 36°C, but we estimate that they are made in heat-shocked cells at an average of 5–10% of the rate of cells grown at 25°C. The predominant proteins synthesized at 36°C are the heat-shock proteins, which are also among the major proteins synthesized at 33°C.

These alterations in the pattern of protein synthesis are accompanied by a dramatic alteration in the distribution of polysomes during heat shock (Lindquist et al. 1975; Fig. 1A). For simplicity, only the patterns of polysomes found in cells grown at 25°C and 36°C for 1 hour are shown in Figure 1A. Extracts of cells grown at 33°C display what can best be described as a superimposition of these polysome profiles. During these studies we have taken care to standardize the growth conditions of the cells and to maximize three factors: (1) the degree of translational repression of 25°C mRNAs (by temperature optimization); (2) the recovery of ribosomes in the postmitochondrial supernatant loaded onto the gradients (by optimizing ionic conditions); and (3) the proportion of ribosomes found in the polysomal region of the gradients (by repeated serum stimulation). The conditions we use (described in the legend to Fig. 1) result in an average of 62% of the ribosomes sedimenting at a rate indicating more than two ribosomes on each message in cells

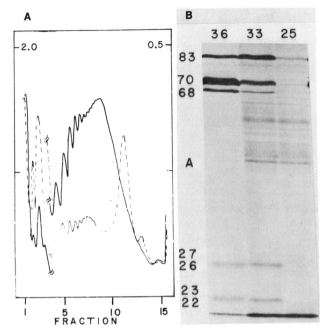

Figure 1
Analysis of protein synthesis in cells. (A) Optical density profiles of postmitochondrial extracts of cells grown at 25°C (———) and 36°C for 1 hour (- - -) analyzed on linear sucrose gradients. There is a fourfold decrease in the scale at the breaks, as indicated on the right margin. (B) Autoradiogram of proteins labeled during a 20-minute pulse with [35S]methionine in cells grown at 25°C, 33°C, and 36°C for 1 hour. The data in this and all other figures were collected on a single batch of extracts, but the results are general (Pardue et al. 1981). Schneider 2-L cells were fed to a density of $6-8 \times 10^6$ cells/ml the night before the experiment, and fed with 0.1 volume of fresh media at 4 hours and 1 hour before the culture was split into three equal aliquots. These were incubated for an additional hour at the indicated temperature. At 45 minutes, 1 ml of cell suspension was removed, and 7.5 μl of [35S]methionine was added to this small culture at 55 minutes. At 75 minutes, the cells were spun out of the small culture and resuspended in 150 μl of sample buffer, frozen, and subsequently analyzed on polyacrylamide gels as described in Mischke and Pardue (1982). Cells were spun out of the large cultures after 60 minutes, washed two times with phosphate-buffered saline, lysed in 0.01 vol (relative to the original culture volume) of Int K buffer (250 mM KCl; 2.5 mM $MgCl_2$; 20 mM HEPES [pH 7.2]; 10 mM EGTA), plus 5% RLS (Storti et al. 1981) and 1% Triton X-100. These extracts were maintained on ice for 15 minutes with occasional vortexing, then spun at 12,000g for 20 minutes. The resulting postmitochondrial supernatant was frozen in liquid N_2, and stored at −70° for further analysis. Aliquots of 400 μl were thawed, adjusted to gradient salts, and analyzed on 0.5−1.5 M sucrose gradients in an SW 27 rotor (gradient buffer: 500 mM KCl; 5 mM MgCl; 20 mM HEPES (pH 7.2); +/− 10 mM EGTA).

grown at 25°C. In cells grown at 33°C and 36°C, 44% and 31%, respectively, of the ribosomes sediment in the polysomal region of the gradients.

LOCATION OF 25°C mRNAs IN HEAT-SHOCKED CELLS

The majority of 25°C mRNAs sediment within sucrose gradients as if they were associated with polysomes, regardless of the temperature at which the cells are grown. This conclusion is based on several lines of evidence, all of which involve making extracts of cells, separating the polysomes from these extracts on sucrose gradients (Fig. 1A), and extracting total RNA from fractions of the gradients using phenol and chloroform.

IN VITRO TRANSLATION OF POLYSOMAL RNAs

We have utilized the ability of the rabbit reticulocyte lysate to translate added mRNA indiscriminately as an assay for functional mRNA in fractions from sucrose gradients (Pardue et al. 1981; Fig. 2). This method has identified three classes of abundant mRNAs based on their sedimentation from extracts of heat-shocked and control cells. The first class, consisting of the heat-shock-specific mRNAs, is found only in cells incubated at elevated temperatures. This class is localized to both the polysomal fractions (4–15, Figs. 1 and 2) and ribonucleoprotein (RNP) fractions (1–4, Figs. 1 and 2) of the sucrose gradients from cells grown at 33°C and 36°C. The second class appears to contain most of the abundant 25°C mRNAs (characterized by RNAs encoding the proteins migrating between actin and the large heat-shock proteins in Figs. 1 and 2). This class also contains the mRNA encoding hsp83, which is translated at all temperatures (Fig. 1). The class-two mRNAs are found in both RNP and polysomal fractions in cells grown at all three temperatures. The sedimentation of the 25°C mRNAs in this class indicates that they are associated with a similar number of ribosomes and that the amount of 25°C mRNAs appears to be similar at the three temperatures (see Δ in Fig. 2 and below). It should be emphasized that these 25°C mRNAs are associated with polysomes in cells grown at 36°C, but their translation products are severely underrepresented in these cells (Fig. 1B). The third class of mRNAs is found in fractions containing RNPs and polysomes in cells grown at 25°C and 33°C, but only in fractions containing RNPs in cells grown at 36°C. This class appears to be a minor subset of the 25°C mRNAs, mostly encoding small polypeptides.

NORTHERN HYBRIDIZATION TO POLYSOMAL RNAs

The presence of RNAs in fractions from the sucrose gradients was also assayed by denaturing the RNA by treatment with glyoxal, separating the RNA on agarose gels, transferring the RNA to nitrocellulose, and hybridizing these filters with cloned DNA probes coding for known

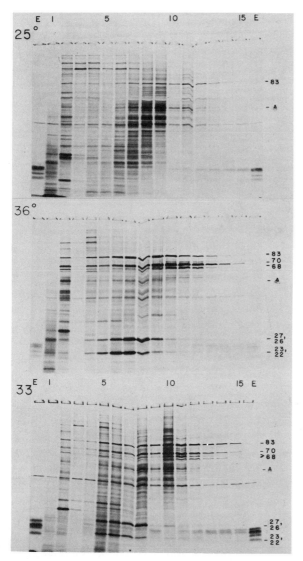

Figure 2
Autoradiographs of the proteins encoded by RNA from polysome gradient fractions. RNA was isolated from fractions of the sucrose gradients shown in Fig. 1A (and a gradient from cells grown at 33°C) by extraction with phenol and chloroform. Equal aliquots of RNA from equal-sized fractions of the gradients (except fractions 1, 2, and 3 in all cases were half-sized aliquots) were translated in the rabbit reticulocyte lysate (Mischke and Pardue 1982). The products were analyzed on SDS polyacrylamide gels (Pardue et al. 1981). All the translations were within the linear range of added RNA concentration versus counts incorporated into protein. The positions of the major heat-shock protein and an unidentified 25°C protein (Δ) are given to the right of the autoradiograms. E, Endogenous protein synthesis.

RNAs. This approach allows the quantitation of a particular RNA of interest with increased sensitivity over the translation assay. By comparison with the assay for functional RNA, this approach would also allow the detection of sequences that have been irreversibly converted to a nontranslatable form. No irreversibly inactivated RNAs of this type were detected.

We have studied the distribution of nine RNAs encoding known proteins and rRNA in polysome gradients of cells grown at 25°C and 36°C in this way. The results confirm the existence of the three classes identified by the translation assays described above. The 36°C mRNAs encoding hsp70, hsp27, hsp26, hsp23, and hsp22 belong to the first class above and are found only in cells grown at 36°C. Three 25°C mRNAs (encoding α- and β-tubulin and cytoplasmic actin, none of which is translated in heat-shocked cells) and the mRNA encoding hsp83 belong to the second class above. For example, α-tubulin mRNA is found on polysomes in cells grown at all three temperatures (Fig. 3A). There is a decrease in the amount of α-tubulin sequences in cells grown at 33°C and 36°C (42% and 48%, respectively), but there is little alteration in the average number of ribosomes associated with each mRNA at the three temperatures (10–13 ribosomes per message, Fig. 3B). Thus, for instance, if each ribosome associated with α-tubulin mRNA in 36°C cells were making α-tubulin molecules at the same rate that it does at 25°C, there would be only a 50% reduction in α-tubulin synthesis. In contrast to this expectation, cells grown at 36°C actually make α-tubulin at 5–10% of the rate of cells grown at 25°C (Fig. 1B). Thus, a ribosome which is associated with α-tubulin mRNA in cells at 36°C is making five- to tenfold less protein than a corresponding ribosome in cells at 25°C. The simplest, but not the only, explanation of these data is that elongation is specifically blocked or drastically reduced on 25°C mRNAs in cells grown at 36°C. Only a single member of the third class of mRNAs has been studied so far. The RNA encoding a small ribosomal protein is found on polysomes only in cells grown at 25°C.

THE EVIDENCE THAT THE 25°C mRNAs ARE ACTUALLY RIBOSOME ASSOCIATED

The association of 25°C mRNAs with polysomes in heat-shocked cells is stable to 0.5 M KCl and 1% Triton X-100, indicating that nonspecific interactions are unlikely to account for their sedimentation properties. We have also mixed RNA labeled with [^{3}H]uridine in cells grown at 25°C with the extracts of heat-shocked and control cells before separating these extracts on polysome gradients. There is no evidence in this type

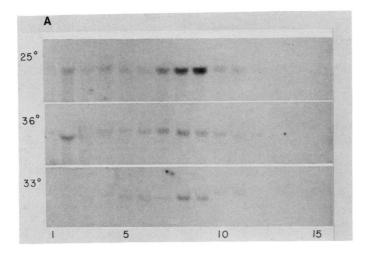

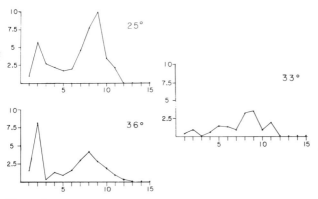

Figure 3

Analysis of α-tubulin RNA in fractions of polysome gradients by hybridization with a cloned gene sequence. (A) Autoradiograms of filters containing transferred RNA probed with [^{32}P]labeled, α-tubulin DNA sequences. (B) Densitometer tracings of the autoradiograms in A. *Ordinate*, arbitrary units of area contained in densitometer tracings of the α-tubulin hybridization. *Abscissa*, fraction number. Fractions of RNA from the samples analyzed in Fig. 2 were denatured by treatment with glyoxal, separated on 0.7% agarose gels, and transferred to nitrocellulose. Nick-translated DNA from a genomic α-tubulin clone was hybridized to these filters. All manipulations as described in Mischke and Pardue (1982). Hybridization intensity was within the linear range of detection (one-fourth of the amount of RNA processed in parallel gives one-fourth of the area under hybridization peaks).

of experiment for nonspecific associations of the labeled RNAs. Finally, in extracts treated with 0.025 M EDTA, all of the RNAs detectable by translation are released from the polysome region into the RNP region of sucrose gradients. Thus, by classical criteria, the 25°C mRNAs are associated with ribosomes in heat-shocked cells.

SUMMARY AND CONCLUSIONS

The translational regulation of protein synthesis in heat-shocked *Drosophila* cells appears at least in part to involve a selective inhibition of elongation on 25°C mRNAs. The evidence for this statement comes from a detailed analysis of the subcellular compartmentalization of mRNAs in heat-shocked and control cells. We have identified a class of mRNAs that contains most of the abundant 25°C mRNAs and is found on polysomes of approximately the same size in cells grown at 25°C, 33°C, and 36°C. However, the translation products of these 25°C mRNAs are only metabolically labeled in cells grown at 25°C and 33°C, indicating that the ribosomes associated with these RNAs in cells at 36°C are not making protein at the normal rate. The simplest, but not the exclusive, interpretation of these data is that there is a specific block of elongation on 25°C mRNAs in heat-shocked cells. This mechanism of translational repression has not been reported in systems studied in detail to date.

ACKNOWLEDGMENTS

This work was supported by a grant from the National Institutes of Health. D.G.B. is a graduate fellow of the Whitaker Health Sciences Fund. We are indebted to D. Mischke for encouragement, the transfer of knowledge, and the tubulin clone.

REFERENCES

McKenzie, S.L., S. Henikoff, and M. Meselson. 1975. The localization of RNA from heat induced polysomes at puff sites in *D. melanogaster. Proc. Natl. Acad. Sci.* **72:** 1117–1121.

Mirault, M.-E., M. Goldschmidt-Clermont, C. Moran, A.P. Arrigo, and A. Tissières. 1977. The effect of heat shock on gene expression in *Drosophila melanogaster. Cold Spring Harbor Symp. Quant. Biol.* **42:** 819–827.

Mischke, D. and M.L. Pardue. 1982. Organization and expression of α-tubulin genes in *Drosophila melanogaster. J. Mol. Biol.* **156:** 449–466.

Pardue, M.L., D.G. Ballinger, and M.P. Scott. 1981. The expression of heat shock genes in *Drosophila melanogaster. ICN-UCLA Symp. Mol. Cell. Biol.* **28:** 415–427.

Storti, R.V., M.P. Scott, A. Rich, and M.L. Pardue. 1980. Translational control of protein in synthesis in response to heat shock in *D. melanogaster* cells. *Cell* **22:** 825–834.

Translation and Turnover of *Drosophila* Heat-shock and Nonheat-shock mRNAs

Christiane Krüger and
Bernd-Joachim Benecke
Lehrstuhl für Biochemie
Ruhr-Universität Bochum
D-463 Bochum, Federal Republic of Germany

The *Drosophila* heat-shock response has been widely used as a model system to study regulation of gene expression in eukaryotes. In addition to the dramatic changes observed at the level of gene activity and transcription (for a review see Ashburner and Bonner 1979), recent investigations indicate a pronounced translational control in *Drosophila* cells upon heat shock (Storti et al. 1980; Krüger and Benecke 1981). This regulation at the level of polypeptide synthesis is achieved by a strong discrimination between heat-shock-specific mRNA and normal cellular mRNA molecules. 25°C mRNA, which is present in heat-shocked cells in unaltered amounts, is no longer translated efficiently in vivo, although the overall translatability of these messages appears unaffected when assayed in vitro in a cell-free system obtained from either rabbit reticulocytes or *Drosophila* cultured cells. A lysate from heat-shocked *Drosophila* cells, however, effectively discriminates the 25°C mRNA, whereas heat-shock-specific messages are not affected, thus reflecting the in vivo situation. Furthermore, it has been shown that this discrimination is not due to a simple competition of heat-shock versus nonheat-shock messages.

In the present report, we have studied the in vitro translation and biological stability of both populations of mRNA under different culture conditions and have investigated the fate of heat-shock-specific mRNA under recovery conditions, i.e., during transition of heat-shocked cells to normal culture conditions at 25°C.

IN VITRO TRANSLATION OF 25°C mRNA

We have studied the effects of supplementation of heat-shock lysates with nonheat-shock lysate fractions and vice versa, in order to understand the mechanism of translational control observed upon heat-shock in *Drosophila* cells. The results of these studies are summarized in Table 1. It appears that there is no soluble factor present in the heat-shock lysate capable of suppressing the in vitro translation of 25°C mRNA in control lysates. On the other hand, heat-shock lysates supplemented with ribosomes isolated from control cells regain the ability of effective in vitro translation of normal cellular messages. Initiation factors were purified from rabbit reticulocyte lysates or from *Drosophila* Kc cells by DEAE-cellulose chromatography. Supplementation of the heat-shock lysate with these initiation factor preparations did not restore the translation of 25°C mRNA. These data demonstrate that the lesion does not take place at the level of initiation factors. The activity of these initiation factor preparations was verified in a nonhemin-treated rabbit reticulocyte lysate. However, they reproducibly reduced the in vitro translation efficiency of *Drosophila* lysates, irrespective of the message used. One might speculate that the *Drosophila* cell-free translation extract is such a delicately balanced system that the addition of protein fractions tends to unbalance the whole system.

Table 1
In Vitro Translation of 25°C mRNA

A. In control lysate with heat-shock components added

Control	100.0%
Control + cytoplasm (control cells)	86.3%
Control + cytoplasm (heat-shocked cells)	86.0%

B. In heat-shock lysate with control components added

Heat-shock lysate	100.0%
Heat-shock lysate + initiation factors (reticulocytes)	92.2%
Heat-shock lysate + initiation factors (*Drosophila*)	95.0%
Heat-shock lysate + control ribosomes	180.2%

In vitro translation of 25°C mRNA was studied in *Drosophila* control lysates supplemented with particle-free cytoplasm of control or heat-shocked cells (A) or in heat-shock lysates supplemented with ribosomes or initiation factors isolated from equal amounts of control cells (B). Isolation of initiation factors was by 0.5 M KCl wash of ribosomes, dialysis, DEAE-cellulose chromatography, and ammonium sulfate precipitation.

STABILITY OF mRNA

Since results obtained in other systems show that modulations in the translation efficiency of eukaryotic mRNA may be accompanied by altered biological stability of these messages (Benecke et al. 1978; Farmer et al. 1978), we have analyzed the lifetime of both heat-shock and nonheat-shock mRNAs in cultured *Drosophila* cells. Normal cellular messages labeled for 14 hours had a turnover with an apparent half-life of about 7 hours (Fig. 1A). At elevated temperature (37°C), the decay rate of the mRNA molecules was markedly increased (Fig. 1B) resulting in a half-life of 2.5−3 hours. When cells were heat-shocked only for shorter periods (for example 1 hr) and were allowed to recover at 25°C, the initial rapid decay of the 25°C mRNA slowed down and coincided with the normal turnover rate (Fig.1B, broken line). Labeled under heat-shock conditions and analyzed during prolonged incubation at 37°C (Fig. 1C), mRNA revealed decay rates essentially identical to those obtained with 25°C mRNA under these conditions. This indicates that heat-shock mRNAs are subjected to the same degradation process that regulates the lifetime of normal cellular messages.

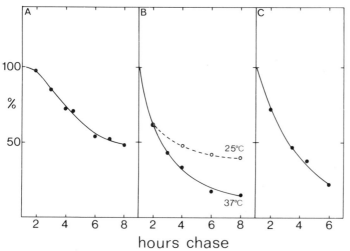

Figure 1

Stability of heat-shock and nonheat-shock mRNA. *Drosophila* Schneider cells were labeled for 14 hours at 25°C (25°C mRNA) or for 1 hour starting 20 minutes after transfer to 37°C (heat-shock [hs] mRNA). At the end of labeling, the cells were transferred to isotope-free medium supplemented with 1 mM unlabeled uridine. At the times indicated, cells were harvested and poly(A)$^+$ RNA was isolated from the cytoplasm, followed by SDS-sucrose gradient analysis. The radioactivity associated with the mRNA peak at 1-hour chase was set as 100%. (*A*) Normal cellular mRNA at 25°C; (*B*) normal cellular mRNA at 37°C (●) or at 25°C after 1-hour heat shock (o); (*C*) heat-shock mRNA at 37°C.

THE SHUT-OFF OF HEAT-SHOCK POLYPEPTIDE SYNTHESIS

A rapid shut-off of heat-shock-specific polypeptide synthesis is observed in cells which recover from heat shock (Fig. 2). After 2 hours at 25°C, a marked decrease in the synthesis of heat-shock proteins followed by an almost complete cessation at about 4 hours was observed. Interestingly, this switch-off does not affect all heat-shock proteins simultaneously, smaller polypeptides being synthesized significantly longer than larger ones. Comparable results were obtained by in vitro translation of poly(A)$^+$ RNA isolated from heat-shocked cells which were allowed to recover for various time periods. Figure 3 shows the in vitro translation products that were obtained in the rabbit reticulocyte lysate with mRNA isolated after 1, 3, and 5 hours at 25°C, respectively. Tne reduced rates of synthesis of individual heat-shock proteins parallel those observed in vivo. Again, the synthesis of larger heat-shock polypeptides is terminated faster than that of the smaller products. Identical data were obtained from in vitro translation experiments with a cell-free system of *Drosophila*, irrespective whether the cells were

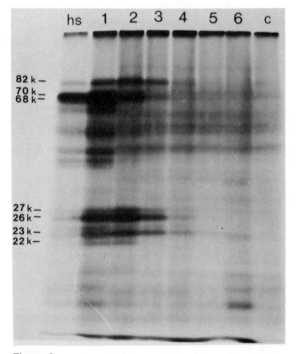

Figure 2
Heat-shock-specific polypeptide synthesis during the transition from 37°C to 25°C. Cells were labeled with [³H]leucine for 30 minutes at 37°C (heat shock [hs]) or at the times indicated after transfer from 37°C to 25°C (1–6 hr) or at 25°C without any heat shock prior to labeling (c). The molecular weight of heat-shock proteins is indicated.

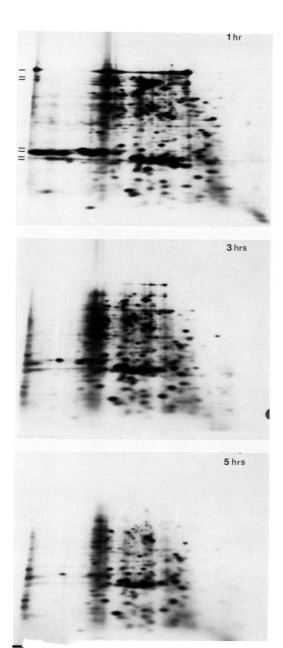

Figure 3
In vitro translation of polysomal mRNA isolated from *Drosophila* cells after various times of recovery. Kc cells subjected to heat-shock for 1 hour were transferred to 25°C and poly(A)$^+$ RNA was isolated after 1, 3, and 5 hours, respectively. Isolation of mRNA, in vitro translation, and analysis of the products by two-dimensional gel electrophoresis was as described by Krüger and Benecke (1981). The bars indicate the positions of the heat-shock proteins as shown in Fig. 2.

heat-shocked or not (not shown). This rapid shut-off of heat-shock-specific protein synthesis does not allow any conclusions as to the presence and/or amounts of heat-shock mRNAs because of the effective translational control involved in the heat-shock response. Therefore, the fate of heat-shock-specific mRNA was analyzed by hybridization of poly(A)[+] RNA labeled under heat-shock conditions and isolated after 1, 3, and 5 hours of chase. The RNA was hybridized to cloned DNA corresponding to the major heat-shock gene coding for the 70-kD polypeptide (this clone was kindly provided by Dr. Tissières, Geneva). Table 2 illustrates the rapid disappearance of heat-shock mRNA in response to recovery conditions. The decay rates of the mRNA analyzed are very similar or identical to the corresponding protein synthesis rates in vivo and in vitro. In contrast to the onset of the heat-shock response, no translational control is involved in the restoration of normal macromolecular metabolism under recovery conditions. Instead, heat-shock mRNA is selectively degraded much faster than cellular mRNAs would disappear under these conditions and even faster than heat-shock mRNA turnover at 37°C.

Accordingly, the regulation at the level of protein synthesis that is observed in response to heat-shock and the transition to normal physiological conditions seem to be mediated by a series of well-coordinated events. This system will be used for investigating in detail the underlying mechanisms of this translational control system on the molecular level and may provide further insight in the regulation of eukaryotic gene expression.

ACKNOWLEDGMENTS

We wish to thank Dr. Tissières for providing the cloned DNA, Dr. H. Trachsel for valuable discussions in connection with the isolation of initiation factors, and the Deutsche Forschungsgemeinschaft for financial support.

Table 2
Hybridization of mRNA to the Major 70-kD Heat-shock Gene

RNA	cpm
25°C mRNA	73
Heat-shock mRNA	1707
Heat-shock mRNA (after 1 hr at 25°C)	478
Heat-shock mRNA (after 3 hr at 25°C)	109
Heat-shock mRNA (after 5 hr at 25°C)	79

Drosophila Schneider cells were labeled with [³H]uridine for 1 hr at 25°C, 37°C, or 37°C followed by incubation for 1, 3, and 5 hr at 25°C under chase conditions. Cytoplasmic poly(A)[+] RNA was hybridized to cloned heat-shock DNA immobilized on nitrocellulose filters for 24 hr at 42°C as described in detail by Kafatos et al. (1979).

REFERENCES

Ashburner, M. and J.J. Bonner. 1979. The induction of gene activity in *Drosophila* by heat shock. *Cell* **17:** 241–254.

Benecke, B.J., A. Ben-Ze'ev, and S. Penman. 1978. The control of mRNA production, translation and turnover in suspended and reattached anchorage-dependent fibroblasts. *Cell* **14:** 931–939.

Farmer, S., A. Ben-Ze'ev, B.J. Benecke, and S. Penman. 1978. Altered translatability of messenger RNA from suspended anchorage-dependent fibroblasts: Reversal upon cell attachment to a surface. *Cell* **15:** 627-637.

Kafatos, F., C.W. Jones, and A. Efstratiadis. 1979. Determination of nucleic acid sequence homologies and relative concentrations by a dot hybridization procedure. *Nucleic Acids Res.* **7:** 1541–1552.

Krüger, C. and B.J. Benecke. 1981. In vitro translation of *Drosophila* heat-shock and non-heat-shock mRNAs in heterologous and homologous cell-free systems. *Cell* **23:** 595–603.

Storti, R.V., M.P. Scott, A. Rich, and M.L. Pardue. 1980. Translational control of protein synthesis in response to heat shock in *D. melanogaster*. *Cell* **22:** 825–834.

Preferential Translation of Heat-shock mRNAs in HeLa Cells

Eileen Hickey and Lee A. Weber
Biology Department
University of South Florida
Tampa, Florida 33620

HeLa cells shifted to 42°C from the normal incubation temperature of 37°C undergo dramatic quantitative and qualitative changes in protein synthesis. First, the overall rate of protein synthesis rapidly declines to 20–30% of that found at 37°C. This is followed by induction of the three size classes of heat-shock polypeptides that are characteristic of vertebrate cells. With our gel systems, the major human heat-shock polypeptides show molecular weights of 27,000 (p27), 70,000 (p70), and 80,000 (p80). The first two size classes of heat-shock polypeptides are resolved into multiple components by two-dimensional electrophoresis (Hickey and Weber 1982).

Synthesis of the heat-shock polypeptides reaches a maximum between 3–5 hours of hyperthermia, and then decreases to a low level by 8 hours at 42°C. This increase and decline in heat-shock polypeptide synthesis reflects the accumulation and subsequent loss of heat-shock polypeptide mRNA sequences. This was initially shown by in vitro translation of cytoplasmic RNA isolated from heat-treated cells, using the reticulocyte lysate (Hickey and Weber 1982). Analysis of the translation products shows that synthesis of each size class of heat-shock polypeptide generally corresponds to the amount of specific translatable heat-shock polypeptide mRNA present at any time. More recently, we have used a cloned DNA fragment (B 8, Craig et al. 1979) containing a *Drosophila* p70 structural gene as a hybridization probe to measure changes in the level of the human mRNA sequence. Total cytoplasmic RNA isolated from control cells and cells incubated at 42°C for increasing periods of time was fractionated on agarose gels under denaturing conditions, transferred to nitrocellulose, and hybridized with the ^{32}P-labeled DNA probe. Autoradiography revealed hybridization with a single 2.6-kb RNA species which is barely detectable in control cell RNA, and which increases and decreases in abundance in parallel with p70 synthesis. These results indicate that the major long-term changes in heat-shock polypeptide synthesis represent changes in the amount of heat-shock mRNAs in the cells.

Despite the large reduction in the overall rate of protein synthesis at 42°C, several lines of evidence argue that most mRNAs present at 37°C (37°C RNAs) remain stable during heat shock. Protein synthesis rapidly recovers to the control rate upon return to 37°C of cells subjected to up to 1 hour of heat shock. This recovery occurs in the absence of new RNA synthesis (McCormick and Penman 1969). Translation of total cytoplasmic RNA from control and heat-shocked cells showed no large changes in the concentration of 37°C mRNAs (Hickey and Weber 1982). Finally, we have prepared a library of cDNA sequences from total polyadenylated RNA from heat-shocked cells. When individual plasmids are hybridized with ^{32}P-labeled cDNA to mRNA from heat-shocked or 37°C cells, the majority of plasmids hybridize equally well with cDNA from both sources. This suggests that the abundance of most 37°C mRNAs does not change during heat shock (E. Hickey and L.A. Weber, unpubl.). The heat-shock mRNAs are, therefore, translated in the presence of a large amount of mRNA coding for 37°C polypeptides, under conditions where the overall rate of translation is severely inhibited. The experiments described below were designed to determine the mechanism by which rapid synthesis of heat-shock polypeptides continues under conditions that apparently inhibit the translation of the majority of cellular mRNAs.

HEAT-SHOCK mRNAS ARE TRANSLATED EFFICIENTLY DESPITE THE INHIBITION OF POLYPEPTIDE CHAIN INITIATION CAUSED BY HEAT SHOCK

HeLa cells undergo a reduction in the size and number of polyribosomes within 10 minutes following transfer to 42°C. This change in ribosome distribution is the result of a lesion in the initiation step of mRNA translation which underlies the inhibition of protein synthesis (McCormick and Penman 1969). By 1 hour of heat shock, the number of large polyribosomes increases, although control levels are not restored (McCormick and Penman 1969; E. Hickey and L.A. Weber, unpubl.). The increase in large polyribosomes corresponds temporally with the onset of heat-shock polypeptide synthesis.

We have analyzed the distribution of specific mRNAs among the polyribosomes by fractionating ribosomal particles on sucrose gradients and translating RNA isolated from each fraction in the reticulocyte lysate. At 37°C, mRNA coding for actin is predominantly found associated with polyribosomes containing from 12–17 ribosomes. Following 90 minutes at 42°C, actin mRNA is most abundant in a fraction containing 7–12 ribosomes. An increase in actin mRNA sedimenting slower than 80S monoribosomes is also noted. Similarly, mRNAs coding for histones are shifted from polyribosomes containing 4–5 ribosomes at 37°C to structures containing 1–2 ribosomes and to the postribosomal supernatant, at 42°C. The heat-shock mRNAs encoding p70 and p80 are found on large polyribosomes containing 12–20 ribosomes at 42°C. Many mRNAs coding for polypeptides as large or larger than heat-shock polypeptides are shifted at 42°C to a smaller size class of polyribosomes than those occupied by heat-shock mRNAs. This suggests that the heat-shock mRNAs initiate translation particularly efficiently, even when the rate of initiation on other mRNAs has been reduced.

When cells incubated at 42°C for 90 minutes are returned to 37°C, the rate of protein synthesis recovers to approximately 50% of the control within the first 30 minutes at the control temperature. As shown in Figure 1, left and center panels, histone (lower arrow) and actin (p43) mRNAs become associated with a larger number of ribosomes during recovery. In contrast, heat-shock p70 and p80 mRNAs are found on essentially the same size polyribosomes at both temperatures. Changes in the rate of initiation, therefore, do not seem to affect translation of heat-shock mRNAs as drastically as they affect other RNAs.

ALTERATIONS IN THE RELATIVE RATE OF INITIATION AND ELONGATION CHANGE THE PATTERN OF POLYPEPTIDE SYNTHESIS

We have investigated further whether the pattern of polypeptide synthesis at 42°C is influenced by differences in the ability of various mRNAs to initiate translation. If this is the case, and 37°C mRNAs are not specifically sequestered in an inactive form, it should be possible to restore these mRNAs to larger polyribosomes at 42°C by reducing the rate of polypeptide chain elongation. As elongation is progressively inhibited, a point is reached where the attachment of ribosomes to mRNA becomes limited by the time required for previously initiated ribosomes to move away from the ribosome binding site. When the rate of elongation rather than the rate of initiation becomes limiting for translation, differences in initiation efficiency between mRNAs play a correspondingly smaller role

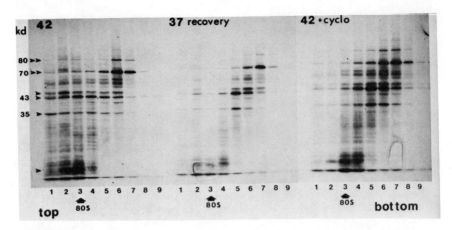

Figure 1
Distribution of specific mRNAs in cytoplasmic extracts during heat shock, recovery, and cycloheximide treatment at 42°C. HeLa cells in suspension culture were labeled with [³H]uridine for 24 hours prior to heat shock to allow for quantitation of RNA recovery. They were then concentrated to 2×10^6 cells/ml in fresh medium, allowed to recover from centrifugation for 30 minutes at 37°C, and then incubated at 42°C for 2 hours. The culture was then divided into three separate flasks. The first was incubated for an additional 30 minutes at 42°C, the second for an additional 30 minutes at 37°C (37° recovery), and the third was incubated for 30 minutes at 42°C with I μg/ml of cycloheximide present during the final 15 minutes. Each culture was then poured over crushed frozen saline. Cytoplasmic extracts were prepared by Dounce homogenization and fractionated on 10–30% sucrose gradients. RNA was prepared for translation from each gradient fraction using a rapid procedure to be published elsewhere. After correcting for differential losses of RNA during preparation, a constant proportion of RNA present in each gradient fraction was translated in the reticulocyte lysate in the presence of [³⁵S]methionine. Translation products were analyzed by SDS gel electrophoresis and fluorography. Note that heat-shock polypeptide p27 is not labeled with methionine.

in determining the pattern of protein synthesis. Drugs such as cycloheximide, which specifically slow elongation, can be used to approach this condition. Conversely, conditions that preferentially inhibit initiation accentuate synthesis of polypeptides encoded by the more efficient mRNAs (Lodish 1974).

In Figure 1 (right panel) it can be seen that treatment of heat-shocked cells at 42°C with a low level of cycloheximide causes a shift of actin, p35, and histone mRNAs to larger polyribosomes, similar to those in the cells allowed to recover at 37°C. Thus, the increase in initiation upon recovery, or the reduction in the elongation rate caused by cycloheximide at 42°C, both cause an increase in the average number of ribosomes associated with 37°C mRNAs. The average number of ribosomes bound to heat-shock mRNAs appears to be affected to a lesser extent. It should also be noted that the majority of the 37°C mRNAs found sedimenting slower than 80S monoribosomes at 42°C become associ-

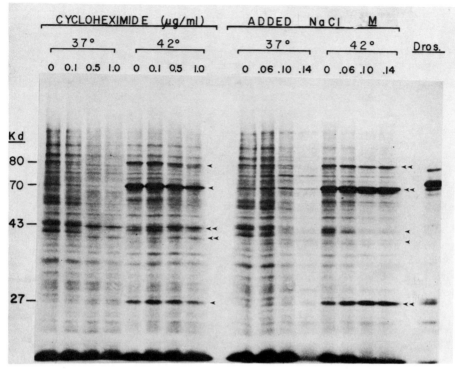

Figure 2
Effect of cycloheximide on hypertonic stress on heat-shock polypeptide synthesis. Cells were incubated at either 37°C or 42°C for 2 hours and then each culture was split into eight separate flasks. Each was then preincubated for 10 minutes with the indicated concentration of cycloheximide or excess NaCl. The cells were then labeled with [³H]leucine for 30 minutes and labeled polypeptides were analyzed by gel electrophoresis and fluorography. Each lane contains approximately the same amount of trichloroacetic acid (TCA)-precipitable radioactivity with the exception of the highest concentration of NaCl at 37°C, where incorporation was severely inhibited. Double arrows indicate polypeptides whose relative synthesis is enhanced by the treatments; single arrows indicate those inhibited.

ated with polyribosomes during cycloheximide treatment, or following return to 37°C. This would suggest that the mRNAs accumulated in this fraction during heat shock represent a consequence of the kinetics of initiation rather than a translationally inactive pool.

Figure 2 shows the differential effects of inhibition of elongation with cycloheximide, or inhibition of initiation with hypertonic stress, on the pattern of polypeptide synthesis. As the rate of elongation is progressively inhibited by increasing concentrations of cycloheximide, synthesis of the heat-shock polypeptides declines relative to the major 37°C polypeptides. When heat-shocked cells are treated with increasing concentrations of excess NaCl, which further inhibits initiation (Saborio et al. 1974), the pattern of polypeptide synthesis shifts more strongly toward heat-shock polypeptide production. Cells incubated at 42°C for 90 minutes and then allowed to recover at 37°C for 30 minutes contain levels of heat-shock mRNA comparable with cells maintained continuously at 42°C. In such recovering cells, we found that the same ratio of heat-shock polypeptide and p43 synthesis seen at 42°C could be generated at 37°C by inhibiting protein synthesis with excess NaCl to the 42°C rate (Hickey and Weber 1982). It appears therefore that differences in the efficiency of initiation between the heat-shock mRNAs and the majority of 37°C mRNAs might be sufficient to explain the apparent preferential translation of heat-shock mRNA at elevated temperatures.

INHIBITION OF INITIATION BY TEMPERATURE DIRECTLY AFFECTS THE PATTERN OF POLYPEPTIDE SYNTHESIS

Heat-shocked HeLa cells allowed to recover at 37°C for 1 hour show increased rates of protein synthesis and contain essentially the same levels of heat-shock mRNAs as were present during heat shock. We have therefore used cells recovering from heat shock to test the effect of inhibition of initiation brought about by increasing temperature on the relative synthesis of the heat-shock polypeptides and actin (p43). Cells were incubated at 42°C for 90 minutes and allowed to recover at 37°C for 30 minutes. The culture was then divided, and separate samples were incubated for 15 minutes in the presence of [^{35}S]methionine at 36°C, 38°C, 40°C, 42°C, and 44°C. Aliquots were taken for analysis of labeled polypeptides by gel electrophoresis, and the distribution of polyribosomes in the remaining cells was determined by sucrose gradient centrifugation. The proportion of total ribosomes present as 80S mono-

Table 1
The Effect of Inhibition of Initiation by Temperature on Relative Synthesis of Heat-shock Polypeptides

Temperature	Percent maximum incorporation[a]	Percent ribosomes in 80S[b]	Percent incorporation into[c] p70 + p80	p43
36°C	100	32	20	8
38°C	89	33	20	9
40°C	93	39	23	8
42°C	48	53	28	7
44°C	14	62	32	5

[a]Incorporation is measured by TCA precipitable cpm of [^{35}S]methionine.
[b]Measured by planimetry of optical density traces of sucrose density gradients.
[c]Measured by gel electrophoresis, fluorography, and densitometry. Exposures were kept within the linear range of the film.

ribosomes was used as an indication of the degree of inhibition of initiation. The results summarized in Table 1 show that at temperatures above 40°C protein synthesis is inhibited. Increasing inhibition is paralleled by reduction of initiation and preferential synthesis of heat-shock polypeptides in a manner analogous to the effects of hypertonic stress. Thus, the general reduction in the rate of initiation brought about by either condition has a lesser effect on the translation of heat-shock mRNAs than on the majority of 37°C mRNAs.

The diversion of the protein synthetic machinery from total cellular mRNA coding for the heat-shock polypeptides is part of the heat-shock response in most organisms (this volume). There may, however, be significant differences among species in the mechanism by which this translational control operates. In yeast, for example, control mRNA has been reported to be degraded at elevated temperatures (Lindquist 1981). However, in *Drosophila,* the dramatic shift toward heat-shock polypeptide synthesis occurs without loss of mRNA. In vitro translation experiments using homologous cell-free systems suggest that the translational machinery is modified during heat shock to selectively translate heat-shock mRNA (Storti et al. 1980). In HeLa cells, no mRNA-specific translational control mechanism needs to be postulated to explain preferential translation of heat-shock mRNAs. Our data suggest that the heat-shock mRNAs possess structural features that confer on them a high intrinsic translational efficiency. Accordingly, translation of heat-shock mRNAs is affected by the lesion in initiation induced by hyperthermia much less than is translation of the majority of cellular mRNAs. The general reduction in the rate of initiation can quantitatively account for the observed inhibition of 37°C mRNA translation. Polyribosome-bound actin mRNA, for example, is translated on the average by half as many ribosomes at 42°C as at 37°C. When the significant increase in the fraction of actin mRNA molecules not associated with ribosomes at the higher temperature is also considered, the three-fold reduction in the absolute rate of actin synthesis observed during heat shock is easily explained. In conclusion, we would like to point out that a similar kinetic explanation for the preferential translation of heat-shock mRNA during heat shock in *Drosophila* has not yet been finally ruled out.

ACKNOWLEDGMENTS

This research was supported by NIH Grant CA26847.

REFERENCES

Craig, E.A., B.J. McCarthy, and S.C. Wadsworth. 1979. Sequence organization of two recombinant plasmids containing genes for the major heat shock-induced protein of *D. melanogaster. Cell* **16:** 575–588.

Hickey, E.D. and L.A. Weber. 1982. Modulation of heat shock polypeptide synthesis in HeLa cells during hyperthermia and recovery. *Biochemistry* **21:** 1513–1521.

Lindquist, S. 1981. Regulation of protein synthesis during heat shock. *Nature* **293:** 311–314.

Lodish, H.F. 1974. Model for the regulation of mRNA translation applied to haemoglobin synthesis. *Nature* **251:** 385–388.

McCormick, W. and S. Penman. 1969. Regulation of protein synthesis in HeLa cells: Translation at elevated temperatures *in vivo. J. Mol. Biol.* **39:** 315–333.

Saborio, J., S-S. Pong, and G. Koch. 1974. Selective and reversible inhibition of initiation of protein synthesis in mammalian cells. *J. Mol. Biol.* **85:** 195–211.

Storti, R., M.P. Scott, A. Rich, and M.L. Pardue. 1980. Translational control of protein synthesis in response to heat shock in *D. melanogaster* cells. *Cell* **22:** 825–834.

Control of Polypeptide Chain Elongation in the Stress Response: A Novel Translational Control

G. Paul Thomas and
Michael B. Mathews
Cold Spring Harbor Laboratory
Cold Spring Harbor, New York 11724

Organisms representative of most taxonomic groups respond to a variety of metabolic insults by the induction of a small suite of characteristic polypeptides — the heat-shock or stress proteins — which are seemingly highly conserved throughout evolution (Kelley and Schlesinger 1978; Ashburner and Bonner 1979). Work in this laboratory has been directed to a description of the mechanistic aspects of the induction of stress proteins (SPs) of HeLa cells upon exposure to amino acid analogs (Thomas et al. 1981; G.P. Thomas and M.B. Mathews; G.P. Thomas et al.; both in prep.). This paper briefly recapitulates some of our findings and then describes some more recent findings regarding translational controls which appear to operate in this inducing system.

Using as inducer the proline analogue L-azetidine-2-carboxylic acid (AzC), profound alterations in both transcription and translation are observed. At the level of protein synthesis, there is a time- and concentration-dependent decrease in the synthesis of the normal background of cellular proteins and concomitant induction of the SPs. At the level of RNA synthesis, the patterns of newly made cytoplasmic RNAs also simplify with time: no rRNA is produced and the only new mRNAs exported from the nucleus are the (probable) SP mRNAs. Thus, the mRNA made within a given period represents decreasing proportions on normal mRNAs and increasing proportions of SP mRNAs. The general parallelism between patterns of RNA and protein synthesis suggests that protein synthesis may become restricted to the translation of newly synthesized mRNAs.

Induction of SP mRNAs and the shut-off of production of normal mRNAs, normal proteins, and rRNA commence simultaneously, after about 2−3 hours. As one might anticipate, induction of the SP mRNAs does not occur if RNA synthesis is blocked during this period. A second, unexpected, finding was that after 2−3 hours in the presence of AzC and absence of RNA synthesis, protein synthesis declined precipitously although preexisting mRNAs remained in translatable form and undiminished amount. A further surprise was that although not active in protein synthesis, the mRNAs encoding the normal polypeptide complement were found to be on polysomes and have polysomal distributions essentially identical to those of untreated cells. That they are integral components of the polysomes is indicated by the EDTA sensitivity of their polysomal associations. The most obvious explanation combining these findings is that protein synthesis upon a class of mRNAs is blocked at the level of elongation, and that there is a second population of polysomes that is active and produces SPs almost exclusively.

What determines which mRNAs are present in these two polysome classes? A comparison of the amounts of "normal" and SP mRNAs present in cells with the levels and patterns of protein synthesized in vivo at the same times (Thomas et al. 1981) leads us to suggest two possibilities. For discussion, we define "old" mRNA as that made before activation of the response (2−3 hours after exposure) and "new" as that made after this time. The first possibility would be that translation on old mRNA is blocked and that only new mRNAs are used. If there were also to be a continuous requirement for "newness," a property that distinguishes the most recently made mRNAs but that is lost with time, the SPs would thereby come to dominate protein synthesis as a consequence of the transcriptional controls. A second possibility would be that the SP mRNAs are discriminated for, and although normal mRNAs are still made at decreasing rates, these are not used. If the second were true, one would expect an initial decline in protein synthesis followed by a progressive recovery and increase paralleling the accumulation of the SP mRNAs, as appears the case for heat shock in *Drosophila* (McKenzie et al. 1975; Ashburner and Bonner 1979; Storti et al. 1980). What is observed with AzC treatment, though, is a gradual decline in overall level of protein synthesis, increasingly dominated by SP synthesis, and reflecting the synthesis, rather than accumulation, of the SP mRNAs. Since no new ribosomes will be made, protein synthesis is limited to

preexisting subunits, the majority of which presumably become sequestered into the inactive class and carry old mRNAs. In the next section,
we outline more recent findings regarding this apparent elongation
control.

DISCRIMINATION BETWEEN mRNAs RESIDES WITH RIBOSOMES

Puromycin is an analog of aminoacyl-tRNA and can be incorporated by
ribosomes both in vivo and in vitro resulting in the release of peptidyl-
tRNAs and, under some conditions, of mRNAs as well (Blobel 1971). If
elongation on normal mRNA were blocked in AzC-treated cells, one
might hope to obtain differential effects of puromycin treatment on active
and inactive polysomes, since even if inactive polysomes bound puromycin, its mode of action requires peptide bond formation and translocation
reactions, presumably not performed (or drastically retarded) by inactive
ribosomes. A preliminary experiment of this type is shown in Figure 1.
Polysomes isolated from AzC-treated HeLa cells were treated, or not,
with puromycin and fractionated by sucrose gradient sedimentation.
Cell-free translation in the reticulocyte lysate was used to assay intact
polysomes direct from gradient fractions or naked RNA isolated from the
same fractions.

Two results emerge. First, the near-exclusive synthesis of stress
proteins that is observed by pulse-labeling intact cells is reproduced
when polysomes are assayed. Even in the presence of all the components required for efficient translation of naked mRNA (as provided by
the reticulocyte lysate), only polysomes bearing SP mRNAs are run off
so that the SPs are almost exclusive products. By contrast, when naked
mRNAs are prepared from the same fractions, the restriction on normal
mRNAs is lifted and normal cellular polypeptides become prominent
products. (Tropomyosin, actin, and intermediate filament proteins are
representatives of the class of protein and are indicated in Fig. 1.)
This result argues that the inhibition of translation on the normal mRNAs
is mediated through protein components of the polysomes, and that
these are retained in the high-salt conditions employed in their isolation.

Second, puromycin has some effect on the sedimentation of both
polysomes and mRNA activities. With untreated polysomes, the bulk of
polysomes sedimented in fractions 4, 5, and 6, as did the mass of
mRNA activities for both normal and stress proteins. Treatment with
puromycin shifted most of the polysomes up the gradient into lighter
regions to peak at fractions 6 and 7, and the majority of mRNA activities
to fractions 7 and 8. We tentatively suggest that there may be two
effects of such puromycin treatment. One is on the active polysomes

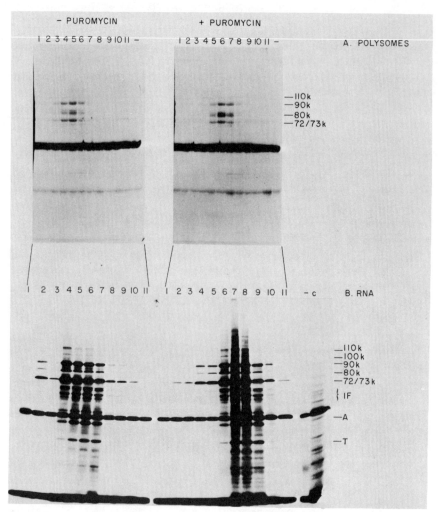

Figure 1

Translation of intact polysomes and of polysomal RNAs from AzC-treated HeLa cells. HeLa cells were exposed to AzC for 16 hours, after which few proteins others than the SPs are being synthesized. Cells were lysed with NP40 and DOC, the postmitochondrial supernatant made 0.5 M with KCl and polysomes pelleted through 1 M and 2 M cushions of sucrose also in high salt. The polysomes were resuspended and one-half incubated with 1 mM puromycin for 15 minutes each at 0° and then 37°C, the other half were similarly treated but did not receive puromycin. The polysomes were then separated in a linear sucrose gradient. Sedimentation was from left to right. Equal aliquots of the fractions were translated either directly or after deproteinization by phenol and chloroform extractions. (Under the conditions employed for translation of polysomes, protein synthesis appears limited to run-off, since the products do not appear affected by any of three specific inhibitors of initiation—edeine, pactamycin, or aurintricarboxylic acid—but is, of course, obliterated by sparsomycin, an inhibitor of elongation.) In both cases, the amounts used were below saturation. Products were analyzed by SDS-gel electrophoresis. (T) Tropomyosin; (A) actin; (IF) intermediate filament proteins.

210

bearing SP mRNAs and results in some transfer to lighter regions, but these retain their translatability: naked translatable mRNAs for the normal or stress protein do not appear to be released, however. A second effect is on the inactive polysomes (bearing some SP mRNA perhaps and greater amounts of normal mRNAs), which effectively releases the mRNA activities to more slowly sedimenting forms that are still not translatable in the reticulocyte lysate unless stripped of their associated proteins. In gradients of both treated and untreated polysomes, there is a fraction of SP mRNAs that sediment as large polysomes but do not appear to be translated as polysomes: their significance is unclear at present.

A direct quantitative comparison between the amounts of SPs synthesized in response to either polysomes or polysomal RNAs cannot be made: this is for one or both of two reasons. Since the synthesis on polysomes is limited to run-off, the specific distribution of ribosomes on polysomes (that is, the average number of ribosomes occupying a given portion of an mRNA), affects the amount of radioactivity that can by incorporated during completion of nascent chains: there are no data regarding the actual distributions of ribosomes on active and inactive polysomes. A second feature that would confuse the picture is the probable existence of some SP mRNAs in inactive polysomes. The above experiment therefore requires some refinement, but the observation that the bulk of normal mRNAs cosediment in inactive form with the active polysomes stands, as does the finding that the restriction on translation can be relieved by deproteinization. Since only the active polysomes appear capable of being run off under these conditions, purification of ribosomal subunits from active polysomes free from those of inactive polysomes becomes feasible. It will be of some interest to compare the polypeptide compositions of active and inactive subunits.

COMPARISONS WITH OTHER SYSTEMS

In both *Drosophila* and HeLa cells, continuous heat shock results in an initial disaggregation of polysomes, followed by a gradual recovery of protein synthesis and polysomes which is dependent on RNA synthesis. It may be that this need for RNA synthesis is analogous to that apparent in analog-treated cells, namely that protein synthesis may become dependent on the production of new mRNA. For *Drosophila,* studies of RNA synthesis under heat-shock conditions would be consistent with this idea as the SP mRNAs are the only RNAs produced and SPs are the only proteins being synthesized (Ashburner and Bonner 1979). Although severe heat shocks cause both induction of SPs and obliteration of synthesis of normal proteins, less severe shocks have less dramatic effects on synthesis of normal proteins. One possibility is that

the apparently absolute requirement for RNA synthesis is a secondary control invoked only under the most severe stress conditions, or that the restrictions on RNA synthesis are less complete but that the requirement for new RNA is retained.

Another effect of heat shock is control of initiation of protein synthesis mediated by either phosphorylation of eIF-2α (see Ernst et al., this volume) and/or by dephosphorylation of ribosomal protein S6 (see Glover, this volume). These features may cause the initial disaggregation of polysomes which would appear to be only transient, since protein synthesis resumes if RNA synthesis is not inhibited. Alternatively, protein synthesis may become independent of the normal controls through the production of modified mRNAs which do not require interactions with eIF-2α or S6.

As regards the fate of the preexisting mRNAs in heat-shocked Drosophila, there have been conflicting reports describing their presence in (Sondermeijer and Lubsen 1978; Kruger and Benecke 1980) or absence from (McKenzie et al. 1975; Storti et al. 1980) polysomes. More recent results from Pardue's laboratory (see Ballinger and Pardue, this volume) suggest that elongation controls of the type outlined above for analog-treated cells may operate in heat-shocked Drosophila in addition to the early effects on polypeptide initiation. Thus, it may be that in both human and insect cells the stress response(s) may proceed by regulation at several levels of gene expression, both at transcription and translation, and that common control mechanisms are utilized by the different organisms.

ACKNOWLEDGMENTS

We wish to thank Susan Danheiser for excellent technical assistance. This work was supported by NIH grant GM 27790.

REFERENCES

Ashburner, M.A. and J.J. Bonner. 1979. The induction of gene activity in Drosophila by heat shock. Cell 17: 241–254.

Blobel, G. 1971. Release, identification and isolation of messenger RNA from mammalian ribosomes. Proc. Natl. Acad. Sci. 68: 832–835.

Kelley, P.M. and M.J. Schlesinger. 1978. The effect of amino acid analogues and heat shock on gene expression in chicken embryo fibroblasts. Cell 15: 1277–1286.

Kruger, C. and B.J. Benecke. 1980. In vitro translation of Drosophila heat shock and non-heat shock mRNAs in heterologous and homologous cell-free systems. Cell 23: 595–603.

McCormick, W. and S. Penman. 1969. Regulation of protein synthesis in HeLa cells: Translation at elevated temperatures. *J. Mol. Biol.* **39:** 315–333.

McKenzie, S.L., S. Henikoff, and M. Meselson. 1975. Localization of RNA from heat-induced polysomes at puff sites in *Drosophila melanogaster. Proc. Natl. Acad. Sci.* **72:** 1117–1121.

Sondermeijer, P.J.A. and N.H. Lubsen. 1978. Heat shock peptides in *Drosophila hydeii* and their synthesis in vitro. *Eur. J. Biochem.* **88:** 331–339.

Storti, R.V., M.P. Scott, A. Rich, and M.L. Pardue. 1980. Translational control of protein synthesis in response to heat shock in *D. melanogaster* cells. *Cell* **22:** 825–834.

Thomas, G.P., W.J. Welch, M.B. Mathews, and J.R. Feramisco. 1981. Molecular and cellular effects of heat shock and related treatments of mamalian tissue culture cells. *Cold Spring Harbor Symp. Quant. Biol.* **46:** 985–996.

Heat Shock, Protein Phosphorylation, and the Control of Translation in Rabbit Reticulocytes, Reticulocyte Lysates, and HeLa Cells

Vivian Ernst, Ellen Zukofsky Baum,
and Pranhitha Reddy
Department of Biochemistry
Brandeis University
Waltham, Massachusetts 02254

Protein synthesis in the enucleated rabbit reticulocyte is controlled by the specific phosphorylation-dephosphorylation of the α-subunit (38 kD) (eIF-2α) of the initiation factor, eIF-2 (for reviews see Clemens 1980; Gross 1980; Levin et al. 1980). eIF-2 binds initiator Met-tRNA$_f$ and GTP to form the ternary complex [eIF-2·Met-tRNA$_f$·GTP] and catalyzes the subsequent binding of Met-tRNA$_f$ to the 40S ribosomal subunit. A second factor, RF, stimulates the binding of both Met-tRNA$_f$ and GTP to eIF-2 and promotes the exchange of GTP for GDP on eIF-2 (Clemens et al. 1982; Siekerka et al. 1982). In heme-deficient reticulocytes or reticulocyte lysates, protein chain initiation becomes rapidly inhibited due to the activation of the heme-regulated, cyclic nucleotide-independent eIF-2α kinase (HRI) from its latent inactive form. Activation of HRI in intact cells or lysates is accompanied by HRI phosphorylation and produces a significant increase in the phosphorylation state of eIF-2α (Ernst et al. 1979; Leroux and London 1982). Recent studies indicate that the specific phosphorylation of eIF-2α by HRI prevents the interaction of eIF-2 with RF, and thereby inhibits both the binding of GTP and Met-tRNA$_f$ to eIF-2 and the GDP/GTP exchange reaction. These and other observations suggest that phosphorylation of eIF-2α inhibits initiation by preventing the normal catalytic recycling of eIF-2 during initiation (Clemens et al. 1982; Siekerka et al. 1982).

A similar inhibition of initiation is observed in intact reticulocytes or reticulocyte lysates incubated at elevated temperatures (40–42°C). This

215

inhibition also displays biphasic kinetics of inhibition and is accompanied by polyribosome disaggregation and depletion of [40S·Met-tRNA$_f$] initiation complexes (Mizuno 1975; Bonanou-Tzedaki and Arnstein 1976). We have examined the function of eIF-2α phosphorylation-dephosphorylation in the mechanism of inhibition induced in lysates incubated at 42°C. We conclude that the phosphorylation of eIF-2α is a primary event in the inhibitory mechanism based on the following observations:

1. Inhibition, which takes place in the presence of optimal concentrations of hemin (10–20 μM), is accompanied by the activation of the heme-regulated eIF-2α kinase, HRI, and increased phosphorylation of lysate eIF-2α;
2. the restoration of protein synthesis to inhibited lysates that occurs when lysates are shifted to normal temperatures (30°C) is accompanied by dephosphorylation of eIF-2α;
3. inhibition is prevented or reversed by cAMP (10 mM), GTP (2 mM), eIF-2, or RF—features that are characteristic of the inhibition caused by eIF-2α kinase activation and the consequent phosphorylation of eIF-2α (for reviews see Clemens 1980; Gross 1980).

Incubation of HeLa cells at 42°C also produces polyribosome disaggregation and inhibition of initiation. Inhibition occurs within 10–15 minutes; however, unlike reticulocytes, prolonged incubation of HeLa cells at 42°C results in the reformation of polyribosomes and restoration of protein synthesis by 60–120 minutes (McCormick and Penman 1969). This recovery of protein synthesis at 42°C is due to a basal level of translation of a class of proteins normally synthesized at 37°C and to the preferential synthesis of two heat-shock-induced proteins (95 and 68 kD). Both the recovery of protein synthesis at 42°C and the synthesis of the 95-kD and 68-kD proteins are inhibited by actinomycin D. These and other preliminary data indicate that the synthesis of the heat-shock proteins is under both transcriptional and translational control. Prolonged heat-shock (2 hr) is accompanied by decreased phosphorylation of five major polypeptides (mw = 50K, 41K, 36K, 27K, and 25K). Studies in progress are aimed at determining (1) the mechanism of the initial inhibition of protein synthesis in HeLa cells at 42°C; (2) the mechanism of the preferential translation of the heat-shock mRNAs at 42°C; and (3) the role of protein phosphorylation-dephosphorylation events in these control mechanisms.

CHARACTERISTICS OF THE INHIBITION OF PROTEIN SYNTHESIS IN RETICULOCYTE LYSATES INCUBATED AT 42°C

Incubation of hemin-supplemented lysates (20 μM hemin) at 42°C results in an initial burst of synthesis for 9–10 minutes, which involves several

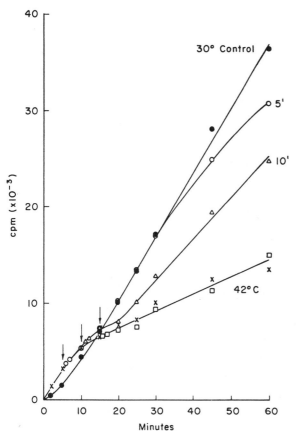

Figure 1
Protein synthesis in reticulocyte lysates at 30°C and 42°C. Assays for protein synthesis in hemin-supplemented (20 μM hemin) lysates were carried out as described previously (Ernst et al. 1978). ($\bullet$) Lysate incubated at 30°C; (X) lysate incubated at 42°C; ($\circ$, $\triangle$, $\square$) lysates incubated at 42°C for 5, 10, 15 minutes, respectively, before shifting to 30°C. Incorporation of [^{14}C]leucine into protein is given as cpm $\times$ 10^{-3} per 5 μl of reaction mixture.

reinitiation cycles, followed by an abrupt decline in rate (Fig. 1). In-hibition is accompanied by polyribosome disaggregation and loss of [40S·Met-tRNA$_f$] initiation complexes, indicating that the lesion is at the level of protein chain initiation (Mizuno 1975; Bonanou-Tzedaki 1976; Bonanou-Tzedaki et al. 1978). The inhibition is reversible within the first 10 minutes of incubation at 42°C; transferring lysates back to normal temperatures (30°C) results in reformation of polyribosomes, and resto-ration of protein synthesis to almost control rates (Fig. 1 and Mizuno 1977). In addition, inhibition of protein synthesis at 42°C is prevented by high levels of cAMP (10 mM), GTP (2 mM) (Mizuno 1977), or by two

Table 1
Effects of cAMP, GTP, RF, and eIF-2 on Inhibition of Reticulocyte Lysate Protein Synthesis at 42°C

Incubation temperature	Additions	Incorporation (cpm × 10⁻³)
Experiment 1		
30°C	—	11.9
42°C	—	4.9
42°C	GTP (2 mM)	14.1
42°C	cAMP (10 mM)	10.7
Experiment 2		
30°C	—	9.3
42°C	—	4.3
42°C	RF	6.2
42°C	eIF-2	11.3
42°C	RF + eIF-2	22.4

Hemin-containing protein synthesis reaction mixtures (25 μl) were incubated at 30°C or 42°C for 30 minutes as described (Ernst et al. 1978). cAMP (10 mM), GTP-Mg^{++} (2 mM), eIF-2 (14 pmoles) and RF (1–2 pmoles) were added at the start of incubation. All values given represent the amount of [^{14}C]leucine incorporated (cpm × 10⁻³) per 5 μl incubation. eIF-2 and RF were purified from reticulocyte lysates. eIF-2 was purified from the ribosomal salt-wash by DEAE cellulose and phosphocellulose (P-11) chromatography and glycerol gradient centrifugation. RF was purified from postribosomal supernatants following the procedure essentially as described by Amesz et al. (1979) (Matz, Levin, and London, in prep.; V. Ernst, data not shown).

initiation factor preparations, RF and eIF-2 (Table 1). These properties of lysates inhibited by heat shock closely resemble those of lysates inhibited by heme-deficiency, double-stranded RNA, or oxidized glutathione (GSSG). The latter three modes of inhibition are all due to the phosphorylation of the α subunit (38 kD) (eIF-2α) of initiation factor eIF-2, which leads to the inactivation of eIF-2 in protein synthesis (for reviews see Clemens 1980; Levin et al. 1980).

[^{32}P]PHOSPHOPROTEIN PROFILES OF INHIBITED LYSATES

The observations described above implicate the phosphorylation of eIF-2α in the mechanism of inhibition of initiation in lysates incubated at 42°C. We directly compared the phosphorylation state of eIF-2α, and other lysate phosphoproteins, in control and inhibited lysates. Lysates were incubated under protein synthesis conditions in the presence of a [^{32}P]ATP and [^{32}P]GTP regenerating system (Ernst et al. 1980; Jackson and Hunt 1980) and lysate [^{32}P]phosphoprotein profiles analyzed by both single and two-dimensional polyacrylamide gel electrophoresis and au-

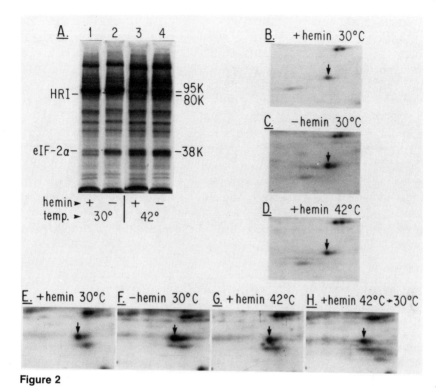

Figure 2

[^{32}P]Phosphoprotein profiles of control and inhibited lysates. Lysate protein synthesis reaction mixtures (25 μl) were incubated for 10 or 30 minutes in the presence of a [^{32}P]ATP and [^{32}P]GTP regenerating system as previously described (Ernst et al. 1980). Lysate [^{32}P]phosphoprotein profiles were analyzed by single (*A*) and two-dimensional (*B-H*) PAGE and autoradiography (O'Farrell 1975; Ernst et al. 1980). (*A*) Single-dimension SDS-PAGE of lysate [^{32}P]phosphoproteins. Incubations were for 10 minutes. Lane 1, hemin-supplemented lysate at 30°C (control); lane 2, heme-deficient lysate at 30°C; lane 3, hemin-supplemented lysate at 42°C; lane 4, heme-deficient lysate at 42°C. The heme-regulated kinase and eIF-2α(P) migrate as 80K and 38K polypeptides in this gel system. (*B-H*) Two-dimensional polyacrylamide gel electrophoresis of lysate [^{32}P]phosphoproteins. Incubation was for 30 minutes. (*B*) Hemin-supplemented lysate at 30°C; (*C*) heme-deficient lysate at 30°C; (*D*) hemin-supplemented lysate at 42°C; (*E*) hemin-supplemented lysate at 30°C; (*F*) heme-deficient lysate at 30°C; (*G*) hemin-supplemented lysate at 42°C; (*H*) hemin-supplemented lysate at 42°C for 10 minutes followed by incubation at 30°C for 20 minutes. The arrows indicate the phosphorylated α-subunit (eIF-2α) (38K, pl 5.4) of eIF-2. (*B-D*) and (*E-H*) represent eIF-2α phosphorylation in two different lysate preparations. [^{32}P]-incorporation into eIF-2α in each panel was as follows: (*B*) 74 cpm; (*C*) 187 cpm; (*D*) 216 cpm; (*E*) 187 cpm; (*F*) 580 cpm; (*G*) 299 cpm; (*H*) 176 cpm.

toradiography (Fig. 2). As described previously, hemin-supplemented lysates undergoing linear rates of protein synthesis display a basal level of eIF-2α phosphorylation (Fig. 2A, lane 1). In contrast, lysates inhibited by heme deficiency (Fig. 2A, lane 2), or heat shock (Fig. 2A, lane 3) show a rapid and extensive increase in the phosphorylation of eIF-2α (38 kD) (Ernst et al. 1979). In heme-deficient but not heat-shock lysates, inhibition is also accompanied by a significant increase in the phosphorylation of the heme-regulated eIF-2α kinase, HRI (80K). Analysis of lysate [^{32}P]phosphoprotein profiles by two-dimensional gel electrophoresis yielded similar results (Fig. 2B, C, and D). Inhibition at 42°C is accompanied by reduced phosphorylation of a 95K polypeptide (Fig. 2A, lane 3); the possible function of this polypeptide in the control of initiation is not known. When lysates incubated at 42°C are shifted to 30°C, the observed restoration of protein synthesis (Fig. 1) is accompanied by the concomitant dephosphorylation of eIF-2α (Fig. 2, G and H). These data strongly implicate the phosphorylation of eIF-2α in the mechanism of inhibition of initiation at 42°C. Both eIF-2α kinase and eIF-2α phosphatase activities of lysates incubated at 42° were studied directly (see below).

ACTIVATION OF THE eIF-2α KINASE, HRI, AT 42°C IN THE POSTRIBOSOMAL SUPERNATANT OF RABBIT RETICULOCYTE LYSATES

Incubation of reticulocyte lysates, or lysate postribosomal supernatants at 42° results in the activation of the heme-regulated eIF-2α kinase, HRI, based on the following criteria:

1. All translational inhibitor activity copurifies with HRI on carboxymethyl Sephadex (CMS) (Bonanou-Tzedaki et al. 1981; V. Ernst et al., unpubl.);
2. the CMS-inhibitor preparation obtained from heat-shock lysates inhibits hemin-supplemented control lysates (30°C) with biphasic kinetics and this inhibition is prevented or reversed by RF, eIF-2, cAMP (10 mM), or GTP (2 mM) (Fig. 3);
3. a two- to three-fold increase in eIF-2α kinase activity is observed in the CMS-inhibitor preparation from heat-shocked lysates compared with similar preparations from control lysates (Table 2);
4. this eIF-2α kinase activity is blocked by antibodies prepared against highly purified HRI (Table 2).

Bonanou-Tzedaki et al. (1981) have also demonstrated that HRI antibodies block the inhibitory activity generated in heat-shock lysates.

eIF-2α PHOSPHATASE ACTIVITY IN LYSATES INCUBATED AT 42°C

The dephosphorylation rates of eIF-α in control and inhibited lysates was monitored by comparing the loss of radio label from exogenously added

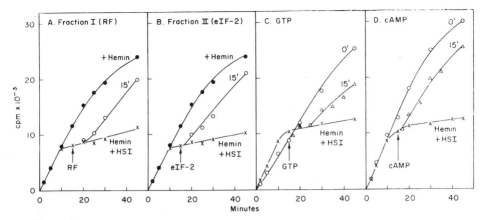

Figure 3

Inhibition of translation by heat-induced inhibitor. Reversal by RF, eIF-2, GTP, and cAMP. Assays for protein synthesis (50 μl) in hemin-supplemented lysates (20 μM hemin) at 30°C were carried out as previously described (Ernst et al. 1978). (●) + hemin control lysate; (X) lysates inhibited by heat-induced translational inhibitor (HSI) (CM Sephadex fraction, 120 μg, see Table 2). (○, Δ) additions at 15 minutes were RF (1–2 pmoles) (*A*); eIF-2 (14 pmoles) (*B*); GTP (2 mM) (*C*); cAMP (10 mM) (*D*). (○) GTP and cAMP (*C* and *D*) added at zero minutes. Incorporation of [¹⁴C]leucine into protein is given as cpm × 10⁻³ per 5 μl reaction mixture.

[³²P]eIF-2, essentially as described by Safer and Jagus (1979). No significant difference in the rate of dephosphorylation of [³²P]eIF-2α in 30°C or 42°C lysates was observed (V. Ernst et al., unpubl.). These data suggest that the increase in eIF-2α phosphorylation that is observed in heat-shock lysates is due mainly to HRI activation, rather than inhibition of phosphatase activity.

PROTEIN SYNTHESIS AND PROTEIN PHOSPHORYLATION IN HELA CELLS EXPOSED TO HEAT SHOCK

Incubation of HeLa cells at elevated tmeperatures (40–45°C) produces an initial inhibition of protein synthesis within 10–15 minutes. Inhibition is accompanied by rapid polyribosome disaggregation, indicating a lesion at some step in protein chain initiation. However, unlike reticulocytes, further incubation of HeLa cells at 42°C results in the reformation of polyribosomes and restoration of protein synthesis to almost control rates by 60–120 minutes (McCormick and Penman 1969; data not shown). This restoration of protein synthesis is accompanied by the de novo synthesis of two major heat-shock polypeptides (95 kD and 68 kD) in addition to a basal synthesis of a class of proteins synthesized at normal, control temperatures (Fig. 4A, lane 6). A comparison of the

Table 2

Effect of HRI Antibody on the eIF-2α Kinase Activated in Lysates at 42°C

Additions to protein kinase assays		$[P_i]$ Incorporation into eIF-2α (cpm)
CM Sephadex fraction	Immune serum	
A. 30°C + hemin	—	630
30°C − hemin	—	1480
42°C + hemin	—	1330
B. 30°C − hemin	—	2230
	+ control serum	1860
	+ anti-HRI serum	420
42°C + hemin	—	2800
	+ control serum	2310
	+ anti-HRI serum	560

Reticulocyte lysates (1 ml) were incubated for 30 minutes under the following conditions: (1) 20 μM hemin at 30°C; (2) without added hemin at 30°C; and (3) plus 20 μM hemin at 42°C. These lysates were then centrifuged at 150,000 g for 1.25 hour to pellet the ribosomes; the postribosomal supernatants (PRS) were collected and chromatographed on carboxymethyl Sephadex (CMS). All the inhibitory activity detected in the PRS obtained from lysates incubated at 30°C minus hemin or 42°C plus hemin was eluted from the column at low salt (a property of the heme-regulated kinase, HRI). The CMS fractions were concentrated by ammonium sulfate precipitation and assayed for inhibitory activity by addition to heme-supplemented control lysates incubated at 30°C (see Fig. 3) (Ernst et al. 1978). Purification of inhibitory activity is about 10-fold at this step. eIF-2α kinase activities of inhibitory fractions were assayed essentially as described previously (Levin et al. 1976). In experiment A, [γ-^{32}P]ATP was added at a specific activity of 20,000 cpm/pmole (80 μM ATP) and reactions (20 μl) contained 1 pmole eIF-2 and 10-12 μg of CMS fraction. Incubation was for 10 minutes at 37°C. In experiment B, incubations contained 25–27 μg CMS fraction, 75 μM ATP; 13,600 cpm × [γ-^{32}P]ATP/pmole, and assays were pretreated for 15 minutes at 0° with control or anti-HRI serum (1 μl) (Petryshyn et al. 1979) as indicated. ^{32}P-Labeled polypeptides were analyzed by SDS-PAGE and autoradiography (Levin et al. 1976). The polypeptide bands (38K) corresponding to eIF-2α were identified, excised, solubilized in H_2O_2, and counted by liquid scintillation spectrometry. All volumes given have been corrected for basal [^{32}P] incorporation (> 100 cpm) in the absence of added eIF-2.

[^{35}S]methionine-labeled polypeptides from cells incubated at 37°C and 42°C by two-dimensional gel electrophoresis (O'Farrell 1975) reveals that at 42°C (1) there is inhibition of the synthesis of a class of proteins normally synthesized at 37°C, and (2) there is a shift in the pI of a 35K polypeptide from pI 5.8 (37°C) to pI 5.6 (42°C) (data not shown). This shift in pI is apparently not due to phosphorylation as demonstrated by analysis of [^{32}P]-labeled cell phosphoproteins in both single dimension SDS-PAGE (Fig. 4B), or in two-dimensional PAGE. However, the prolonged incubation (2 hr) of cells at 42°C is accompanied by the marked

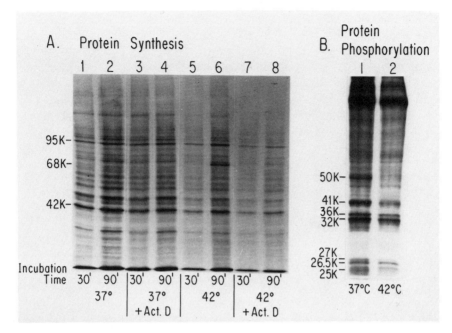

Figure 4
HeLa cell polypeptide synthesis and phosphorylation at 37°C and 42°C. (A) HeLa cell
polypeptide synthesis at 37°C and 42°C. HeLa cells were grown in Joklik medium supplemented with 10% calf serum to a density of 5×10^5 cells/ml. The cells were collected,
resuspended at a concentration of 1×10^6 cells/ml in fresh medium and incubated at 37°C for
1 hour. [^{35}S]Methionine (40 μCi/ml) was added to the culture which was then divided into four
aliquots and incubated at 37°C or 42°C in the presence and absence of 5 μg/ml actinomycin D
as indicated in the figure. [^{35}S]Methionine-labeled polypeptides from control (37°C) (lanes 1–
4) or heat-shock (42°C) (lanes 5–8) HeLa cells were separated in SDS-PAGE and analyzed
by fluorography. Labeling with [^{35}S]methionine was for 30 minutes (lanes 1, 3, 5, and 7) or 90
minutes (lanes 2, 4, 6, and 8). Total cell extracts from 5×10^4 were loaded in each lane. (B)
[^{32}P]phosphoprotein profiles of heat-shocked HeLa cells. Cells were grown in Dulbecco's
phosphate-free medium for 3 hours essentially as described in Fig. 3. Cells were transferred to
fresh medium containing [^{32}P]orthophosphate (0.5 mCi/ml) and incubated at 37°C or 42°C for
2 hours as indicated. Total cellular proteins were separated in SDS-PAGE and [^{32}P]phospho-
proteins identified by autoradiography. Lanes 1 and 2, [^{32}P]phosphoproteins of control 37°C
cells and heat-shocked cells, respectively. Total cell extracts from approximately 4×10^4 cells
were loaded in each lane.

dephosphorylation of four major polypeptides (50K, 41K, 36K, 32K) as
well as a group of low-molecular-weight proteins (23–27K) (Fig. 4B).
   Both the restoration of protein synthesis, and the specific translation of
the 95K and 68K heat-shock proteins are inhibited by actinomycin D
(Fig. 4A, lane 8). In vitro translation of poly(A)$^+$ mRNA from control
(37°C) and heat-shock cells (42°C, 120 min) in a microccocal nuclease-
treated, mRNA-dependent reticulocyte lysate indicates that the synthesis

of the 95K and 68K polypeptides is only directed by poly(A)$^+$ mRNA from heat-shocked cells. The 95K and 68K polypeptides are located predominantly in the soluble fraction of cells, although low levels of these polypeptides are associated with the cytoskeleton, ribosomes, and nuclei (V. Ernst et al., unpubl.).

Studies in progress are aimed at determining (1) the mechanism of inhibition of initiation in HeLa cells during the early stages of heat shock, (2) the mechanism by which the heat-induced 95K and 68K polypeptides are preferentially translated at 42°C, and (3) the function of the phosphorylation of eIF-2α and other cellular proteins in these control mechanisms.

ACKNOWLEDGMENTS

We are grateful to Drs. D. H. Levin and B. Matz (MIT) for providing us with highly purified preparations of eIF-2 and RF. We would also like to thank Dr. R. Petryshyn (MIT) for anti-HRI serum, and Ms. H. Youssofian for invaluable technical assistance. This work was supported by NSF grant PCM-8022837 awarded to V. Ernst and Biomedical Research Support Grant (6796) awarded to Brandeis University.

REFERENCES

Amesz, H., H. Goumans, T. Haubrich-Morree, H.O. Voorma, and R. Benne. 1979. Purification and characterization of a protein factor that reverses the inhibition of protein synthesis by the heme-regulated translational inhibitor in rabbit reticulocyte lysates. *Eur. J. Biochem* **98:** 513.

Bonanou-Tzedaki, S. and H.R.V. Arnstein. 1976. Subcellular localization of a lesion in protein synthesis in rabbit reticulocytes incubated at elevated temperatures. *Eur. J. Biochem.* **61:** 397.

Bonanou-Tzedaki, S., M.K. Sohl, and H.R.V. Arnstein. 1981. Regulation of protein synthesis in reticulocyte lysates. Characterization of the inhibitor generated in the post-ribosomal supernatant by heating at 44°C. *Eur. J. Biochem.* **114:** 69.

Bonanou-Tzedaki, S., K.E. Smith, B.A. Sheeran, and H.R.V. Arnstein. 1978. Reduced formation of initiation complexes between Met-tRNA$_f$ and 40S ribosomal subunits in rabbit reticulocyte lysates incubated at elevated temperatures. *Eur. J. Biochem.* **84:** 601.

Clemens, M.J. 1980. Translational control of protein synthesis in erythroid cells. In *Biochemistry of cellular regulation 1, gene expression* (ed. M.J. Clemens), pp. 128–141. CRC Press Inc., Boca Ratan, Florida.

Clemens, M.J., V.M. Pain, S.-T. Wong, and E. Henshaw. 1982. Phosphorylation inhibits guanine nucleotide exchange on eukaryotic initiation factor 2. *Nature* **296:** 93.

Ernst, V., D.H. Levin, I.M. London. 1978. Evidence that glucose-6-phosphate regulates protein synthesis in reticulocyte lysates. *J. Biol. Chem.* **253:** 7163.

_____. 1979. In situ phosphorylation of the α-subunit of eukaryotic initiation factor two in reticulocyte lysates inhibited by heme-deficient double-stranded RNA, oxidized glutathione, or the heme-regulated protein kinase. *Proc. Natl. Acad. Sci.* **76:** 2118.

Ernst, V., D.H. Levin, A. Leroux, and I.M. London. 1980. Site-specific phosphorylation of the α-subunit of eukaryotic initiation factor eIF-2 by the heme-regulated and double-stranded RNA-activated eIF-2α kinases from rabbit reticulocyte lysates. *Proc. Natl. Acad. Sci.* **77:** 1286.

Gross, M. 1980. The control of protein synthesis by hemin in rabbit reticulocytes. *Mol. Cell. Biochem.* **31:** 25–36.

Leroux, A. and I.M. London. 1982. Regulation of protein synthesis by phosphorylation of initiation factor 2α in intact reticulocytes and reticulocyte lysates. *Proc. Natl. Acad. Sci.* **79:** 2147.

Levin, D.H., R.S. Ranu, V. Ernst, and I.M. London. 1976. Regulation of protein synthesis in reticulocyte lysates: Phosphorylation of methionyl-tRNA$_f$ binding factor by protein kinase activity of the translational inhibitor isolated from heme-deficiency lysates. *Proc. Natl. Acad. Sci.* **73:** 3112.

Levin, D., V. Ernst, A. Leroux, P. Petryshyn, R. Fagard, and I.M. London. 1980. Regulation of eukaryotic protein synthesis by the phosphorylation of initiation factor eIF-2. In *Protein phosphorylation and bioregulation,* FMI-EMBO Workshop Basel (ed. G. Thomas, E.J. Podesta, and J. Gordon), pp. 128–141. S. Karger, Basel, Switzerland.

McCormick, W. and S. Penman. 1969. Regulation of protein synthesis in HeLa cells: Translation at elevated temperatures. *J. Mol. Biol.* **39:** 315.

Mizuno, S. 1975. Temperature sensitivity of protein synthesis initiation in the reticulocyte lysate system. Reduced formation of the 40S ribosomal subunit Met-tRNA$_f$ complex at an elevated temperature. *Biochim. Biophys. Acta* **414:** 273.

_____. 1977. Temperature sensitivity of protein synthesis initiation. Inactivation of a ribosomal factor by an inhibitor formed at elevated temperatures. *Arch. Biochem. Biophys.* **179:** 289.

O'Farrell, P.H. 1975. High resolution two-dimensional electrophoresis of proteins. *J. Biol. Chem.* **250:** 4007.

Petryshyn, P., H. Trachsel, and I.M. London. 1979. Regulation of protein synthesis in reticulocyte lysates: Immune serum inhibits heme-regulated protein kinase activity and differentiates heme-regulated protein kinase from double-stranded RNA-induced protein kinase. *Proc. Natl. Acad. Sci.* **76:** 1575.

Safer, B. and R. Jagus. 1979. Control of eIF-2 phosphatase activity in rabbit reticulocyte lysate. *Proc. Natl. Acad. Sci.* **76:** 1094.

Siekerka, J., L. Mauser, and S. Ochoa. 1982. Mechanism of polypeptide initiation in eukaryotes and its control by the phosphorylation of the α-subunit of initiation factor 2. *Proc. Natl. Acad. Sci.* **79:** 2537.

Heat-shock Effects on Protein Phosphorylation in *Drosophila*

Claiborne V. C. Glover
Department of Biochemistry
Stanford University School of Medicine
Stanford, California 94305

Heat shock is an attractive system for studying the role of protein phosphorylation in gene expression. In *Drosophila,* heat shock results in dramatic changes in the pattern of protein synthesis. Translation of most normal (25°C) proteins is rapidly suppressed, and synthesis of a specific set of heat-shock proteins is subsequently induced. These effects are mediated by controls acting at both the transcriptional and translational levels. Because of the magnitude of the changes involved, the heat-shock system offers an excellent opportunity to detect protein phosphorylation events associated with transcription and translation. This is particularly true of the general reduction in synthesis of most normal 25°C proteins because this reduction involves a large decrease in the percentage of ribosomes in polysomes as well as a large reduction in the percentage of active genes.

Heat-shock-induced changes in the phosphorylation of acid-soluble proteins are described in this paper. This group of proteins has been chosen for an initial study because it includes the basic ribosomal proteins, histones, and the high-mobility-group (HMG) proteins. Heat-shock effects on translation might involve altered phosphorylation of ribosomal proteins; similarly, heat-shock-induced changes in transcription might involve altered phosphorylation of histones and/or HMG proteins.

DISCUSSION

Heat shock of *Drosophila* tissue culture cells labeled in vivo with $^{32}P_i$ generates reproducible changes in the labeling of specific acid-soluble proteins (Fig. 1). Of the six most prominent changes (designated 1–6 in

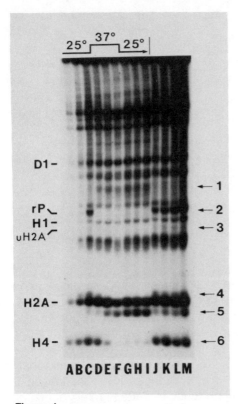

Figure 1
Effect of heat shock on phosphorylation of acid-soluble proteins in vivo. A 25°C culture of *Drosophila* Kc tissue culture cells was labeled in vivo with $^{32}P_i$ at zero time and allowed to incorporate label for 2 hours at 25°C. The culture was then divided into two aliquots. One aliquot (control) was incubated an additional 4 hours at 25°C; the other was subjected to a 2-hour temperature shock at 37°C and then returned to 25°C for 2 hours. Labeled precursor was present throughout the experiment. Acid-soluble proteins were extracted from whole cells and analyzed by acid-urea gel electrophoresis as described (Glover 1982). An autoradiograph of the gel is shown. Electrophoresis is from top to bottom. (A, B, C) 20, 60, and 120 minutes after zero time; (D, E, F) 20, 60, and 120 minutes after temperature shift-up; (G, H, I) 20, 60, and 120 minutes after shift-down. (J, K, L, and M) Control samples corresponding in time to lanes D, E, F, and I, respectively. Labeled bands corresponding to histones H1, H2A, and H4, chromosomal protein D1, and a ribosomal protein (rP) are indicated (Glover 1982). uH2A represents ubiquinated H2A. Six labeled proteins whose phosphorylation is significantly affected by heat shock are indicated by arrows in the right margin.

Fig. 1), four involve decreased labeling with ^{32}P (bands 2, 3, 4, and 6) while two involve increased labeling (bands 1 and 5). None of these changes is significantly reversed by a 2-hour recovery period at 25°C (Fig. 1, G−I). In an attempt to assess the significance of heat shock-induced changes in protein phosphorylation, cell fractionation and other techniques have been employed to identify these proteins.

Dephosphorylation of a Ribosomal Protein

The most striking change in Figure 1 is the rapid and complete dephosphorylation of band 2. Identification of band 2 as a ribosomal protein (rP) has previously been described in detail (Glover 1982) and is based primarily on cell fractionation experiments (e.g., Fig. 2). This basic ribosomal protein is a component of cytoplasmic rather than mitochondrial ribosomes and has a molecular weight of 32 kD. It is the only ribosomal protein that is significantly labeled by ^{32}P$_i$ in vivo in *Drosophila*. These characteristics strongly resemble those of ribosomal protein S6 described in a variety of higher eukaryotes (Wool 1979). Estimates of the actual percentage of phosphorylated rP can be obtained from stained patterns in two-dimensional gels (e.g., Fig. 3A and C). In cells growing at 25°C, as much as 40% of rP has been found in the phosphorylated state. In contrast, rP is completely dephosphorylated in heat-shocked cells. The kinetics of dephosphorylation following a shift-up to 37°C are quite rapid (see Fig. 1), dephosphorylation being complete within about 10 minutes (Glover 1982). Heat-shock effects on the remaining bands designated in Figure 1 occur with slower kinetics (minutes to hours).

Effects on Histone Phosphorylation

Bands 3, 4, and 6 (Fig. 1) are all associated with nuclei (Figs. 2 and 3). This, combined with their low molecular weight and acid solubility, suggests that they are histones. Band 6 is very probably monophosphorylated H4 because of its exact comigration with singly modified H4 (Fig. 3). Band 4, which migrates in the streak associated with H3, may be either phosphorylated H3 or a phosphorylated form of an H3 or other histone variant. Unfortunately, percentage phosphorylation cannot be calculated for these proteins because of inadequate resolution and/or the presence of histone acetylation which also affects first-dimension mobility.

Band 3 has tentatively been identified as ubiquinated H2A (uH2A) on the basis of its electrophoretic mobility in two-dimensional gels (Levinger and Varshavsky 1982). It is accompanied (Fig. 3E) by a second protein having the expected mobility of uH2B. The fact that uH2A is phosphory-

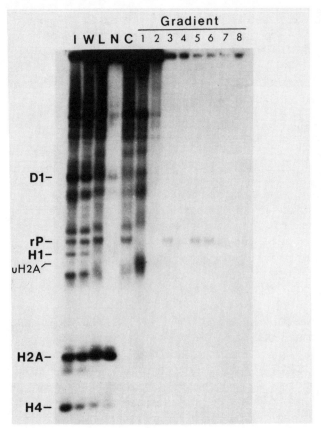

Figure 2
Cell fractionation. *Drosophila* Kc cells labeled for 5 hours at 25°C with $^{32}P_i$ were harvested, washed, detergent lysed, and fractionated into nuclei and cytosol. A portion of the cytosol was further fractionated by sedimentation in a sucrose gradient (Glover 1982). Acid-soluble proteins were extracted from each fraction and analyzed by gel electrophoresis and autoradiography as in Fig. 1. I, intact cells; W, washed cells; L, lysed cells; N, nuclei; C, cytosol; lanes 1–8, sucrose gradient fractions (sedimentation from left to right; lane 3 corresponds to monosomes, lanes 5–7 to polysomes).

lated whereas uH2B is not (Fig. 3F) is consistent with the state of phosphorylation of H2A and H2B themselves. As mentioned above, band 3 (or uH2A) appears to dephosphorylate in response to heat shock. However, examination of the stained gels of Figure 3, A and C, reveals that the apparent dephosphorylation in this case may have another explanation, because the stained spot corresponding to this protein is also reduced in intensity following heat shock. Thus, this protein may be either degraded or, perhaps more likely, deubiquinated.

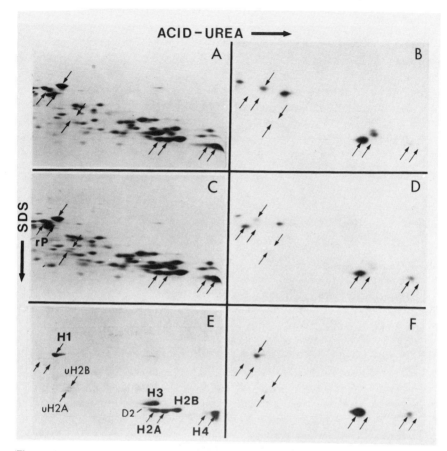

ACID-UREA ⟶

Figure 3

Two-dimensional gel electrophoresis. *Drosophila* Kc cells were labeled in vivo with $^{32}P_i$ at either 25°C or 37°C. Acid-soluble proteins were then extracted from whole cells or from isolated nuclei and analyzed by two-dimensional gel electrophoresis as described (Glover 1982). Only the bottom half of the first-dimension gel was subjected to electrophoresis in the second dimension. (*A, C, E*) Stained gels; (*B, D, F*) autoradiograph. (*A* and *B*) Whole-cell extracts from heat-shocked cells; (*C* and *D*) whole-cell extracts from 25°C cells; (*E* and *F*) nuclear extracts from 25°C cells. Arrows in each panel indicate corresponding mobilities. Histones H1, H2A, H2B, H3, and H4, histone variant D2, and the phosphorylated ribosomal protein (rP) are indicated. uH2A and uH2B represent ubiquinated H2A and H2B, respectively. Most of the remaining spots in *A* and *C* are ribosomal proteins (Glover 1982). H2A, H4, and rP are each designated by a pair of arrows because phosphorylation substantially reduces the mobility of these proteins in the first dimension. Spots numbered 2–6 in *D* correspond to bands 2–6 in Fig. 1. Band 1 is not included on the gel.

Interestingly, the same behavior is associated with the nonphosphorylated uH2B.

Other Effects

Nothing is known about the two proteins whose phosphorylation appears to increase following heat shock (Fig. 1, bands 1 and 5) except that neither is nuclear. Analysis of appropriate two-dimensional gels reveals additional acid-soluble and insoluble proteins whose phosphorylation is affected by heat shock (not shown). These also have yet to be identified. Finally, of the remaining acid-soluble proteins which have been identified, three are phosphorylated: histones H1 and H2A and chromosomal protein D1. Heat shock has no discernible effect on the phosphorylation of any of these proteins.

SUMMARY AND CONCLUSIONS

Although the functional significance of the various phosphorylation changes described above remains to be determined, data derived from other systems suggest a number of interesting possibilities.

Increased phosphorylation of ribosomal protein S6 is correlated with increased translation in most systems, including serum stimulation of tissue culture cells (Wool 1979) and activation of sea urchin eggs by fertilization (Ballinger and Hunt 1981). Thus, an attractive hypothesis is that phosphorylation of rP in *Drosophila* is required for efficient translation of most normal 25°C mRNAs (but not heat-shock mRNAs) and that dephosphorylation of this protein is responsible for the rapid breakdown of polysomes following heat shock. This suggestion is supported by the fact that ribosome dephosphorylation and polysome breakdown occur at similar rates and is consistent with available data from in vitro translation studies (Scott and Pardue 1981). If true, this hypothesis implies a significant role for protein phosphorylation changes in induction of at least one aspect of the heat-shock response. A simple model involves transduction of an external stimulus (heat shock) to generate an intracellular signal which affects the activity of either a protein kinase or phosphatase. This in turn leads to dephosphorylation of a specific ribosomal protein, thereby inhibiting translation of normal 25°C mRNAs.

Proposals concerning the function of histone phosphorylation have been derived primarily from cell-cyle studies (Gurley et al. 1981). In most systems, H3 phosphorylation is observed only at mitosis and is thought to be associated in some way with chromosome condensation. Phosphorylation of H4 appears to be minimal in many systems. The low levels that do occur have been attributed to cytoplasmic phosphorylation events involved in transport and deposition of newly synthesized H4 (Gurley et al. 1981). These data might argue that the gradual dephosphorylation of H4 and/or H3 following heat shock in *Drosophila* results from a disruption of normal progress through the cell cycle. However, because so little is known about the function of histone phosphorylation in *Drosophila,* it cannot be ruled out that these decreases in histone phosphorylation may be associated with other events such as the general reduction in transcription which accompanies heat shock.

An association between uH2A and active chromatin has recently been observed in *Drosophila* (Levinger and Varshavsky 1982). In addition, both uH2A and uH2B are absent from transcriptionally inactive metaphase chromosomes (Wu et al. 1981). Hence, the possibility that heat shock causes a substantial decrease in the total amount of both uH2A and uH2B is intriguing. Together, these observations suggest that loss of uH2A and uH2B might be associated with suppression of transcription during heat shock.

Many of the observations described in this paper are preliminary. However, the finding that heat shock causes substantial changes in protein phosphorylation offers an important opportunity to acquire more information about the functional significance of phosphorylation. Further progress will require: (1) confirming and extending the identification of the phosphoproteins affected by heat shock, (2) determining the actual change in percentage of phosphorylation for these proteins, as has been done for rP, and (3) most importantly, demonstrating causal relationships between specific phosphorylation events and particular heat-shock-induced effects using either in vitro or genetic techniques.

ACKNOWLEDGMENTS

I would like to acknowledge the encouragement and support of Dr. Douglas Brutlag in whose laboratory this work was done. I am also grateful to Dr. David Hogness for use of facilities and to Ken Relloma for technical assistance. This work was supported by a postdoctoral fellowship award from the National Institutes of Health and by National Institutes of Health grant GM 28079.

REFERENCES

Ballinger, D.G. and T. Hunt. 1981. Fertilization of sea urchin eggs is accompanied by 40S ribosomal subunit phosphorylation. *Dev. Biol.* **87:** 277–285.

Glover, C.V.C. 1982. Heat shock induces rapid dephosphorylation of a ribosomal protein in *Drosophila. Proc. Natl. Acad. Sci.* **79:** 1781–1785.

Gurley, L.R., J.A. D'Anna, M.S. Halleck, M.S. Barham, R.A. Walters, J.J. Jett, and R.A. Tobey. 1981. Relationship between histone phosphorylation and cell proliferation. In "Protein phosphorylation" (ed. O.M. Rosen and E.G. Krebs). *Cold Spring Harbor Conf. Cell Proliferation* **8:** 1073–1094.

Levinger, L. and A. Varshavsky. 1982. Selective arrangement of ubiquinated and D1 protein-containing nucleosomes within the *Drosophila* genome. *Cell* **28:** 375–385.

Scott, M.P. and M.L. Pardue. 1981. Translational control in lysates of *Drosophila melanogaster* cells. *Proc. Natl. Acad. Sci.* **78:** 3353–3357.

Wool, I.G. 1979. The structure and function of eukaryotic ribosomes. *Annu. Rev. Biochem.* **48:** 719–754.

Wu, R.S., K.W. Kohn, and W.M. Bonner. 1981. Metabolism of ubiquinated histones. *J. Biol. Chem.* **256:** 5916–5920.

Changes in Protein Phosphorylation and Histone H2b Disposition in Heat Shock in *Drosophila*

**Marilyn M. Sanders,
Dona Feeney-Triemer, Anne S. Olsen,
and James Farrell-Towt**
*Department of Pharmacology
UMDNJ-Rutgers Medical School
Piscataway, New Jersey 08854*

The heat shock response is induced within a few minutes of the transfer of *Drosophila* cells from 25° to 37°C. Synthesis of new RNA or protein as a mechanism of induction has been ruled out by experiments in the presence of RNA and protein synthesis inhibitors. Since changes in protein posttranslational modification seemed a possible mechanism of induction, we investigated changes in protein phosphorylation in heat shock by two-dimensional gel electrophoresis of solubilized whole cells. This method of analysis showed that one basic polypeptide, which we have identified as a ribosomal protein, is dephosphorylated in heat shock. A parallel experiment utilized [^{35}S]methionine to label heat-shock polypeptides. This experiment showed that histone H2b is a heat-shock polypeptide in *Drosophila* and that histone synthesis becomes noncoordinate in heat shock (Sanders 1981). Histone synthesis has been thought to be tightly coordinately regulated and coupled to DNA synthesis. The perturbation of histone synthesis in heat shock presents an opportunity to investigate (1) the disposition of an identified heat-shock polypeptide of known function and (2) the consequences of perturbed histone synthesis.

EFFECT OF HEAT SHOCK ON TOTAL CELL PROTEIN PHOSPHORYLATION

Drosophila K_c cells, labeled with $^{32}P_i$ phosphate for 3 hours, were subjected to heat shock or control conditions for varying periods of time.

The cells were solubilized for two-dimensional gel electrophoresis in two different systems which together display acidic and neutral as well as basic nucleic acid-binding polypeptides (O'Farrell 1975; Sanders et al. 1980). Analysis of the autoradiograms from paired heat-shocked and control samples showed that only one major polypeptide changed significantly in phosphate content in heat shock. This polypeptide had a molecular weight of about 30 kD, was very basic, and was 40% dephosphorylated in 5 minutes and 80–100% dephosphorylated in 1 hour of heat shock. Because the dephosphorylation was so specific and the time course coincided with the time course of induction of the response, it seemed possible that this change was involved in early regulatory events in heat shock.

A subcellular fractionation experiment showed that the polypeptide copurified with the ribosomes (Fig. 1, A and B). It was not washed off with 0.5 M KCl, suggesting that it is an integral part of the ribosome and not a loosely associated factor. Similar observations have also been reported by Glover (1982). Attempts to separate the ribosomal subunits by conventional in vitro procedures for eukaryotic systems resulted in disintegration of the subunits. However, we were able to separate the ribosomal subunits in good yield by in vivo incubation of the cells in 0.1 mM puromycin for 2–3 hours (A. M. Fallon, pers. comm.). Figure 1, C–F, shows the staining patterns and autoradiograms of the 60S and 40S subunits. A comparison of the staining patterns shows that the subunits are not appreciably contaminated with each other. The autoradiograms show that the phosphopolypeptide is present in the 40S subunit. The molecular weight, subunit location, and ability of this polypeptide to be phosphorylated suggest it is analogous to the mammalian ribosomal protein S6 (Wool 1979).

ACCUMULATION OF HISTONE H2B IN HEAT SHOCK

Heat-shock histone H2b (H2b^{hs}) was identified by its electrophoretic properties and by peptide-mapping experiments (Sanders 1981). In the first hour of heat shock, it is made at 3–4 times its normal rate of synthesis while synthesis of H3, H4, and H1 is decreased from two- to more than tenfold (Sanders 1981). Histone H2a probably continues to be made at its normal control rate in heat shock.

The noncoordinate nature of histone synthesis in heat shock raised several questions. First, does DNA synthesis continue in heat shock? Second, does extra H2b accumulate in heat shock or is old histone H2b degraded at a faster rate to keep the total cell content of histone H2b normal? And finally, what is the fate of the heat shock H2b in the cell?

Incorporation of [³H]thymidine into DNA stops when the cells are

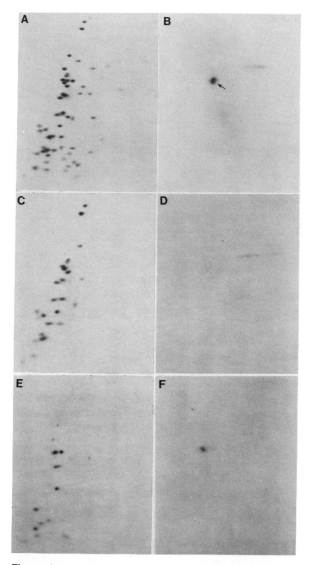

Figure 1

Phosphoprotein present in *Drosophila* ribosomes and ribosomal subunits. Purified ribosomes from ^{32}P-labeled cells grown at 25°C were solubilized and subjected to two-dimensional PAGE (Sanders et al. 1980). The charge fractionation is in the horizontal dimension and size fractionation in the vertical dimension. Total ribosomes stained gel (*A*) and autoradiogram (*B*). The phosphopeptide marked with the arrow in *B* is dephosphorylated in heat shock. The ribosomal subunits were separated on sucrose gradients after puromycin treatment of ^{32}P-labeled cells as described in the text. 60S subunit stained gel (*C*); 60S autoradiogram (*D*); 40S stained gel (*E*); and 40S autoradiogram (*F*). The phosphopeptide dephosphorylated in heat shock is substantially enriched in the 40S subunit.

transferred from 25°C to 37°C. At the same time, incorporation of labeled amino acid into H2b^{hs} continues at a linear rate for 6 hours in heat shock. The rate of synthesis of H2b^{hs} is such that approximately 50% of the cells' total complement of histone H2b should accumulate in 6 hours if old histone H2b is not degraded during heat shock. The stability of old histone H2b in heat shock was tested by a label and chase experiment, with [^{35}S]methionine incorporated at 25°C and chased in parallel cultures with cold methionine at 37°C or 25°C. Quantitation of autoradiograms from two-dimensional gel electrophoresis of solubilized whole cells from the heat shock and control chase samples showed that old histone H2b was stable in 2 hours and 6 hours of heat shock.

A second label and chase experiment showed that ^{35}S-labeled H2b^{hs} synthesized in heat shock was as stable in 24 hours of recovery at 25°C in the presence of cold methionine as was labeled histone H2b synthesized at 25°C for the same period of time. A third experiment was carried out with cold methionine present in heat shock and [^{35}S]methionine present in recovery. This experiment showed that histone H2b synthesis in recovery was typical of that for other heat-shock polypeptides. The cells resumed H2b synthesis at a normal rate after several hours of recovery, during which time H2b was synthesized at a rate intermediate between the normal and heat-shock rates. These experiments indicate that a net excess of histone H2b over other histones and DNA accumulates in heat shock.

LOCATION OF H2b^{hs} IN THE CELLS

Subcellular fractionation experiments showed that nearly all the histone H2b^{hs} was associated with nuclei. An irreproducible fraction, ranging from undetectable amounts to 20% of the H2b^{hs}, was present in soluble form in the postnuclear supernatant fraction. When the nuclei were treated with staphylococcal nuclease, more than 95% of the DNA was solubilized as nucleosomes in 0.6 M NaCl. Figure 2 shows that histone H2b^{hs} (and an equivalent amount of histone H2a made in heat shock) is released with nucleosomes from nuclease-digested nuclei (Fig. 2, lanes c−e). The remainder is bound to the residue remaining after the nucleosomes have been removed (Fig. 2, lane f). The nucleosomes containing histone H2b^{hs} released from these nuclei sediment normally as 11S particles on sucrose gradients. Their density in CsCl is indistinguishable from the density of nucleosomes from control cells. These data indicate that the H2b^{hs} present in nucleosomes has assembled as a component of the normal nucleosome octamer. An altered particle density would have been obtained if H2b^{hs} was associated with chromatin in another way, for example as a ninth histone on the nucleosome. The facts that only H2b^{hs} is left over in the chromatin-depleted residue and that ap-

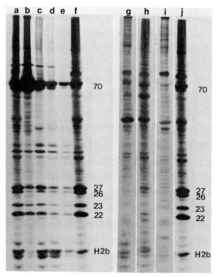

Figure 2

Protein staining and labeling patterns in nuclear subfractions from heat-shocked and control cells. Nuclei were pelleted from homogenates of ^{35}S-labeled heat-shocked or control cells (Sanders 1981). Following digestion with staphylococcal nuclease, nuclei were subfractionated with increasing concentrations of NaCl, and the subfractions were electrophoresed on discontinuous SDS acrylamide gels and subjected to autoradiography. All lanes shown are from the same experiment and gel. Autoradiograms of nuclear subfractions from heat-shocked cells (lanes a–f). Lane a, total nuclear pellet; b, soluble nuclear proteins; c, d, e, nucleosome subfractions released stepwise by 0.2, 0.3, and 0.6 M NaCl, respectively; f, residue after chromatin is removed. Lanes g, h, stained gels from the chromatin-depleted nuclear residues from cells labeled at 25°C and 37°C, respectively; i, j, autoradiograms from the 25°C and 37°C chromatin-depleted residues, respectively. Position of migration of major heat-shock polypeptides and histone H2b are marked.

proximately equal amounts of H2b[hs] and H2a labeled in heat·shock are incorporated into nucleosomes suggest that H2b can only be assembled into nucleosomes with H2a. These two histones interact in solution in the absence of DNA to form a stable dimer (Isenberg 1979).

Figure 2 also shows that nonchromatin-associated H2b[hs] fractionates with the chromatin-depleted nuclear residue. H2b[hs] is not dissociated from this residue by 2 M NaCl and thus must not be in a complex with DNA. Figure 2, lanes g–j, show that changes occur in this residue in heat shock. The nuclear pellet visibly increases in size, and the staining pattern of the heat-shocked residue compared with control (Fig. 2, lanes g and h) shows that the heat-shocked residue accumulates several prominent polypeptides as well as an increase in background staining. The autoradiogram in lane j shows that most of the mass accumulated is not labeled in heat shock, but substantial amounts of heat-shock polypeptides are present in this residue.

The striking increase in the amount of protein present in the chroma-tin-depleted nuclear residue in heat shock suggested that a morphologi-cal change in the nuclei should be detected in heat shock. Figure 3, A and B, shows that a nuclear pellet from control cells contains mainly nuclei that are identified by fluorescence in the presence of the DNA-binding dye, 33258 Hoechst. In contrast, the nuclear pellet from heat-shocked cells (Fig. 3, C and D) contains a large number of structures in addition to nuclei; some of these structures are considerably larger than nuclei. The increase in number of nonnuclear structures relative to nuclei is a minimum of fourfold when fluorescent and total particles larger than half the diameter of nuclei are counted in adjacent microscope fields. The nuclei and the nonnuclear structures have been partially separated by centrifugation. From these preliminary investigations it appears that the nonnuclear structures contain the staining polypeptides that increase in amount in heat shock in the nuclear pellet. Histone H2b[hs] is associat-ed only with the nuclei and some, if not all, of the other heat-shock polypeptides are bound to the nonnuclear structures.

COMMENT

These experiments have shown that a 30,000-dalton ribosomal phos-phoprotein becomes dephosphorylated early in the heat-shock response, an observation similar to that reported by Glover (1982). We have further shown that this phosphoprotein is located on the 40S ribosomal subunit. This polypeptide shows a number of properties in common with the mammalian 40S subunit phosphoprotein S6, including changes in phos-phate content when protein synthesis is perturbed. It is possible that this heat-shock dephosphorylation is a significant regulatory step in the shutdown of normal protein synthesis seen early in the response. This correlation is purely circumstantial at this time, but can be tested using the *Drosophila* in vitro translation system. In vitro translation experi-ments have shown already that the control point is associated with ribosomes (Kruger and Benecke 1981; Scott and Pardue 1981).

Experiments designed to determine the disposition of histone H2b[hs] in heat-shocked cells have shown that an excess of H2b[hs] over other histones and DNA accumulates in heat shock. Some of the H2b[hs] exchanges with old H2b in chromatin, and this amount appears not to exceed the amount of H2a synthesized and incorporated into nucleo-somes in heat shock. Although noncoordinate histone synthesis in so-matic cells has not been reported previously, this exchange suggests that mechanisms for histone exchange in the absence of DNA synthesis exist. No currently measurable change in chromatin properties accompa-

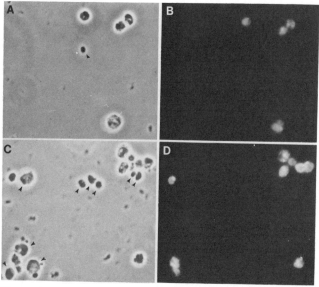

Figure 3
Particles present in the nuclear fraction of heat-shocked and control cells. Aliquots of homogenates of *Drosophila* cells incubated at 25°C (control) or 37°C (heat shock) for 1 hour were centrifuged onto coverslips under conditions analogous to those used to pellet nuclei. The coverslips were inverted on a solution of 1 μg/ml 33258 Hoechst in homogenization buffer (Sanders 1981) and viewed in the microscope with either phase or fluorescent illumination. Only nuclei fluoresce with 33258 Hoechst. (*A* and *B*) Particles cosedimenting with nuclei from control cells with phase and fluorescence, respectively; (*C* and *D*) particles from heat-shocked cells, with phase and fluorescence, respectively. Nonnuclear particles in *A* and *C* are marked by arrows.

nies the exchange but this does not rule out the possibility that some aspect of chromatin structure is remodeled by it.

The H2b^hs that is not chromatin-bound is associated with the residue remaining after chromatin is removed from nuclei. This residue increases in mass in heat shock and accumulates several prominently staining protein bands in addition to a substantial fraction of labeled heat-shock polypeptides. Morphological observations show that a nonnuclear structure that copurifies with nuclei assembles in heat shock. The nonnuclear structure has a very simple protein composition and is probably a component of the cytoskeleton; some heat shock polypeptides are associated with it while H2b^hs is associated with the nuclear matrix or ghost. A further description of the nature of the association of H2b^hs and the other heat-shock polypeptides with nuclear matrix and the nonnuclear structure awaits complete separation of these particles.

ACKNOWLEDGMENTS

The excellent technical assistance of Julia T. Hsu and Barbara Brief is gratefully acknowledged. This work was supported by Public Health Service grant No. AG 02066.

REFERENCES

Glover, C.V.C. 1982. Heat shock induces dephosphorylation of a ribosomal protein in *Drosophila. Proc. Natl. Acad. Sci.* **79:** 1781–1785.

Isenberg, I. 1979. Histones. *Ann. Rev. Biochem.* **48:** 159–191.

Kruger, C. and B.-J. Benecke. 1981. In vitro translation of *Drosophila* heat shock and non-heat-shock mRNAs in heterologous and homologous cell-free systems. *Cell* **23:** 595–603.

O'Farrell, P.H. 1975. High resolution two-dimensional electrophoresis of proteins. *J. Biol. Chem.* **250:** 4007–4021.

Sanders, M.M. 1981. Identification of histone H2b as a heat shock protein in *Drosophila. J. Cell Biol.* **91:** 579–583.

Sanders, M.M., V.E. Groppi, Jr., and E.T. Browning. 1980. Resolution of basic cellular proteins including histone variants by two-dimensional gel electrophoresis: Evaluation of lysine to arginine ratios and phosphorylation. *Anal. Biochem.* **103:** 157–165.

Scott, M.P. and M.L. Pardue. 1981. Translational control of lysates of *Drosophila melanogaster* cells. *Proc. Natl. Acad. Sci.* **78:** 3353–3357.

Wool, I.G. 1979. The structure and function of eukaryotic ribosomes. *Ann. Rev. Biochem.* **48:** 719–754.

Properties of Three Major Chicken Heat-shock Proteins and Their Antibodies

Milton J. Schlesinger,
Philip M. Kelley*, Guiseppe Aliperti**,
and Carol Malfer
Department of Microbiology and Immunology
Washington University School of Medicine
St. Louis, Missouri 63110

About 3 years ago, we made the serendipitous observation that primary cultures of chicken embryo fibroblasts synthesized a new set of mRNAs and proteins after brief exposures to temperatures 3° to 4° above their normal range. The experiments leading to these results were preceded by studies showing that certain amino acid analogs, when added to cell cultures lacking the normal cognate amino acids, changed rather dramatically the cell's protein synthesis pattern (Kelley and Schlesinger 1978). For the chicken cell system, the proteins induced by the analogs and heat shock were identical and some of them seemed analogous to proteins made in heat-shocked *Drosophila*. In addition, this response to high temperature and amino acid analogs was detected in several mammalian cell lines. Other laboratories have shown that a similar set of proteins is induced in chicken cells by a variety of agents such as arsenite, heavy metals, sulfhydryl oxidants, and even viruses.

Based on data reported by a number of investigators, it appears that cells from virtually every organism possess a genetic program for responding to stress, and heat-shock proteins are one set of phenotypes of this program. In this report, we describe properties of three major chicken heat-shock proteins with subunit $M_r = 89,000$, $70,000$, and $24,000$ and summarize studies with polyclonal antibodies obtained from rabbits injected with homogeneous preparations of the chicken heat-shock protein subunit polypeptides. Details of the experimental procedures are described elsewhere (Kelley and Schlesinger 1982).

*Present address: Department of Genetics, University of California, Berkeley, California 94720; **Department of Medicine, St. Mary's Hospital, St. Louis, MO.

PURIFICATION OF CHICKEN HEAT-SHOCK PROTEINS

Prior to purification, we determined which subcellular fraction was most enriched for a chicken heat-shock protein. Lysates of heat-shocked chicken cells, some labeled with [^{35}S]methionine, were prepared by various procedures (i.e., sonication, detergents, homogenization in low- or high-salt buffers) and centrifuged to remove particulate material (nuclei and membranes). Proteins in the various fractions were analyzed by SDS-PAGE and autoradiography. All three major chicken heat-shock proteins could be recovered in good yields from a cell cytoplasmic fraction (i.e., the supernatant fraction after 90 minutes centrifugation at 100,000g). Thus, in their native state these proteins exist predominantly as soluble proteins, although significant amounts of chicken hsp24 and hsp70 were present in the nuclear fraction. Very little chicken heat-shock proteins were associated tightly with membranes or secreted from cells. The buffers employed for these cell extractions included protease inhibitors, i.e., phenylmethanesulfonylfluoride and kallikrein, since extracts of heat-shocked cells are unusually rich in protease activity.

The first step in our purification scheme consisted of a size separation carried out on a Sepharose 6B column. Fractions were analyzed by gel electrophoresis and autoradiography. All three proteins fractionated as oligomers or in complex with other proteins. Most of the chicken hsp89 had an M_r ~550,000; chicken hsp70 had an M_r ~165,000, and chicken hsp24 had an M_r ~185,000. Large amounts of the latter protein eluted with the void volume of the column and subsequent studies confirmed that the chicken hsp24 readily aggregates. A homogeneous preparation of this protein was obtained by pooling appropriate fractions from the column, dialyzing, concentrating, and separating the material by preparative SDS-PAGE. Chicken hsp70 and hsp89 were further fractionated by ion-exchange chromatography on DEAE cellulose and finally subjected to gel electrophoresis. Homogeneous preparations of SDS-denatured proteins were prepared in complete Freund's adjuvant and injected into rabbits. From high-titer sera, IgG fractions were prepared and shown by radioimmune precipitation to react with the proteins induced by heat shock and arsenite (Fig. 1). There was very little protein that reacted with anti-chicken hsp24 IgG in nonstressed chicken fibroblasts although cross-reactive antigens were found in other chicken cells (see below). Anti-chicken hsp70 IgG precipitated two proteins (M_r = 68 kD and 70 kD) from nonstressed cells and increased amount of the latter from stressed cells. Anti-chicken hsp89 IgG precipitated equivalent amounts of protein from normal and stressed cells. Some proteins coprecipitated with the chicken heat-shock proteins but we attribute most of this to nonspecific "sticking" of the more highly radioactive proteins in the extract (i.e. actin, tubulin, and the chicken heat-shock proteins).

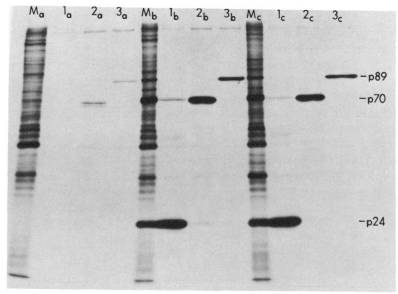

Figure 1
Specificity of antibodies raised against chicken heat-shock proteins. [^{35}S]methionine-labeled cell extracts from untreated CEF (a), CEF heat-shocked for 1 hour at 45°C (b), and CEF treated for 3 hours with 50 μM sodium arsenite (c) were prepared. (M) Total extract. Lane 1, protein precipitated with anti-chicken hsp24; lane 2, protein precipitated with anti-chicken hsp70; and lane 3, protein precipitated with anti-chicken hsp89. Experimental details are in Kelley and Schlesinger (1982).

STUDIES WITH ANTIBODIES TO CHICKEN HEAT-SHOCK PROTEINS

The antibodies have been used in a series of solid-phase radioimmune assays based on a technique referred to as the "Western blot." In this procedure, extracts of nonradioactive protein (~100 μg) are separated by SDS-PAGE and the proteins electrophoretically transferred (blotted) to diazotized paper. The paper is sequentially treated with antibodies and iodinated (^{125}I) protein A. An example of the technique is shown in Figure 2 where anti-chicken hsp70 and hsp89 were used as probes and found to react with proteins of similar molecular weights from human and mouse tissue culture cells and normal and heat-shocked *Drosophila* embryos. Results of these analyses showed that (1) all embryonic and adult chicken tissues as well as yeast, dinoflagellates, slime molds, corn seedling roots, worms, frogs, flies, rodents, and human have a protein of M_r ~70,000 that crossreacts with anti-chicken hsp70, (2) all embryonic and adult chicken tissues and flies, frogs, rodents, and humans have a

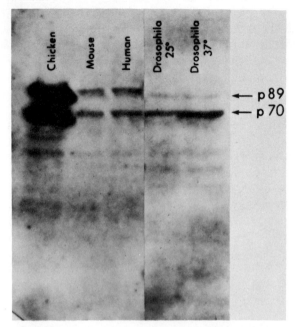

Figure 2
Autoradiogram of a "Western blot." About 80 μg of protein were initially separated in a 10% SDS-polyacrylamide slab gel. Experimental details are in Kelley and Schlesinger (1982).

protein of M_r ~85–90,000 that crossreacts with anti-chicken hsp89, (3) all chicken muscle tissues have a protein of M_r= 22,000 that cross-reacts with but is not antigenically identical with chicken hsp24. A cross-reacting protein of M_r ~24,000 was found in chicken embryo tissues after heating the intact embryo. The "Western blot" technique offers a sensitive method (~10 ng of antigen are detectable) for analyzing levels of stress proteins. We are currently examining animal tissues and tissue culture cells subjected to a variety of conditions for presence of these proteins.

Antibodies to chicken hsp89 have been used to investigate complexes between the chicken hsp89 and the Rous sarcoma pp60^{src} phosphoki-nase (see Yonemoto et al., this volume). We used these antibodies as well as the others to determine the state of phosphorylation of the chicken heat-shock proteins. Extracts of heat-shocked and normal cells that had been incubated with carrier-free [^{32}P]phosphate for 2 hours were immediately boiled in 2% SDS and 5% β-mercaptoethanol and then analyzed by radioimmune precipitation. All three chicken heat-shock proteins were phosphorylated; chicken hsp89 was heavily phos-phorylated on serine residues in both normal and heat-shocked cells;

some of the chicken hsp70 from normal and stressed cells contained ^{32}P; low levels of ^{32}P were in chicken hsp24 in heat-shocked cells but the antigenically related 22-kD protein was heavily phosphorylated. The significance of these results is not clear at this time.

We have also prepared monolayers of chicken embryo fibroblasts for immunofluorescent staining with the antibodies. No antigens were on the cell surface, but in cells fixed with methanol to allow penetration of the antibodies widespread intense staining with all antibodies was observed. Very little staining was seen over the cell nucleus; anti-chicken hsp89 showed a reticulated, cytoplasmic pattern and anti-chicken hsp70 and hsp24 stained stress fibers as well as the cytoplasm (Fig. 3). Only anti-chicken hsp24 showed strong differences between normal and heat-shocked cells. In 5-day, well-formed fused chicken embryonic myotubes, the anti-chicken hsp24 showed strong general staining compared with anti-chicken hsp70 and hsp89. Binding of anti-chicken hsp24 antigen to myofibrils was detected in cells permeabilized with detergent and then fixed. From these observations we conclude that chicken hsp89 is strictly cytoplasmic and "non-structural," whereas chicken hsp70 and hsp24 are probably distributed in both nuclear and cytoplasmic compartments and interact with the cell's cytoskeleton.

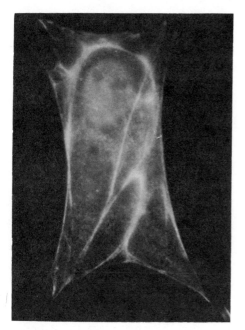

Figure 3
Immunofluorescence staining of chicken embryo fibroblast fixed with methanol and reacted with anti-chicken hsp70. Rhodamine-labeled goat anti-rabbit IgG was used for detecting fluorescence. The pattern, although unusual, was not atypical.

REGULATION OF CHICKEN HEAT-SHOCK PROTEIN SYNTHESIS

We have reported elsewhere our evidence that induction of chicken heat-shock protein synthesis is regulated primarily at the level of transcription (Kelley and Schlesinger 1978; Kelley et al. 1980). More recent studies using in vitro protein synthesis as an assay of levels of mRNA show that chicken heat-shock mRNAs are unstable ($T_{1/2} \cong 3$ hours) when heat-shocked cells are returned to their normal physiological temperature (Schlesinger et al. 1982). However, if either actinomycin D or cycloheximide are added to heat-shocked cells immediately after they are returned to the lower normal temperature, then chicken heat-shock mRNAs are stable. These results need to be confirmed by more direct measurements of mRNA levels. We have isolated a chicken cell genomic clone containing a chicken hsp24 gene and subcloned a 5.7-kb *Eco*RI fragment into a pBR322 variant plasmid. This material should be an effective probe for chicken hsp24 mRNA measurements.

Based on our preliminary data, we postulate that heat shock of avian cells—in a manner probably analogous to *Drosophila* cells — shuts off most mRNA transcription and turns on transcription of a few selected genes. The chicken heat-shock mRNAs compete with the preexisting mRNA for ribosomes and eventually dominate the cells' polysomes. Although there may well be properties of the chicken heat-shock mRNAs that allow them to be selectively translated at the high temperature, we believe that it is the excessive amounts of these mRNAs that permit them to outcompete normal cell mRNAs. Normal cell mRNAs, once removed from polysomes, are degraded. This is distinct from the *Drosophila* cell where there is good evidence for selective translation of heat-shock mRNAs and retention of normal cell mRNAs. The recovery of normal mRNA synthesis and shut-off of heat-shock mRNA appears more complex. We find that if chicken cells are stressed at 43°C instead of 45°C, chicken hsp70 synthesis stops after 2 hours. These data are consistent with the hypothesis put forward by Lindquist (see DiDomenico and Lindquist, this volume) that chicken hsp70 autoregulates its synthesis and that levels of this protein are correlated with the stressed state of a cell. For total recovery, mRNA and protein synthesis are necessary after cells return to normal physiological temperatures. The newly transcribed normal mRNAs replace from polysomes the chicken heat-shock mRNAs which are then degraded. It is, presumably, in this manner that homeostasis prevails.

SUMMARY AND CONCLUSION

Studies of the three major heat-shock proteins in chicken cell cultures have revealed the following:

1. Chicken hsp24 can be isolated as an oligomer of M_r ~185,000 but higher-molecular-weight aggregates readily form. At least two species are detected after isoelectric focusing and low levels of a phosphory-lated form occur. Cross-reacting protein with subunit $M_r = 22,000$ is present in substantial amounts in normal embryonic and adult chicken muscle cells. No cross-reacting material has been detected in normal mammalian tissue culture cells. High levels of chicken hsp24 mRNA are present in heat-shocked chicken fibroblasts and virtually none is found in the normal cell. A cDNA probe made from chicken hsp24 mRNA provided the means for isolating and cloning a chicken hsp24 gene.

2. Chicken hsp70 can be isolated as a dimer of M_r ~165,000 and occurs as multiple species, some of which are phosphorylated. Cross-reacting proteins of similar subunit molecular weight are found in all chicken tissues and in widely divergent species from yeast to humans. This protein interacts with cytoskeletal structures in the cytoplasm and possibly in the cell nucleus as well.

3. Chicken hsp89 can be found in higher-molecular-weight aggregates, occurs in several forms separable by isoelectric focusing, and is phosphorylated on serine residues. Cross-reacting proteins occur in normal chicken tissues and eukaryotic organisms ranging from flies to humans. This protein is localized to the cell cytoplasm.

The function of these three proteins has not yet been defined and represents one of the major unsolved problems in understanding the heat-shock response. We have focused on the chicken hsp24 protein—and hope to determine how it is related to the cross-reacting 22,000-mw normal muscle protein. Of particular interest is the organization of the genes for these two proteins and how their expression is regulated—since the former is transcribed only upon stress whereas the latter is transcribed at all times in a specific, differentiated cell.

ACKNOWLEDGMENTS

Much of the research described was supported by grants from the Public Health Service National Institute of Health (1 R01 GM2803, 5P 30 CA 16217 and AM 20579).

REFERENCES

Kelley, P.M. and M.J. Schlesinger. 1978. The effect of amino acid analogues and heat shock on gene expression in chicken embryo fibroblasts. *Cell* **15:** 1277–1286.

————. 1982. Antibodies to two major chicken heat shock proteins cross react with similar proteins in widely divergent species. *Mol. Cell Biol.* **2:** 267–274.

Kelley, P.M., G. Aliperti, and M.J. Schlesinger. 1980. *In vitro* synthesis of heat shock proteins by mRNAs from chicken embryo fibroblasts. *J. Biol. Chem.* **255:** 3230–3233.

Schlesinger, M.J., G. Aliperti, and P.M. Kelley. 1982. The response of cells to heat shock. *Trends Biochem. Sci.* (in press).

Modifications of the 70-kD Heat-shock Protein of *Drosophila melanogaster*

**Ruggero Caizzi, Corrado Caggese,
and Ferruccio Ritossa**
*Institute of Genetics
University of Bari
Bari, Italy*

It is a frequent feature of many eukaryotic polypeptides to undergo modifications before acquiring enzymatic activity. A series of observations on the enzyme glutamine synthetase 1 (GS1), which, under definite conditions, responds to heat-shock treatments (Scalenghe and Ritossa 1976), led us to search for some derivative of the major heat-shock polypeptide in *Drosophila*.

DISCUSSION

The major heat-shock polypeptide —hsp70 — has a molecular weight of 70 kD and is coded for by about five genes located at regions 87A7 and 87C1 (Henikoff and Meselson 1977; Ish-Horowicz et al. 1977; Craig et al. 1979; Ish-Horowicz et al. 1979). Two-dimensional chromatography and fingerprint analysis of the [35S]methionine tryptic peptides of this protein have been presented (Mirault et al. 1978; Ish-Horowicz et al. 1979) (see Fig. 2A). We observed that after heat shock minor polypeptides of 66 kD, 64 kD, and 43 kD (apparent) on SDS polyacrylamide gels were present in [35S]methionine-labeled extracts obtained in the usual way. Longer exposures were needed for a good visualization of these minor bands in normal extracts. After tryptic digestion and two-dimensional chromatography, these polypeptides showed patterns that were similar to those of hsp70. We also observed that extracts of heat-shocked salivary glands, when prepared in buffer containing 0.5% sodium deoxycholate and incubated at 4°C for several hours, showed polypeptide patterns that were considerably modified with respect to their unincubated controls. The most remarkable change was the progressive disappearance of the 70-kD band and the concomitant appearance of three bands of apparent 43 kD, 32 kD, and 30 kD (Fig. 1), the most prominent being usually the 30-kD band. Fingerprints and two-dimensional chromatographic patterns obtained from these three bands after tryptic digestion of [^{35}S]methionine-labeled polypeptides are similar or identical to those of the hsp70 (Fig. 2, C and D). This observation suggests that either the 70-kD protein is being degraded upon incubation to the 43-, 32-, and 30-kD bands or that the latter reflect altered configurations

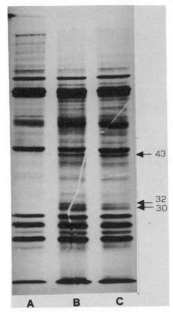

Figure 1
[^{35}S]methionine-labeled heat-shock polypeptide patterns before and after incubation in 0.5% sodium deoxycholate. Third instar larvae raised at 23°C were heat-shocked at 37°C for 20 minutes. Salivary glands were excised and incubated for 1 hour at 23°C with [^{35}S]methionine, exactly as described previously (Caggese et al. 1979). The glands were homogenized with 0.01 M sodium phosphate buffer (pH 7.5), 0.015 M NaCl, and 0.5% sodium deoxycholate (100 μl buffer per 10 pairs of glands). Homogenates were centrifuged for 20 minutes at 12,000g at 0°C. Samples were precipitated with trichloroacetic acid (10%) for 30 minutes on ice. The precipitate was collected, washed with ethanol and ether, and dried. The pellet was dissolved and analyzed on 10% SDS-polyacrylamide gels as detailed by Caggese et al. (1979). (A) [^{35}S]methionine heat-shock polypeptide pattern of an homogenate precipitated immediately after preparation (3.5 × 10^5 cpm). (B) [^{35}S]methionine heat-shock protein pattern after incubation of the homogenate for 12 hours at 4°C (6 × 10^5 cpm). (C) Exactly as lane B but ribonuclease was added to the extract (0.5 mg/ml). Exposure was for 24 hours.

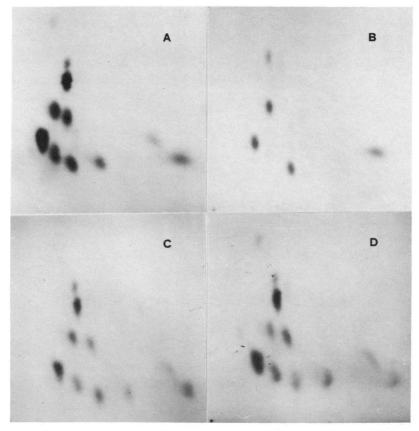

Figure 2
Fingerprints of various heat-shock polypeptides obtained before or after incubation of the extracts with sodium deoxycholate at 4°C. Heat-shock polypeptides were prepared as described in Fig. 1. They were separated on 10% SDS-polyacrylamide gels. After autoradiography, the labeled bands were cut and proteins eluted electrophoretically into dialysis bags. The eluted fractions were precipitated with trichloroacetic acid, washed with ethanol and ether, and then dried. After oxidation with performic acid, they were digested with trypsin and fingerprinted as detailed by Ish-Horowicz et al. (1979b). (*A*) hsp70. (*B*) hsp40 from flies carrying the *Df(3R)kar*[D2] deletion. (*C*) 32-kD polypeptide obtained after incubation for 12 hours at 4°C with sodium deoxycholate. (*D*) 30-kD (apparent) polypeptide obtained after incubation as for *C*.

of the 70-kD protein so that it migrates in SDS polyacrylamide gels with different mobilities but without loss of amino acids.

The nucleotide sequences of three of the genes coding for the hsp70 have been determined (Ingolia et al. 1980). From these sequences, the amino acid sequence of the protein has been determined and the size and number of the [^{35}S]methionine-containing peptides can be predicted for to faster-migrating bands, we should be able to detect loss of peptides in tryptic digests of these bands. A control for this can be found in a 40-kD heat-shock polypeptide produced in stocks carrying the $Df(3R)kar^{D2}$ deletion (Fig. 3). This deletion effectively removes the C terminal portion of one of the genes of the hsp70 (Ish-Horowicz and Pinchin 1980). In this case it is known that part of the 70-kD protein is missing, which results in there being fewer peptides following tryptic digestion (Caggese et al. 1979).

According to Ingolia et al. (1980) the hsp70 has 641 amino acid residues and contains 13 methionines. The distribution of these methionine residues relative to lysines and arginines is such that tryptic digestion would produce 13 methionine-containing peptides. After two-dimensional chromatography the hsp70 resolves into the expected 13 spots whereas after fingerprint analysis only 11 methionine spots are resolved (Fig. 2A). However, two of these spots are more heavily labeled than the others, and we think that each of them represents two unresolved spots.

The 40-kD fragment obtained from Drosophila stocks carrying the $Df(3R)kar^{D2}$ deletion resolves into six spots after two-dimensional chromatography and into five spots after fingerprinting (Fig. 2B). One of these five spots is an apparent doublet. The 40-kD fragment is expected to contain 366 amino acid residues. The expected number of methionine-containing

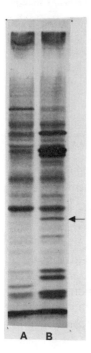

Figure 3
Polypeptide patterns obtained from larvae heterozygous for the $Df(3R)kar^{D2}$ deletion. (A) Uninduced control; (B) heat-shock pattern. Arrow indicates hsp40.

A   B

peptides after tryptic digestion of this fragment is six. The number of methionine-containing peptides is in accord with the expectation and suggests that transcription of the abnormal message produced by the 70-kD gene cut by the *Df(3R)kar*[D2] deletion does not proceed in open frame beyond the breakpoint to include a methionine codon. In contrast, the 43-kD, 32-kD, and 30-kD (apparent) peptides show the full set of methionine peptides, similar to that characteristic of the hsp70. The 32-kD band has 11 spots after fingerprinting, but one spot has a different position with respect to the spots obtained from the hsp70 (Fig. 2C). Also, the two-dimensional chromatogram of the 43-kD polypeptide exhibits some unusual spots while exhibiting most of the pattern of the 70-kD protein (not shown). The fact that each of the proteins with apparent molecular weights 43 kD, 32 kD, and 30 kD exhibits tryptic digests that are similar or indistinguishable from those of the hsp70 suggests that their formation by proteolytic cleavage of the 70-kD protein is highly improbable. Other observations also argue against proteolytic cleavage: (1) two-dimensional gel analysis shows that all the labeled polypeptides observed after incubation with sodium deoxycholate at 4°C are also present in untreated controls, although longer exposures are needed to visualize some of them (i.e., those at apparent 43 kD, 32 kD, and 30 kD positions); (2) the pattern produced by Coomassie Blue staining of the preparations is apparently unchanged whether the extracts are analyzed immediately or after incubation at 4°C; (3) the addition of up to 1 mg/ml protein does not prevent the transformation; (4) a 5- to 20-fold dilution does not prevent the transformation; (5) the electrophoretic mobility and the fingerprints of other heat-shock proteins (84 and 27 kD) are unaffected under the conditions that lead to transformation of the hsp70.

Analysis of the products of tryptic digestion of the 43-kD, 32-kD, and 30-kD polypeptides produced during incubation at 4°C in the presence of sodium deoxycholate shows that they contain all the peptide fragments characteristic of unmodified hsp70. This suggests that the hsp70 can be induced to assume different tertiary structures in vitro that lead to changes in electrophoretic mobility without altering its primary structure. We note that our analysis employs conditions that break disulfide bridges.

CONCLUSIONS

The data presented here indicate that polypeptides that migrate faster, but at definite positions, than hsp70 on SDS polyacrylamide gels might contain the entire or almost entire amino acid sequence of the hsp70. It is of interest to us that two such forms (i.e., those at the 64-kD and 43-kD positions) apparently comigrate with the subunits of the enzyme GS1, which, under definite conditions, positively responds to heat-shock treatments. Also, the subunits of GS1 can be transformed into polypeptides of minor apparent molecular weight after definite treatments that exclude proteolytic clea vage, while under other conditions these subunits can be transformed into a polypeptide of apparent molecular weight close to 70 kD (F. Ritossa, unpubl.). Clearly, all these coincidences remain to be worked out. The observation that GS1 of *Drosophila* has or at least is reproducibly associated with a particular RNA-polymer component (R. Caizzi and F. Ritossa, unpubl.) indicates an unprecedented complexity of this enzyme. Preliminary studies of such RNA polymers suggest their possible involvement in a novel kind of cross-linking mechanism.

The apparent similarities between the transformed heat-shock polypeptides as described here and the subunits of GS1 might be pure coincidence. However, their study might represent a method for determining some functional aspect of the heat-shock system in *Drosophila*.

REFERENCES

Caggese, C., R. Caizzi, M. Morea, F. Scalenghe, and F. Ritossa. 1979. Mutation generating a fragment of the major heat shock-inducible polypeptide in *Drosophila melanogaster. Proc. Natl. Acad. Sci.* **76:** 2385–2389.

Craig, E.A., B.J. McCarthy, and S.C. Wadsworth. 1979. Sequence organization of two recombinant plasmids containing genes for the major heat shock-induced protein of *D. melanogaster. Cell* **16:** 575–588.

Henikoff, S. and M. Meselson. 1977. Transcription at two heat shock loci in *Drosophila. Cell* **12:** 441–451.

Ingolia, T.D., E.G. Craig, and B.J. McCarthy. 1980. Sequences of three copies of the gene for the major *Drosophila* heat shock induced protein and their flanking regions. *Cell* **21:** 669–679.

Ish-Horowicz, D. and S.M. Pinchin. 1980. Genomic organization of the 87A7 and 87C1 heat induced loci in *D. melanogaster. J. Mol. Biol.* **142:** 231–245.

Ish-Horowicz, D., J.J. Holden, and W.J. Gehring. 1977. Deletions of two heat activated loci in *Drosophila melanogaster* and their effects on heat-induced protein synthesis. *Cell* **12:** 643–652.

Ish-Horowicz, D., S.M. Pinchin, P. Schedl, S. Artavanis-Tsakonas, and M.E. Mirault. 1979a. Genetic and molecular analysis of the 87A7 and 87C1 heat induced loci in *D. melanogaster. Cell* **18:** 1351–1358.

Ish-Horowicz, D., S.M. Pinchin, J. Gausz, H. Gyurkovics, G. Bencze, M. Goldschmidt-Clermont, and J.J. Holden. 1979b. Deletion mapping of two *Drosophila melanogaster* loci that code for the 70.000 d heat-induced protein. *Cell* **17:** 565–574.

Mirault, M.E., M. Goldschmidt-Clermont, L. Moran, A.P. Arrigo, and A. Tissières. 1978. The effect of heat shock on gene expression in *Drosophila melanogaster. Cold Spring Harbor Symp. Quant. Biol.* **42:** 819–827.

Scalenghe, F. and F. Ritossa. 1976. Controllo dell'attivita' genica in *Drosophila. Il puff al locus ebony e la Glutamina sintetasi 1. Atti Accad. Naz. Lincei.* **13:** 439–528.

The Mammalian Stress Proteins

William J. Welch, James I. Garrels,
and James R. Feramisco
Cold Spring Harbor Laboratory
Cold Spring Harbor, New York 11724

Exposure of mammalian tissue culture cells to elevated growth temperatures results in specific changes in the patterns of proteins being synthesized. Similar to that which has been described for *Drosophila melanogaster* (Tissières et al. 1974; Lewis et al. 1975; McKenzie et al. 1975; and reviewed by Ashburner and Bonner 1979), heat-shock treatment of mammalian cells is characterized by the synthesis and accumulation of a small number of polypeptides concomitant with a decreased production of the normal complement of other cellular proteins. It is now recognized that various other treatments besides heat shock give rise to the same response in eukaryotic cells. For example, exposure of cells to certain drugs (Hightower 1980), transition series metals (Levinson et al. 1980), and various amino acid analogs (Kelley and Schlesinger 1978; Hightower and Smith 1978; Hightower 1980; Hightower and White 1981; Thomas et al. 1982), to name just a few, induces the same set (or a subset) of proteins as does the heat-shock treatment. Hence, this generalized response to perturbations in the normal growth environment of cells is more aptly referred to as the "stress response" and accordingly the proteins induced termed the "stress proteins."

Although it has been established, in general, which proteins are induced during the response, their location and function in the cell still is not as yet fully understood. As to their intracellular location, it does appear that some of the stress proteins migrate to the nucleus following their synthesis in the cytoplasm (Mitchell and Lipps 1975; Vincent and Tanguay 1979; Arrigo et al. 1980; Velazquez et al. 1980; Levinger and Varshavsky 1981). As to the possible function of these proteins, considerable evidence presented in this volume has shown that the induction of the stress response confers a degree of protection to the cell upon subsequent stress situations and that such protection is contingent upon the prior synthesis of the stress proteins.

As an initial step in understanding their role in the stress response, we have purified three of the stress proteins produced in HeLa cells. In addition, rabbit polyclonal antibodies have been raised against four of the mammalian stress proteins and a monoclonal antibody produced against one of the proteins (see Lin et al., this volume). It is hoped that the examination of the properties of the purified proteins and the determination of their intracellular location by immunological techniques will facilitate the analysis of the function of the stress proteins.

THE MAMMALIAN STRESS PROTEINS

The stress proteins induced in a number of different mammalian cell types appear similar when analyzed by one-dimensional SDS-polyacrylamide gels. The levels of six proteins with apparent molecular masses of 72, 73, 80, 90, 100, and 110 kD appear elevated in a variety of cells grown under stress (e.g., baby hamster kidney, rat embryo fibroblasts, gerbil fibroma cells, and HeLa cells). These stress-induced proteins are illustrated in Figure 1. Gerbil fibroma cells and HeLa cells were pulse-labeled with [^{35}S]methionine after growth for 8 hours at either 37°C (lanes A and D, respectively), 43°C (lanes C and F, respectively), or at 37°C in the presence of an amino acid analog of proline, L-azetidine-2-carboxylic acid (AzC) (lanes B and E, respectively), and the polypeptides examined on a 10% SDS-polyacrylamide gel. In general, the responses of both cell types to the different stress situations seem to be similar. A comparison of the polypeptides synthesized in the AzC-treated cells with those produced in the heat-shock-treated cells shows, however, that the relative induction of each individual stress protein is dependent upon the particular agent used to elicit the response (e.g., compare the levels of the 72K and 73K proteins induced by either AzC or heat shock). Furthermore, it should be noted that the reduction of normal protein synthesis in the stressed cell is also dependent upon the duration of the stress treatment. For example, cells incubated in 5 mM AzC for periods of greater than 12 hours synthesize the stress proteins and little else. Heat treatment for a similar period of time causes a similar induction of the stress proteins; but, unlike the AzC treatment, does not result in such a dramatic cessation of normal protein synthesis.

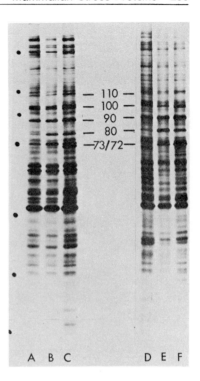

Figure 1
Polypeptide composition of normal and stressed cells. Gerbil fibroma cells (lanes A–C) and HeLa cells (lanes D–F) were pulse-labeled with [^{35}S]methionine after 8 hours of growth at either 37°C (lanes A and D), 43°C (lanes C and F), or at 37°C in the presence of 5 mM AzC, an amino acid analog of proline (lanes B and E). The cells were solubilized in electrophoresis sample buffer and the polypeptides separated on a 10% SDS-polyacrylamide gel. Shown is an autoradiograph of the gel. Molecular-weight markers are indicated by spots on left of the figure and are the same as in Fig. 3.

The proteins induced during the stress response are more clearly identified when analyzed by two-dimensional gel electrophoresis. Figure 2 shows the [^{35}S]methionine pulse-labeled proteins synthesized in normal (A) or stressed HeLa cells incubated in 5 mM AzC (B). Similarly, in a second experiment, the [^{35}S]methionine pulse-labeled proteins synthesized in HeLa cells grown at 37°C (C) or in HeLa cells grown at 43°C for 8 hours (D) are shown. In both experiments, the stressed cells are seen to synthesize higher amounts of the stress proteins. These are designated, with respect to their molecular weight and major isoelectric charge isomer, as the 72K/pI 5.6, 73K/pI 5.5 and 6.3, 80K/pI 5.2, 90K/pI 5.2, and 100K/pI 5.0 proteins. The 110K stress protein, induced somewhat transiently from experiment to experiment, is not indicated here but appears to have a pI of approximately 5.4 (Thomas et al. 1982). Careful inspection of the protein patterns in the stressed cells (B and D) versus those in the normal cells (A and C) reveals that all of the stress proteins with the exception of the 73K/pI 6.3 species appear to be present in the normal uninduced cells. The analysis by two-dimensional gels indicates that a number of the stress proteins appear heterogenous, particularly the 90K, 72K, and 73K proteins. Such charge heterogeneity with respect to the 90K protein appears to be due at least in part to extensive

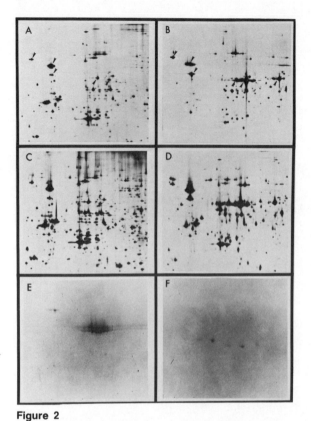

Figure 2
Two-dimensional gel analysis of normal and stressed HeLa cells and of the purified 72K, 73K, and 90K stress proteins. HeLa cells grown at 37°C (A) or grown in the presence of 5 mM AzC (B) were pulse-labeled with [^{35}S]methionine, the cells solubilized, and the proteins analyzed by isoelectric focusing in a 5–7 pH gradient (acid end on the left) followed by electrophoresis into a 10% SDS-polyacrylamide gel. The stress proteins (100K, 90K, 80K, 73K, and 72K) are indicated by arrows. In a second experiment, HeLa cells grown at either 37°C (C) or at 43°C for 8 hours (panel D) were similarly pulse-labeled with [^{35}S]methionine and the polypeptides analyzed as described above. (A–D) Autoradiographs of the gels. In the bottom two panels are shown the Coomassie Blue-stained purified 90K/100K proteins (E) and the 72K/73K proteins (F) analyzed as described above.

phosphorylation. In addition, labeling with [^{33}P]H$_3$PO$_4$ has shown that the 80K and 100K are also apparently phosphoproteins. To date, however, we have been unable to detect phosphorylation of either the 72K or 73K stress proteins. A summary of this data, as well as some of the physical properties of the purified 72K, 73K, and 90K proteins (described below), is presented in Table 1.

TABLE 1
Properties of the Mammalian Stress Proteins

Stress protein	Molecular weight as determined by SDS-PAGE (daltons)	pI of major charge isomer as determined by IEF[a]	Stokes' radius (Å)	Sedimentation coefficient (10⁻¹³ sec)	Phosphoprotein	"Native" molecular weight (daltons)
72K	72,000	5.6	42.6[b]	4.3[b]	−	−
73K	73,000	5.5 and 6.3			−	73,800
80K	80,000	5.2	N.D.[c]	N.D.	+	N.D.
90K	90,000	5.2	69[d]	5.8[d]	+	−
100K	100,000	5.0			+	165,000
110K	110,000	5.4	N.D.	N.D.	N.D.	N.D.

[a]Data presented in Thomas et al. (1982).
[b]The 72K and 73K proteins copurified; this value was determined for the mixture.
[c]N.D., not determined.
[d]The 90K and 100K proteins copurified; this value was determined for the mixture.

PURIFICATION OF THE 72K AND 73K STRESS PROTEINS

As the next step in their analysis, we have purified the three major stress proteins (the 72K, 73K, and 90K proteins) produced in HeLa cells. Additionally, we have partially purified the bulk of the 100K and 80K proteins from the detergent-extracted particulate fraction of these cells. A detailed description of these procedures will be reported elsewhere. Briefly, HeLa cells incubated in the presence of 5 mM AzC were lysed by Dounce homogenization in hypotonic buffer and the lysate centrifuged at 20,000g. The majority of 90K as well as considerable amounts of the 72K and 73K proteins were released into the soluble phase following the disruption of the cells. Conversely, the 80K and 100K proteins, for the most part, were found in the low-speed pellet. For their purification, the supernatant containing the 72K, 73K, and 90K proteins was applied to a DEAE-52 column and the proteins eluted with a linear gradient of NaCl. The peak fractions containing the 90K protein and the 72K and 73K proteins were collected and the proteins further purified by both ion exchange and gel filtration chromatography. The 72K and 73K proteins (major charge isomers of 5.6 and 5.5, respectively) were observed to copurify through five different column chromatographic steps as well as to cosediment through sucrose gradients. One-dimensional peptide maps of the individually iodinated purified 72K and 73K proteins revealed that the two proteins shared many common peptides. The 72K and 73K protein mixture had an apparent native molecular mass of 73.8 kD as determined by gel filtration and sucrose sedimentation analysis. The 90K stress protein was observed both to copurify and to cosediment through sucrose gradients with small amounts of the 100K stress protein. Unlike the 72K and 73K proteins, however, the one-dimensional peptide maps indicated that 90K and 100K were unrelated polypeptides. An apparent native molecular mass of 165 kD was determined for the 90K and 100K protein mixture.

To determine if the purified 72K, 73K, and 90K proteins as well as the 100K protein copurifying with 90K were the same as the corresponding stress proteins made in vivo, a comparison of their patterns on two-dimensional gels with the pattern of [35S]methionine-labeled polypeptides synthesized in HeLa cells exposed to AzC was made. The Coomassie Blue-stained purified 90K and 100K proteins analyzed on a 5−7 isoelectric focusing gradient in the first dimension followed by electrophoresis on a 10% SDS-polyacrylamide slab gel is shown in Figure 2E. The purified 72K/73K protein mixture similarly analyzed and stained by Coomassie Blue is shown in Figure 2F. When the purified proteins were mixed with the [35S]methionine-labeled proteins synthesized in AzC-treated HeLa cells and the samples analyzed as described above, the purified proteins were observed to migrate identically with their in vivo labeled counterparts (data not shown). These results indicated that the proteins purified are indeed the stress-induced 72K, 73K, and 90K proteins and that the proteins are not significantly altered during their purification.

The copurification of minor amounts of 100K with 90K is interesting since the majority of the 100K protein remains with the particulate fraction following the 20,000g centrifugation of the Dounced cell lysate. Furthermore, as is decribed by Lin et al. (this volume), we have observed that the 100K protein appears to be present predominantly in or near the Golgi apparatus in a number of different cell types examined to date and under certain conditions, associated with the nucleus. In light of this copurification of the Golgi-associated 100K protein with 90K, it is worth mentioning here that a number of laboratories have observed that a portion of the transforming protein of Rous sarcoma virus, pp60^{src}, appears to exist as part of a complex with both a 50K protein and the 90K stress protein (Sefton et al 1978; Opperman et al. 1981; Brugge et al. 1981). In addition, the putative transforming protein(s) of PRCII avian sarcoma virus, p105 and p110, appears to exist in a complex with the same 50K cellular protein and the avian 90K stress protein as well (Adkins et al. 1982). The significance of these various protein complexes containing the 90K stress protein is not clear at this time.

PREPARATION AND PARTIAL CHARACTERIZATION OF POLYCLONAL ANTIBODIES DIRECTED AGAINST THE MAMMALIAN STRESS PROTEINS

Having purified the 72K, 73K, and 90K proteins and partially purifying the 80K and 100K stress proteins, we prepared rabbit antibodies against these proteins. Immunoprecipitation from [^{35}S]methionine-labeled HeLa cells grown under stress (Fig. 3, total proteins are shown in lane A) with the 72K/73K antiserum (Fig. 3, lane B), the 80K antiserum (Fig. 3, lane C), and the 90K antiserum (Fig. 3, lane D) specifically brought down polypeptides of the appropriate molecular weights. To determine the exact nature of the isoelectric variants of the stress proteins being recognized by these antibodies, the immunoprecipitates are currently being analyzed by two-dimensional gel electrophoresis.

Preliminary direct immunofluorescence analysis of the intracellular location of the 72K/73K stress proteins has been obtained using our polyclonal antibody. Although not shown here, staining of normal gerbil fibroma cells, HeLa cells, baby hamster kidney cells, and rat embryo fibroblasts with the 72K/73K antibody reveals both a cytoplasmic and nuclear distribution of the antigens. Heat shock or AzC treatment of the cells results in an increased nuclear staining as well as some enhancement of the cytoplasmic staining. Such a predominant nuclear location of the 72K/73K proteins in the stressed mammalian cells may be analogous to that which has been observed for at least one of the heat-shock

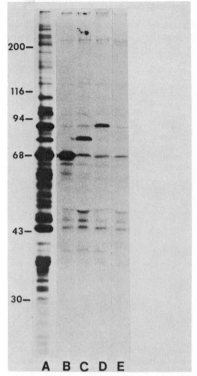

Figure 3
Specificity of antibodies directed against the mammalian stress proteins. Rabbit antisera directed against the 72K/73K, 80K, and 90K stress proteins were produced as described in the text. Immunoprecipitation from [35S]methionine pulse-labeled HeLa cells grown under stress (lane A) with the 72K/73K antiserum (lane B), the 80K antiserum (lane C), and the 90K antiserum (lane D) was done in the presence of 1% deoxycholate, 1% Triton X-100, and 0.1% SDS. The antibody-antigen complex was precipitated by the addition of fixed *Staphylococcus aureus* (IgG-Sorb). In lane E is an immunoprecipitation with no added antibody. Shown is an autoradiograph of a 10% SDS-polyacrylamide gel. Molecular-weight markers are indicated on the left.

proteins in *Drosophila* (Mitchell and Lipps 1975; Arrigo et al. 1980; Velazquez et al. 1980; Levinger and Varshavsky 1981) and in *Chironomus tentans* (Vincent and Tanguay 1979). Although intriguing with respect to the altered transcriptional activity of stressed cells, the significance of a nuclear location of the 72K/73K proteins remains to be determined.

SUMMARY

Mammalian cells grown under stress synthesize and accumulate a small number of proteins with apparent molecular masses of 72, 73, 80, 90,

100, and 110 kD. Fractionation of stressed HeLa cells resulted in the solubilization of all of the 90K as well as considerable amounts of the 72K and 73K proteins; the majority of the 80K and 100K proteins remained with the particulate fractions. Purification of the 72K, 73K, and 90K proteins was accomplished by ion exchange and gel filtration chromatography. 72K and 73K copurified, cosedimented through sucrose gradients, and were found to be related proteins as determined by one-dimensional peptide maps. The 90K protein copurified with small amounts of the 100K protein and the two proteins cosedimented in sucrose gradients. Immunofluorescence studies showed that the 72K/73K proteins were located both in the cytoplasm and in the nucleus, with the nuclear staining increasing upon heat-shock treatment of the cells.

ACKNOWLEDGMENTS

We gratefully acknowledge J. D. Watson for his interest and support of this work. Special thanks to G. P. Thomas for stimulating discussions. We thank B. McLaughlin, P. Renna, and M. Szadkowski for their technical assistance. This work was supported by a NIH grant (GM-28277) to J.R.F. and a Muscular Dystrophy Postdoctoral Fellowship to W.J.W.

REFERENCES

Adkins, B., T. Hunter, and B.M. Sefton. 1982. The transforming proteins of PRCII virus and Rous sarcoma virus form a complex with the same two cellular phosphoproteins. *J. Virol.* (in press).

Arrigo, A.P., S. Fakan, and A. Tissières. 1980. Localization of the heat shock-induced proteins in *Drosphila melanogaster* tissue culture cells. *Dev. Biol.* **78:** 86–103.

Ashburner, M. and J.J. Bonner. 1979. The induction of gene activity in *Drosophila* by heat shock. *Cell* **17:** 241–254.

Brugge, J.S., E. Erickson, and R.L. Erickson. 1981. The specific interaction of the Rous sarcoma virus transforming protein, pp60[src], with two cellular proteins. *Cell* **25:** 363–372.

Hightower, L.E. 1980. Cultured animal cells exposed to amino acid analogues or puromycin rapidly synthesize several polypeptides. *J. Cell. Physiol.* **102:** 407–427.

Hightower, L.E. and M.D. Smith. 1978. Effects of canavanine on protein metabolism in Newcastle disease virus-infected and uninfected chicken embryo cells. In *Negative strand viruses and the host cell* (ed. B.W.J. Mahy and R.O. Barry), pp. 395–405. Academic Press, London.

Hightower, L.E. and F.P. White. 1981. Cellular responses to stress: Comparison of a family of 71–73 kilodalton proteins rapidly synthesized in rat tissue slices and canavanine-treated cells in culture. *J. Cell. Physiol.* **108:** 261–275.

Kelley, P.M. and M.J. Schlesinger. 1978. The effect of amino acid analogues and heat shock on gene expression in chicken embryo fibroblasts. *Cell* **15:** 1277–1286.

Levinger, L. and A. Varshavsky. 1981. Heat-shock proteins of *Drosophila* are associated with nuclease-resistant, high-salt-resistant nuclear structures. *J. Cell Biol.* **90:** 793–796.

Levinson, W., H. Oppermann, and J. Jackson. 1980. Transition series metals and sulfhydryl reagents induce the synthesis of four proteins in eukaryotic cells. *Biochim. Biophys. Acta* **606:** 170–180.

Lewis, M., P.J. Helmsing, and M. Ashburner. 1975. Parallel changes in puffing activity and patterns of protein synthesis in salivary glands of *Drosophila*. *Proc. Natl. Acad. Sci.* **72:** 3604–3608.

McKenzie, S.L., S. Henikoff, and M. Meselson. 1977. Localization of RNA from heat-induced polysomes at puff sites in *Drosophila melanogaster. Proc. Natl. Acad. Sci.* **72:** 1117–1121.

Mitchell, H.K. and L.S. Lipps. 1975. Rapidly labeled proteins on the salivary gland chromosomes of *Drosophila melanogaster. Biochem. Genet.* **13:** 585–602.

Oppermann, H., A.O. Levinson, L. Levintow, H.E. Varmus, J.M. Bishop, and S. Kawai. 1981. Two cellular proteins that immunoprecipitate with the transforming protein of Rous sarcoma virus. *Virology* **113:** 736–751.

Sefton, B.M., K. Beemon, and T. Hunter. 1978. Comparison of the expression of the *src* gene of Rous sarcoma virus in vitro and in vivo. *J. Virol.* **28:** 957–971.

Thomas, G.P., W.J. Welch, M.B. Mathews, and J.R. Feramisco. 1982. Molecular and cellular effects of heat shock and related treatments of mammalian tissue culture cells. *Cold Spring Harbor Symp. Quant. Biol.* **46:** 985–996.

Tissières, A., H.K. Mitchell, and V.M. Tracey. 1974. Protein synthesis in salivary glands of *Drosophila melanogaster:* Relation to chromosome puffs. *J. Mol. Biol.* **84:** 389–398.

Velazquez, J.M., B.J. DiDomenico, and S. Lindquist. 1980. Intracellular localization of heat shock proteins in *Drosophila*. *Cell* **20:** 679-689.

Vincent, M. and R.M. Tanguay. 1979. Heat shock induced proteins present in the cell nucleus of *Chironomus tentans* salivary gland. *Nature* **281:** 501-503.

The Association of the 100-kD Heat-shock Protein with the Golgi Apparatus

Jim Jung-Ching Lin,
William J. Welch, James I. Garrels,
and James R. Feramisco
Cold Spring Harbor Laboratory
Cold Spring Harbor, New York 11724

The functions and intracellular locations of the heat-shock proteins (hsp) have become a topic of major interest in cell biology over the past few years, most likely because of the widespread occurrence of these proteins in many types of normal cells (as well as in heat-shocked cells) and the numerous means by which these proteins (or a subset thereof) can become induced. Initial studies of the intracellular localization of one of the heat-shock proteins have been carried out by autoradiographic analysis of *Drosophila* cells (Mitchell and Lipps 1975; Velazquez et al. 1980). It was found that major heat-shock proteins (most likely hsp70) were highly concentrated in the nuclei upon heat shock and that their transport to the nucleus occurred very rapidly. Vincent and Tanguay (1979) also reported that hsp34 was enriched in the nucleolar preparation of *Chironomus tentans*. Furthermore, Arrigo et al. (1980) reported that the *Drosophila* 68K and 70K heat-shock proteins as well as the four smaller heat-shock proteins are present in the nucleus after heat-shock treatment and that they return to the cytoplasm when the cells are returned to normal growth temperatures. Finally, by biochemical analysis, hsp70, 68, 28, 26, and 22 of *Drosophila* cells were found to be tightly associated with nuclease-resistant, salt-resistant nuclear structures (Levinger and Varshavsky 1981). These experiments showing a nuclear location of at least a few of the heat-shock proteins have led to the speculation that heat-shock proteins may be involved in transcriptional regulation.

267

In this report we have used a monoclonal antibody, JLJ5a (Lin and Queally 1982) and a polyclonal rabbit antiserum directed against the hsp100 of cultured mammalian cells to study the intracellular location of this protein. In these cells (e.g., HeLa, baby hamster kidney, gerbil fibroma, rat L6 myoblasts), the major heat-shock proteins have apparent molecular masses of 72, 73, 80, 90, 100, and 110 kD (Thomas et al. 1982; Welch and Feramisco, this volume). By using indirect immunofluorescence, we have found that the 100-kD protein is mainly localized in the Golgi apparatus of tissue culture cells. In addition, a nuclear localization of the 100-kD protein was also found in a subpopulation of cells. Upon heat treatment, the number of cells with a nuclear localization of the 100-kD protein increased substantially.

IMMUNOPRECIPITATION OF HSP100 FROM CULTURED MAMMALIAN CELLS

Two-dimensional gel analysis of the proteins synthesized in normal and heat-shocked L6 myoblasts shows the induction of the heat-shock proteins in these cells (Fig. 1). The myoblasts were grown at 37°C (panel A) or at 42°C (panel B) for 4 hours and labeled with [^{35}S]methionine for the last hour. The labeled proteins were resolved on two-dimensional gels using pH 5–7 ampholyte gradients in the first dimension and 10% SDS-polyacrylamide gels in the second dimension (Garrels 1979). hsp 100, 90, 80, and 72 can be identified in such cells.

To demonstrate the specificity of the monoclonal antibody JLJ5a for hsp100, an immunoprecipitation reaction is shown (carried out as described before [Lin and Queally 1982]). [^{35}S]methionine-labeled proteins from L6 myoblasts, gerbil fibromas, Rat-1 fibroblasts, and L6 myotubes were incubated with the antibody and the resultant immunoprecipitates were analyzed by two-dimensional gel electrophoresis. In the case of L6 myoblasts, the immunoprecipitate analysis indicating the reaction of the antibody with hsp100 is shown in Figure 1 (panel D). Experiments with the other cell types gave similar results. (In earlier studies with this antibody [Lin and Queally 1982], we reported that the antigen had an apparent M_r of 110 kD; however, the updated apparent M_r of the antigen should be 100 kD). In addition, we have further characterized hsp100 as being apparently one of the phosphoproteins of the cell in that it can be radioactively labeled in vivo by incubation of the cells in [^{33}P]orthophosphate (Fig. 1C).

THE DISTRIBUTION OF HSP100 IN NORMAL CELLS

By using indirect immunofluorescence (using the procedures of Blose [1979]) with the JLJ5a monoclonal antibody, we have studied the intra-

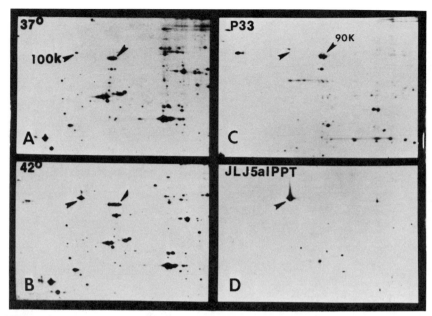

Figure 1
Two-dimensional gel electrophoresis of hsp100. A comparison of the electrophoretic mobilities of hsp100 from [^{35}S]methionine-labeled L6 myoblasts grown at 37°C (*A*) or at 42°C (*B*); [^{33}P]orthophosphate-labeled L6 myoblasts grown at 37°C (*C*); and the immuno-precipitate from [^{35}S]methionine-labeled L6 myoblasts by JLJ5a monoclonal antibody (*D*). The first-dimension (isoelectric focusing) gels contained pH 5–7 ampholytes and the second dimensions were 10% SDS–polyacrylamide gels. The antibody specifically immun-oprecipitates hsp100. The [^{33}P]orthophosphate-labeling pattern suggests that the hsp100 is a phosphoprotein.

cellular localization of the 100-kD dalton protein. A wide variety of tissue culture cell types have been examined and have given essentially the same results. Both normal cells and heat-shocked cells have been studied, since hsp100 is present in both cases (see Fig. 1, for example). The results of such an immunofluorescence experiment using gerbil fibroma cells grown at the normal temperature is shown in Figure 2 (A,C, and E). Bright fluorescence with a lamellarlike structure was seen near the perinuclear region of the cells. In addition, nuclear fluorescence was found in a subpopulation of the cells (~15% of the cells). It is not as yet possible to determine if the nuclear distribution is intra- or extranu-clear. The perinuclear distribution of hsp100 corresponds to the Golgi apparatus of the cell. This has been shown previously by Lin and Queally (1982) by using the following criteria: (1) in double-staining experiments, the localization of the 100-kD protein by the JLJ5a anti-body was coincident with the reaction products of thiamine pyrophos-

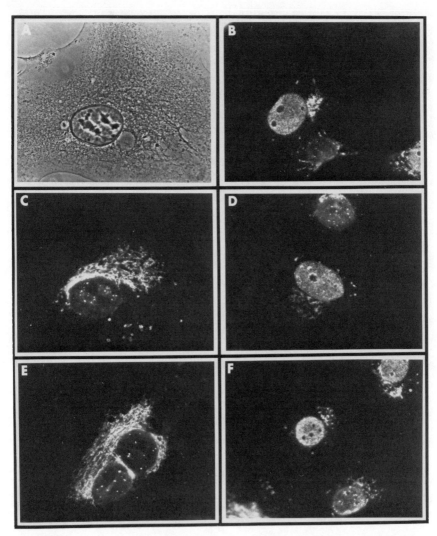

Figure 2

Localization of hsp100 in mammalian cells. Indirect immunofluorescence of gerbil fibroma cells cultured at the normal temperature (37°C) (*A,C,E*) or at higher temperature (43°C) (*B,D,F*), stained by the JLJ5a monoclonal antibody. With heat shock, the Golgi staining pattern changed somewhat to a less organized one and the nuclear staining increased. (*A*) Phase-contrast micrograph; (*B–F*) immunofluorescence micrographs. *A* and *C* show the same field.

phatase, a well-established enzyme marker of the Golgi apparatus (Novikoff and Goldfischer 1961); and (2) treatment of the cells with Colcemid or monensin, drugs known to disrupt and fragment the Golgi apparatus (Robbins and Gonatas 1964; Moskalewski et al. 1976; Tarta-koff and Vassalli 1977; Thyberg et al. 1977; Ledger et al. 1980), induced the rearrangement of 100-kD protein from the normal perinuclear location into vacuoles dispersed throughout the cytoplasm. Rabbit polyclonal antibody raised against purified hsp100 from HeLa cells gave similar staining patterns. Therefore, we concluded that the hsp100 is primarily associated with the Golgi apparatus and in some cases is also present in or near the nucleus in normal cells.

LOCALIZATION OF HSP100 IN HEAT-SHOCKED CELLS

Gerbil fibroma cells were subjected to heat treatment (43°C, 10 hr) and analyzed by immunofluorescence for the distribution of hsp100. To summarize these data, the proportion of cells showing the nuclear staining increased and the Golgi staining appeared to be somewhat disrupted (Fig. 2, B, D, and F). In some of the cells showing the nuclear distribution, the Golgi apparatus staining could not be detected by JLJ5a antibody. Similar types of results were obtained in the experiments with HeLa cells and L6 myoblasts. The time required for heat treatment of the cells to increase the percentage of the cells showing the nuclear staining varied from cell line to cell line. In the case of HeLa cells, for example, a substantial increase in number of cells with nuclear staining could be detected in as little as 2 hours after shifting the cells to the higher temperature. Virtually all of these cells showed both the nuclear staining and the Golgi apparatus staining.

Based upon the immunoprecipitation analysis (e.g., Fig. 1) of the specificity of the monoclonal antibody JLJ5a, it is likely that the same protein is giving rise to both the Golgi region and nuclear staining by immunofluorescence assays. Monoclonal antibodies, however, can have complicating features in these types of experiments such as the presence of more than one species of immunoglobulin or a cross-reaction with a common antigenic site on multiple proteins that are unrelated (Lane and Hoeffler 1980; Dulbecco et al. 1981; Pruss et al. 1981; Lin 1982). We have ruled out the former potential problem since the secreted antibody contained only a single species of light chain as determined by two-dimensional gel analysis (not shown). The latter problem seemingly can be ruled out since the immunoprecipitation analysis shows only a single antigen. A polyclonal rabbit antiserum made against the hsp100 clearly stains the Golgi apparatus in normal cells but has not as yet been used to localize the protein in heat-shocked cells. These experiments, which will be most useful in confirming or denying the possibility that the monoclonal antibody recognizes more than one protein, are currently in progress in our laboratory.

SUMMARY

Indirect immunofluorescence has been used to determine the intracellular location of hsp100 of mammalian cells. The majority of the protein was found to be localized in or near the Golgi apparatus in all the mammalian cell lines tested thus far. In normal, unstressed cells, about 15% of the cells showed a nuclear distribution of the protein as well. Whether this nuclear distribution represents the intranuclear presence of hsp100 or the association of the protein with the outer nuclear surface is not as yet clear. Immunoelectron microscopy most likely will be necessary to settle this question. In heat-shocked or stressed cells, a general increase in the number of cells showing the nuclear distribution occurred with many of the cells showing a slightly disrupted Golgi staining. The functional significance of these distributions of hsp100 is not known at this time; however, recent experiments in our laboratory and in others (see Hightower and White, this volume) have indicated that glucose deprivation leads to the increased synthesis of hsp100. Combining these results with those defining the intracellular location of this protein, the possibility exists that the hsp100 may be involved in the metabolic or catabolic reactions of the Golgi apparatus.

ACKNOWLEDGMENTS

We are indebted to J.D. Watson for his enthusiastic support of this work. We thank B. McLaughlin, P. Renna, and M. Szadkowski for their technical assistance. Support for this work was provided by a Muscular Dystrophy Association grant to J.J-C. L., a NIH grant to J.R.F. (GM 28277), and a Muscular Dystrophy Association Postdoctoral Fellowship to W.J.W.

REFERENCES

Arrigo, A.P., S. Fakan, and A. Tissières. 1980. Localization of the heat shock-induced proteins in *Drosophila melanogaster* tissue culture cells. *Dev. Biol.* **78:** 86–103.

Blose, S.H. 1979. Ten-nanometer filaments and mitosis: Maintenance of structural continuity in dividing endothelial cells. *Proc. Natl. Acad. Sci.* **76:** 3372–3376.

Dulbecco, R., M. Unger, M. Bologna, H. Battifora, P. Syka, and S. Okada. 1981. Cross-reactivity between Thy-1 and a component of intermediate filaments demonstrated using a monoclonal antibody. *Nature* **292:** 772–774.

Garrels, J.I. 1979. Two-dimensional gel electrophoresis and computer analysis of proteins synthesized by clonal cell lines. *J. Biol. Chem.* **254:** 7961–7977.

Lane, D.P. and W.K. Hoeffler. 1980. SV40 large T shares an antigenic determinant with a cellular protein of molecular weight 68,000. *Nature* **288:** 167–170.

Ledger, P.W., N. Uchida, and M.L. Tanzer. 1980. Immunocytochemical localization of procollagen and fibronectin in human fibroblasts: Effects of the monovalent ionophore, monensin. *J. Cell Biol.* **87:** 663–671.

Levinger, L. and A. Varshavsky. 1981. Heat-shock proteins of *Drosophila* are associated with nuclease-resistant, high-salt-resistant nuclear structures. *J. Cell Biol.* **90:** 793–796.

Lin, J.J.C. 1982. Mapping structural proteins of cultured cells by monoclonal antibodies. *Cold Spring Harbor Symp. Quant. Biol.* **46:** 769–783.

Lin, J.J.C. and S.A. Queally. 1982. A monoclonal antibody that recognizes Golgi-associated protein of cultured fibroblast cells. *J. Cell Biol.* **92:** 108–112.

Mitchell, H.K. and L.S. Lipps. 1975. Rapidly labeled proteins on the salivary gland chromosomes of *Drosophila melanogaster. Biochem. Genet.* **13:** 585–602.

Moskalewski, S., J. Thyberg, and U. Friberg. 1976. In vitro influence of colchicine on the Golgi complex in A- and B-cells of guinea pig pancreatic islets. *J. Ultrastruct. Res.* **54:** 304–317.

Novikoff, A.B. and S. Goldfischer. 1961. Nucleoside diphosphatase activity in Golgi apparatus and its usefulness for cytological studies. *Proc. Natl. Acad. Sci.* **47:** 802–810.

Pruss, R.M., R. Mirsky, M.C. Raff, R. Thorpe, A.J. Dowding, and B.H. Anderton. 1981. All classes of intermediate filaments share a common antigenic determinant defined by a monoclonal antibody. *Cell* **27:** 419–428.

Robbins, E. and N.K. Gonatas. 1964. Histochemical and ultrastructural studies on HeLa cell cultures exposed to spindle inhibitors with special reference to the interphase cell. *J. Histochem. Cytochem.* **12:** 704–711.

Tartakoff, A.M. and P. Vassalli. 1977. Plasma cell immunoglobulin secretion. *J. Exp. Med.* **146:** 1332–1345.

Thomas, G.P., W.J. Welch, M.B. Mathews, and J.R. Feramisco. 1982. Molecular and cellular effects of heat shock and related treatments of mammalian tissue culture cells. *Cold Spring Harbor Symp. Quant. Biol.* **46:** 985–996.

Thyberg, J., S. Nilsson, S. Moskalewski, and A. Hinek. 1977. Effects of colchicine on the Golgi complex and lysosomal system of chondrocytes in monolayer culture. An electron microscopic study. *Cytobiologie* **15:** 175–191.

Velazquez, J.M., B.J. DiDomenico, and S. Lindquist. 1980. Intracellular localization of heat shock proteins in *Drosophila. Cell* **20:** 679–689.

Vincent, M. and R.M. Tanguay. 1979. Heat shock induced proteins present in the cell nucleus of *Chironomus tentans* salivary gland. *Nature* **281:** 501–503.

Disruption of the Vimentin Cytoskeleton May Play a Role in Heat-shock Response

Harald Biessmann*, Falko-Gunter Falkner†,
Harald Saumweber‡, and Marika F. Walter§

*Department of Genetics
University of California
Davis, California 95616

†Department of Physiological Chemistry
University of Munich
8 Munich, West Germany

‡Department of Biochemistry and Biophysics
University of California
San Francisco, California 94143

§Department of Zoology
University of California
Davis, California 95616

The effect of heat shock on protein synthesis is quite dramatic. Among the first detectable changes is the disintegration of polysomes engaged in mRNA translation, and this occurs prior to the onset of heat-shock mRNA translation. We found that intracellular protein rearrangements accompany these changes (Falkner and Biessmann 1980) and that two major cytoplasmic proteins of molecular weights 46,000 and 40,000 appear to change cellular compartments briefly after raising the temperature of *Drosophila* Kc cells to 37°C. The 46K protein is immunologically related to vimentin, a component of intermediate-sized (10 nm thick) cytoskeletal filaments in vertebrate cells (Falkner et al. 1981). In addition, the monoclonal antibody Ah6/5/9, directed against the 46K *Drosophila* vimentin analog, stains cytoskeletal meshworks in a variety of vertebrate cells such as baby hamster kidney (BHK), HeLa, chicken, and *Xenopus* fibroblasts, as well as in the protozoan *Paramecium* (unpubl.). The function of this highly conserved protein is still unknown, but the intermediate filament cytoskeleton may serve as attachment structure for polysomes in the cytoplasm (Cervera et al. 1981). In this paper we describe the effects of heat shock and valinomycin on this cytoskeleton in *Drosophila* tissue culture cells and salivary glands.

INTRACELLULAR REDISTRIBUTION OF TWO CYTOPLASMIC PROTEINS AFTER HEAT SHOCK

When Kc cells are heat-shocked, a redistribution of two preexisting cytoplasmic proteins of molecular weights 46,000 and 40,000 takes place and can be visualized on stained SDS-polyacrylamide gels (Fig. 1a). These two proteins, which are normally found in the microsomal fraction of the cell, almost quantitatively appear in the nuclear fraction after a 5-minute heat shock. A monoclonal antibody Ah6/5/9 (IgM) which was produced against the 46K protein (Falkner et al. 1981) recognizes both proteins, indicating that they are immunologically related. This antibody was used to detect the intracellular distribution of these

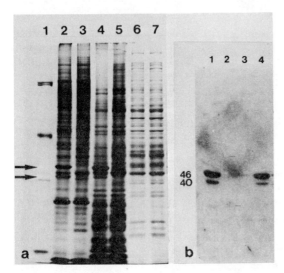

Figure 1
(a) SDS-polyacrylamide gel, stained with Coomassie Blue, of proteins extracted from Kc cells before and after a 5-minute heat shock. Size markers in lane 1 have molecular weights of 165K, 155K, 68K, 35K, and 21.5K. Lane 2, proteins of the nuclear fraction after heat shock and, lane 3, from control cells; lane 4, proteins from the microsomal fraction from control cells and, lane 5, from heat-shocked cells; lane 6, soluble proteins from control and, lane 7, from heat-shocked cells. Arrows indicate the 46K and 40K proteins. (b) Autoradiograph of proteins blotted onto nitrocellulose filter, first reacted with Ah6/5/9 and then with ^{125}I-labeled rabbit anti-mouse Fabγ-fragments. SDS-extractable proteins from nuclear fraction of heat-shocked (lane 1) and control (25°C) cells (lane 2); microsomal fraction of heat-shocked (lane 3) and control cells (lane 4). (Reprinted, with permission, from Falkner et al. 1981.)

proteins before and after heat shock. This was achieved by reacting Ah6/5/9 to protein extracts from different cell fractions which had been separated on a polyacrylamide gel and transferred to nitrocellulose filter (Fig. 1b). The results clearly show the shift of the 46K and 40K proteins from the microsomal compartment to the nuclear fraction after heat shock.

Even though the 40K protein is found in extracts from cells isolated in the presence of SDS and phenylmethylsulfonylfluoride, it is not clear whether it is a unique polypeptide or a degradation product of the 46K protein. We find that the 46K protein is extremely sensitive to proteolysis, and gives rise to even smaller but distinct degradation products (Falkner and Biessmann 1980 and unpubl.)

A VIMENTINLIKE PROTEIN IN *DROSOPHILA*

We believe that this 46K *Drosophila* protein is ortholog to vimentin found in vertebrates because it carries the same antigenic determinant as vimentin enriched from BHK cells (Falkner et al. 1981). Furthermore, it is insoluble in 1% Triton X-100 and can be enriched from *Drosophila* Kc cells by following a typical vimentin isolation procedure (unpubl.). This suggests that much of the molecule, in addition to the few amino acids recognized by Ah6/5/9, is shared by the *Drosophila* 46K protein and vimentin. Finally, indirect immunofluorescence and localization of the antigen by electron microscopy confirm that the 46K *Drosophila* protein forms a cytoplasmic filamentous structure.

EFFECT OF HEAT SHOCK AND VALINOMYCIN ON CYTOSKELETON

The effect of heat shock and valinomycin on the cytoskeleton in Kc cells and in salivary glands is shown in Figure 2. Flattened out Kc cells, grown for a few hours at low density on coverslips, show a very fine cytoplasmic meshwork that is stained with Ah6/5/9. It appears to be denser in the perinuclear region but stretches far into all cellular processes. Heat shock results in a rapid collapse of these structures, and the material forms quite massive aggregates at or around the nuclei.

Valinomycin, which induces all heat shock puffs at a concentration of 10^{-5} M (J.J. Bonner, pers. comm.), shows a similar but less drastic effect on the intermediate filament cytoskeleton. These effects are best studied in tissue culture cells because, like vertebrate cells in culture, they seem to overproduce vimentin. But the intracellular redistribution of the 46K protein as visualized by our antibody could also be observed in isolated cells of slightly squashed salivary glands (Fig. 2). By protein blotting we could show that salivary glands also contain the 46K protein (unpubl.).

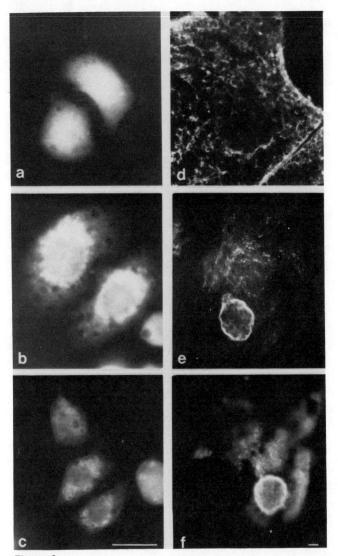

Figure 2
Indirect immunofluorescence staining of Kc cells grown on coverslips (a, b, c) and of salivary gland cells (d, e, f). Preparations were fixed in 4% formaldehyde and treated with methanol and acetone. Control (25°C) cells (a, d), after 10-minute heat shock at 37°C (b, e), and after 90 minutes in medium containing 10^{-5} M valinomycin (c, f). Bars indicate 10 μm. (d and e reprinted, with permission, from Falkner et al. 1981.)

ELECTRON MICROSCOPY

The cytoplasmic structures formed by the 46K protein in Kc cells were identified by electron microscopy. Figure 3a shows an electron micrograph of Triton-extracted Kc cells which were incubated with Ah6/5/9 and ferritin-conjugated goat anti-mouse IgG. In control cells, the ferritin label was associated with filamentous structures of about 10 nm thickness. The nuclear membrane and mitochondria remained unlabeled. After heat shock, large aggregates of ferritin-labeled material appeared at the nuclei of many cells and the cytoplasmic label at the filaments decreased significantly (Fig. 3b).

VIMENTIN CYTOSKELETON MAY BE THE CYTOPLASMIC HEAT-SHOCK TARGET

It has been shown that different organisms respond to heat shock in a similar way. In a mammalian cell type (BHK) and in *Drosophila* cells, we could show that heat shock disrupts the vimentin cytoskeleton (Falkner et al. 1981). Even cilates that also exhibit a heat-shock response contain a cytoskeleton that stains with Ah6/5/9 (unpubl.). We, therefore, suggest that the intermediate filament cytoskeleton may be the common target for heat shock in very distantly related species.

Like valinomycin, two uncouplers of oxidative phosphorylation have recently been shown to disintegrate the tubulin and the vimentin cytoskeletons (Maro and Bornens 1982). Since mitochondria are possibly associated with microtubules and/or intermediate filaments (Heggeness et al. 1978), it is conceivable that these drugs that affect mitochrondrial metabolism may disrupt directly or indirectly the cytoskeletal framework of the cell. However, to our knowledge, their effect on polysomes has not yet been investigated.

The observation that heat shock destroys the vimentin cytoskeleton and thus causes the polysomes to disintegrate may explain recent results of in vitro translation experiments (Krüger and Bennecke 1981; Scott and Pardue 1981). If we assume that heat-shock mRNAs are translated on noncytoskeleton-bound polysomes, whereas normal (25°C) mRNAs require cytoskeleton-bound polysomes, the specificity of a 37°C cell lysate to translate heat-shock mRNAs but not 25°C mRNAs in vitro could be explained by the absence of cytoskeleton-bound polysomes in this lysate.

CONCLUSION

We have demonstrated the presence of a vimentinlike cytoskeleton in *Drosophila* using a monoclonal antibody directed against a major cytoplasmic 46K protein. This cytoskeleton exhibits characteristic features of the vertebrate intermediate filament system. In *Drosophila* and BHK

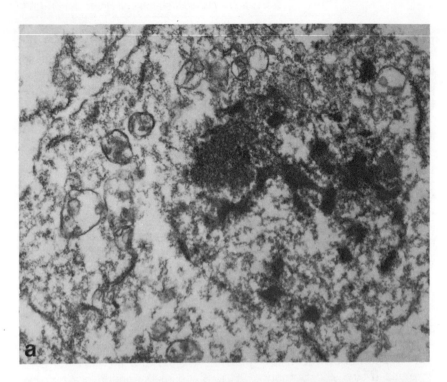

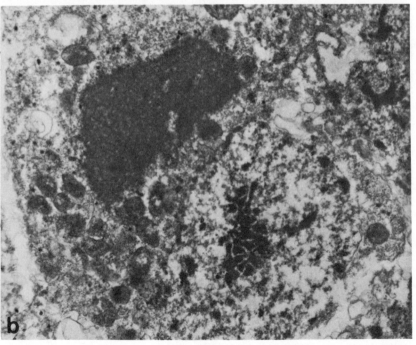

cells, this vimentin cytoskeleton disintegrates after heat shock and aggregates at the cell nucleus. In *Drosophila* we have shown that the ionophore valinomycin, which induces heat shock puffs, has the same effect on the cytoskeleton as heat shock. We discussed the possibility that the disaggregation of polysomes and preferential translation of heat-shock mRNAs in lysates from previously heat-shocked cells may be due to the disruption of this cytoskeleton. Inhibitors of mitochondrial metabolism may also exhibit direct or indirect effects on the cytoskeletal framework. The presence of intermediate filament systems in evolutionarily very distant species like ciliates and vertebrates, both of which show a specific heat-shock response, is a further indication that the cytoskeleton may be the cytoplasmic target of heat shock.

REFERENCES

Cervera, M., G. Dreyfuss, and S. Penmann. 1981. Messenger RNA is translated when associated with the cytoskeletal framework in normal and VSV-infected HeLa cells. *Cell* **23**: 113–120.

Falkner, F.G. and H. Biessmann. 1980. Nuclear proteins in *Drosophila melanogaster* cells after heat shock and their binding to homologous DNA. *Nucleic Acids Res.* **8**: 943–955.

Falkner, F.G., H. Saumweber, and H. Biessmann. 1981. Two *Drosophila melanogaster* proteins related to intermediate filament proteins of vertebrate cells. *J. Cell Biol.* **91**: 175–183.

Heggeness, M.H., M. Simon, and S.J. Singer. 1978. Association of mitochondria with microtubules in cultured cells. *Proc. Natl. Acad. Sci.* **75**: 3863–3866.

Krüger, C. and B.J. Bennecke. 1981. *In vitro* translation of *Drosophila* heat shock and non-heat shock mRNAs in heterologous and homologous cell-free systems. *Cell* **23**: 595–603.

Maro, B. and M. Bornens. 1982. Reorganization of HeLa cell cytoskeleton induced by an uncoupler of oxidative phosphorylation. *Nature* **295**: 334–336.

Scott, M. and M.L. Pardue. 1981. Translational control in lysates of *Drosophila melanogaster* cells. *Proc. Natl. Acad. Sci.* **78**: 3353–3357.

Figure 3
Electron micrograph of thin sections of Kc cells grown at 25°C (*a*) and after a 30-minute heat shock at 37°C (*b*). Cells were extracted with 1% Triton X-100 in 0.14 M NaCl, 10 mM, Tris-HCl, (pH 7.4) for 10 minutes at 4°C. After fixation with 4% formaldehyde and dehydration with methanol and acetone, cells were incubated with Ah6/5/9 and subsequently with ferritin-conjugated goat-anti-mouse IgG. Magnification (*a*) 10,250×; (*b*) 17,500×.

The Human Heat-shock Proteins: Their Induction and Possible Intracellular Functions

Roy H. Burdon
Department of Biochemistry
University of Glasgow
Glasgow G12 8QQ, Scotland

The induction of heat-shock proteins as a common response to thermal stress has been demonstrated in a wide range of organisms including bacteria, plants, and animals. However, the details of molecular mechanisms whereby this is achieved vary slightly from organism to organism. For instance, translation control mechanisms seem less important in the human response than in *Drosophila*. Not surprisingly, there has been considerable speculation with regard to the physiological significance of heat-shock proteins. It has been suggested that they might have a role in the recovery of cellular homeostasis after thermal stress, possibly being involved in processes that regulate hexose transport and/or metabolism. They have also been implicated in the development of "thermotolerance," which would allow cells to survive thermal treatments that would otherwise be lethal.

In this report, the particular features of the heat-shock response in human cells is described together with experiments that indicate roles for heat-shock proteins in (1) the modulation of plasma membrane Na^+K^+-ATPase activity and (2) the generation of thermotolerance. This latter aspect is of clinical importance since thermotolerance can be an impediment to the use of hyperthermia in cancer therapy.

THE MOLECULAR BIOLOGY OF HUMAN
HEAT-SHOCK PROTEIN INDUCTION

Brief hyperthermic treatment of HeLa cells (45°C; 10 min) followed by a return to the normal culture temperature of 37°C for 2 hours results in the increased production of groups of heat-shock proteins at 100 kD, 72–74 kD, and 37 kD (Slater et al. 1981) (Table 1). The 72–74-kD group comprises at least seven polypeptides whereas the 100-kD and 37-kD groups can be resolved into two polypeptides each. The level of synthesis of other cellular proteins at this stage, however, was similar to the levels encountered in control cells (i.e., translational control is not obvious at this stage).

Maximal induction of heat-shock proteins is achieved if the initial hyperthermic trigger is at 45–46°C for between 5–10 minutes (Burdon et al. 1982) and the development period at 37°C is for 2 hours. After 2 hours, the rate of heat-shock protein synthesis declines rapidly (Slater et al. 1981). Heat-shock protein synthesis can nevertheless also be induced in HeLa cells under conditions of continuous heat treatment provided the temperature is not raised above 43°C (Burdon et al. 1982).

During the development phase at 37°C after an initial hyperthermic trigger, the use of actinomycin D, hydroxyurea, and azacytidine suggests transcription but neither DNA replication nor demethylation to be required for hsp gene expression in HeLa cells. Also during the development phase, the disrupted polysomes are reassembled. This process is complete after 2 hours at 37°C. A further feature of the development phase is the accumulation in the cytoplasm at 1–2 hours of specific heat-shock protein in RNAs. In vitro translation experiments (Kioussis et al. 1981) show that the 72–74-kD group is made up of seven primary

Table 1
Human Cell (HeLa) Heat-shock Proteins

Size (kD)	Number in group	Primary translation products	mRNA (cDNA) clones	mRNA size (kb)	Pre mRNA? size (kb)
100	2	one			
72–74	7	α			
		α^1			
		β	pHS6, pHS3	1.9	15.8
		γ	pHS2	6.3	15.8
		δ	pHS6	1.9	15.8
		ϵ	pHS6	1.9	15.8
		ζ			
37	2	one			

translation products (α, α^1, β, δ, γ, ϵ, and ζ). In the case of the 100-kD and 37-kD groups, there is only one primary translation product in each case (Table 1). In all the groups, the translation products arise from both polyadenylated (poly[A]$^+$) and nonpolyadenylated (poly[A]$^-$) in mRNAs. Both the poly(A)$^+$ and poly(A)$^-$ forms accumulate simultaneously in the cytoplasm at 37°C and remain there for at least 6 hours despite the fact that their translation has virtually ceased (Kioussis et al. 1981). Although they still appear to be associated with polysomes at this time, some translational control mechanism must block their expression at this stage.

cDNA sequences to some of the 72−74-kD HeLa cell heat-shock proteins have been cloned in pBR322 at the *Pst*I site (Cato et al. 1981). pHS2 contains a cDNA sequence that hybridizes to mRNA coding for the HeLa cell γ-hsp, pHS3 a sequence that hybridizes to the β-hsp, and pHS6 a sequence that cross-hybridizes with the mRNAs that code for the β-, δ-, and ϵ-hsp (see Table 1).

When cytoplasmic RNA from heat-shocked HeLa cells is electrophoresed and blotted onto DBM paper by the Northern technique and then probed for sequences complementary to these recombinant plasmids, it emerges that the mRNAs for the 72−74-kD heat shock proteins differ widely in their size. Whilst the mRNAs for the β-, δ-, and ϵ-hsp are coded for by mRNAs of an appropriate size (1.9 kb), the γ-hsp arises from an mRNA three times that size (6.3 kb). A search for putative nuclear precursors to these mRNAs indicated high-molecular-weight candidates at 15.8 kb which accumulate at 1−2 hours of development after the initial heat shock, again supporting the notion that transcriptional control is important in the initial response.

The cDNA clones are now being used to isolate genomic clones in order to permit an evaluation of factors which serve to trigger the expression of hsp genes. Besides hyperthermia, several other agents serve as inducers (see Table 2). These include sulfhydryl reagents (e.g., arsenite, cadmium) and chelating agents (8-hydroxyquinoline).

THE INTRACELLULAR FUNCTION OF HUMAN HEAT-SHOCK PROTEINS

Intracellular Location

Although 30% of the 72−74-kD group are found in the nuclear fraction, 70% remain in the cytoplasm even after a pulse-chase. The 37 kD-group also occurs in the nucleus and cytoplasm whereas the 100-kD group appears exclusively cytoplasmic. The DNA-binding capacity of the 72−74-kD proteins was assessed by protein blotting but this was negative. However a role in regulation of transcription cannot be ruled out.

Table 2
HeLa Cell Heat-shock Protein Induction

Inducers	Noninducers
Hyperthermia	dibutyryl cAMP
Brief (10-minute 45°C, development 37°C)	dimethysulfoxide
Continuous ($< 43°$)	sodium butyrate
Sodium arsenite	potassium cyanide
Cadmium sulfate	sodium azide
Copper sulfate	antimycin
Zinc sulfate	oligomycin
8-Hydroxyquinoline	2-deoxyglucose
	5-thioglucose
	glucosamine
	tunicamycin
	glucose starvation
	EMC virus
	interferon

Modulation of Na$^+$K$^+$-ATPase

A rate-limiting reaction in glycolysis has been suggested to be the hydrolysis of ATP to ADP and P$_i$, and in certain tumor cells the responsible catalyst is believed to be the Na$^+$K$^+$-ATPase of the plasma membrane (see Burdon and Cutmore 1982). Brief hyperthermia results in considerable loss of HeLa cell Na$^+$K$^+$-ATPase activity (80% after 10 min at 45°C). However, subsequent incubation of these cells at 37°C for 2 hours results in a partial recovery of activity that can be blocked by actinomycin D or cycloheximide. Since the time of maximal recovery of activity at 37°C is similar to the time of maximal heat-shock protein synthesis after hyperthermic treatment, it is possible that heat-shock proteins may be involved in this partial recovery or repair of enzyme activity. As already mentioned, agents other than heat, such as sodium arsenite, can elicit heat-shock protein induction in HeLa cells (Table 2). Treatment of HeLa cells with 5×10^{-5} M sodium arsenite for 2–3 hours leads not only to increased synthesis of HeLa cell heat-shock proteins but also to elevated levels of Na$^+$K$^+$-ATPase. This increase in ATPase activity can be prevented by inclusion of actinomycin D or cycloheximide in the culture medium along with the sodium arsenite, implying specific gene expression in the alteration of ATPase activity rather than a direct effect of arsenite itself (Burdon and Cutmore 1982). These data lead to hypothesis that heat-shock proteins may be involved in the in vivo modulation of Na^+K^+-ATPase activity. Whether this is achieved directly by association or indirectly by, say, covalent modification is not yet known. However, whereas in control cells the K_m is 7.7 mM, this is reduced to 4.4 mM in cells treated with arsenite (a 1.3-fold increase in V_{max} is also observed). What does appear unlikely however from ouabain-binding studies is that the increase in ATPase activity is due to an increase in the number of α-subunits of Na^+K^+-ATPase (similar in size to the 100-kD heat-shock proteins).

Hexose Transport

Since it was possible that the effects of heat-shock proteins on Na^+K^+-ATPase could influence hexose transport through electrogenic sodium symport systems, the effect of hypothermia and arsenite treatment on [^{3}H]deoxyglucose uptake by HeLa cells was examined. However no evidence was obtained to indicate a clear role for heat-shock proteins in hexose transport.

Generation of Thermotolerance

A practical problem in the clinical application of hyperthermia in cancer therapy is the development of thermotolerance. This is usually defined as the reduced slope of heat survival curves induced by prior heat conditioning. It can usually be induced in cultured human cells by a brief treatment at high temperature (45°C) followed by a development period at 37°C. It can also be induced by continuous heat treatment of cells at < 43°C. Thus, the conditions required for the generation of thermotolerance and heat-shock protein induction are broadly similar. Are heat-shock proteins involved in the generation of thermotolerance and if so how? A possibility was that alterations in Na^+K^+-ATPase activity brought about by heat-shock proteins might yield an enzyme activity with increased thermal resistance. However this was not the case. Thus, although the observed effects on ATPase may be concerned with the repair or recovery from thermal damage, they do not appear to be directly relevant to the generation of thermotolerance.

Recent data from this laboratory now indicate the generation of thermotolerance to be a complex process probably involving the participation of heat-shock proteins as well as functional mitochondria. Examination of cell growth kinetics in response to various heat treatments indicate that human cells without functional mitochondria have a reduced thermal resistance after prior heat treatment (or arsenite treatment). On the other hand, the expected thermotolerance can be induced by prior heat treatment (or arsenite treatment) in cells with normal mitochondria.

SUMMARY

The induction of heat-shock proteins in human cells (HeLa) has broad similarities with the situation in other organisms. However the initial induction appears to be regulated primarily at the level of transcription although some translation control is evident. Nine primary gene products are produced from mRNAs that exist in both poly(A)$^+$ and poly(A)$^-$ forms. Induction does not depend solely on temperature, as other chemical agents will suffice. Such alternative agents have been used to demonstrate the possible involvement of human heat-shock proteins in the modulation of plasma membrane Na$^+$K$^+$-ATPase. This effect, however, does not appear to be directly relevant to the generation of thermotolerance. Instead, it appears that thermotolerance is a complex process likely to involve both heat-shock proteins and functional mitochondria.

ACKNOWLEDGMENTS

This work would not have been possible without the active and expert participation of Dr. Andy Cato, Dr. Adrian Slater, Miss Joanna Kioussis, Mrs. Gillian Sillar, Miss Caroline Cutmore, Miss Ruth Fulton, Mr. M. McMahon, and Mrs. Vera Gill.

REFERENCES

Burdon, R.H. and C.M.M. Cutmore. 1982. Human shock gene expression and the modulation of plasma membrane Na$^+$K$^+$-ATPase activity. *FEBS Letters* **140:** 45–48.

Burdon, R.H., A. Slater, M. McMahon, and A.C.B. Cato. 1982. Hyperthermia and the heat shock proteins of HeLa cells. *Br. J. Cancer* (in press).

Cato, A.C.B., G.M. Sillar, J. Kioussis, and R.H. Burdon. 1981. Molecular cloning of cDNA sequences coding for the major (β, γ, δ and ϵ) heat shock polypeptides of HeLa cells. *Gene* **16:** 27–34.

Kioussis, J., A.C.B. Cato, A. Slater, and R.H. Burdon. 1981. Polypeptides encoded by polyadenylated and non-polyadenylated messenger RNAs from normal and heat shocked HeLa cells. *Nucleic Acids Res.* **9:** 5203–5214.

Slater, A., A.C.B. Cato, G.M. Sillar, J. Kioussis, and R.H. Burdon. 1981. The pattern of protein synthesis induced by heat shock of HeLa cells, *Eur. J. Biochem.* **117:** 341–346.

An Analysis of the Interaction of the Rous Sarcoma Virus Transforming Protein, pp60^{src}, with a Major Heat-shock Protein

Wes Yonemoto, Leah A. Lipsich,
Diane Darrow, and Joan S. Brugge
Department of Microbiology
State University of New York at Stony Brook
Stony Brook, New York 11794

Oncogenic transformation by Rous sarcoma virus (RSV) is mediated by a single viral-encoded polypeptide, pp60^{src}, which has been shown to contain phosphotransferase activity specific for tyrosine residues (Bishop and Varmus 1982). Since the level of total cellular phosphotyrosine increases after transformation by RSV (Sefton et al. 1980), it has been postulated that tyrosine phosphorylation of cellular proteins is directly involved in the events associated with RSV-induced transformation (Bishop and Varmus 1982). Therefore, an initial step to elucidation of the mechanism of Rous transformation is the identification and characterization of the substrates of pp60^{src}-mediated kinase activity as well as those proteins that interact with pp60^{src} and serve to regulate its functional activity.

One protein known to associate with pp60^{src} in a protein complex, has been shown to be identical to one of the proteins whose synthesis is stimulated by a mild heat shock, amino acid analogs, and certain chemical agents (Oppermann 1981a). Although the function of this protein complex is unknown, investigations of the interaction of pp60^{src} with this heat-shock protein may help to characterize the normal cellular function of a "stress" protein as well as aid in the elucidation of the mechanism of transformation by RSV.

THE pp50 : pp60^{src} : pp90 COMPLEX

Two cellular proteins of M_r 90,000 (pp90) and 50,000 (pp50) are specifically coimmunoprecipitated from lysates of RSV-transformed cells using antiserum against pp60^{src}. This coimmunoprecipitation of pp50, pp60^{src}, and pp90 is due to the association of these three proteins into a protein complex (Brugge et al. 1981; Oppermann et al. 1981b). The complexation (between pp60^{src}, pp50, and pp90) can be demonstrated using velocity sedimentation analysis of RSV-transformed cell lysates as shown in Figure 1. The majority of pp60^{src} molecules sedimented as a monomer with a peak in fractions 18–20. However, a small fraction of the pp60^{src} molecules cosedimented and coimmunoprecipated with pp50 and pp90 in a higher-molecular-weight region of the gradient. This evidence, together with previous demonstrations that pp60^{src}, pp50, and pp90 are not structurally nor antigenically related (Hunter and Sefton 1980; Brugge et al. 1981), indicates that these three proteins are associated into a protein complex.

Analysis of the phosphorylation state of the pp60^{src} molecules from the monomer and complex regions of the gradient revealed that the free form of pp60^{src} contains both phosphoserine and phosphotyrosine while those molecules of pp60^{src} that are bound to pp50 and pp90 contain

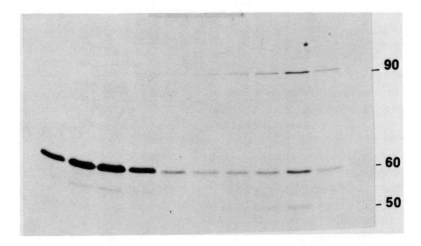

Figure 1
Velocity sedimentation analysis of RSV-transformed CEF. SRA-RSV-transformed CEF were labeled with [^{35}S]methionine lysed in RIPA buffer, and sedimented on glycerol gradients. Alternating fractions were immunoprecipitated with TBR serum as previously described (Brugge et al. 1981). The sedimentation is shown from left to right and fractions 4–26 are displayed. A bovine serum albumin molecular weight marker fractionated with a peak in fractions 18–20.

phosphoserine and very depressed levels of phosphotyrosine (Brugge et al. 1981). In addition, when fractions across the glycerol gradient were assayed by the immune complex phosphotransferase reaction, it was found that only the monomer form of pp60^{src} was able to phosphorylate IgG (Brugge et al. 1981). These results suggest that pp90 and pp50 specifically interact with pp60^{src} molecules that lack phosphotyrosine and that this complex is inactive in phosphotransferase activity.

IDENTIFICATION OF pp90, A HEAT-SHOCK PROTEIN, AND pp50 IN UNINFECTED CELLS

pp90 has been found to be an abundant protein of both avian and mammalian cells and represents approximately 0.5% of the total cellular protein (Brugge et al. 1981). Phosphopeptide and phosphoamino acid analyses of pp90 have indicated that the protein contains multiple phosphoserine sites (Brugge et al. 1981). Thus far, no differences have been detected between the minor fraction of pp90 bound to pp60^{src} and the majority of pp90, which is free (W. Yonemoto and J. S. Brugge, unpubl.). Oppermann and coworkers (1981a) have identified pp90 as one of the heat-shock proteins. The association between pp90, pp60^{src}, and pp50 does not appear to be altered by heat shock or treatment with arsenite (Oppermann et al. 1981a).

pp50 was identified in uninfected chick embryo fibroblasts (CEF) on two-dimensional gels by comparing the migration of immunoprecipitated pp50 with the migration of ^{32}P-labeled CEF proteins (Brugge and Darrow 1982; Gilmore et al. 1982). Partial peptide mapping and phosphoamino acid analysis showed that the pp60^{src}-associated pp50 contained both phosphoserine and phosphotyrosine while the form of pp50 from uninfected CEF contained only phosphoserine. Analysis of the pp50 protein from RSV-transformed CEF on two-dimensional gels resolved the protein into two spots, a phosphoserine-phosphotyrosine-containing species similar to that in the TBR immunoprecipitate, and a phosphoserine-containing species indistinguishable from that found in uninfected CEF (Brugge and Darrow 1982). The analysis of CEF infected with mutants of RSV that have a deletion of the *src* gene, contain only the phosphoserine form of pp50. These results suggest that the phosphorylation of pp50 on phosphotyrosine might be mediated by pp60^{src}.

pp90 AND pp50 INTERACT WITH THE TRANSFORMING PROTEINS OF FSV AND Y73

We have also examined whether the transforming proteins of two independently isolated avian sarcoma viruses are also associated into a complex with pp50 and pp90. Rous, Fujinami (FSV), and Yamaguchi (Y73) sarcoma viruses have unique transforming genes (Bishop and Varmus 1982), yet the protein products of each gene possess tyrosine-specific protein kinase activity (Bishop and Varmus 1982). CEF infected with either FSV or Y73 were analyzed by velocity sedimentation and immunoprecipitation with monoclonal antibody against pp90. It has been demonstrated previously that $pp60^{src}$ and pp50 can be coimmunopreci-pitated from lysates of RSV-transformed CEF using this antibody (Brugge et al. 1981). Analysis of cells infected with either FSV or Y73 showed that the majority of pp90 sedimented as a monomer, however in the higher-molecular-weight leading edge of pp90, the coimmunoprecipi-tation of $pp140^{fps}$, the transforming protein of FSV, was detected with pp90 and pp50 and in a separate experiment, $pp94^{yes}$, the transforming protein of Y73, was coimmunoprecipitated with pp90 and pp50 (Lipsich et al. 1982). Recently, other investigators (Kitamura et al. 1982; M. Shibuya and H. Hanafusa, pers. comm.) have sequenced the onc genes of FSV and Y73 and compared their predicted amino acid sequence with the predicted amino acid sequence of $pp60^{src}$. Despite little nucleotide homology, both $pp140^{fps}$ and $pp94^{yes}$ showed extensive regions of amino acid sequence homology with $pp60^{src}$, especially in the carboxyl half of $pp60^{src}$. Therefore, pp90 and pp50 may bind to a common site on the transforming proteins of RSV, FSV, and Y73 and the interaction of these proteins may be involved in the expression of functional activity of tyrosine-specific protein kinases.

KINETICS OF ASSOCIATION OF THE COMPLEX

In order to investigate the turnover of the $pp50 : pp60^{src} : pp90$ com-plex, Schmidt-Ruppin A (SRA) RSV-transformed CEF were labeled for 15 minutes with [^{35}S]methionine and harvested immediately (Fig. 2A) or chased for 30 minutes (Fig. 2B) or 180 minutes (Fig. 2C) with an excess of cold methionine. Lysates of these cells were then subjected to velocity sedimentation analysis as described in Figure 1. The majority of the newly synthesized $pp60^{src}$ was detected in the region of the gradient where the $pp50 : pp60^{src} : pp90$ complex sediments (Fig. 2A). Within a 30-minute chase period, approximately two-thirds of the pulse-labeled $pp60^{src}$ sedimented as a monomer (Fig. 2B). After a 3-hour chase, approximately 95% of the pulse-labeled $pp60^{src}$ was found in the low-molecular-weight region of the gradient (Fig. 2C). These results indicate that the newly synthesized $pp60^{src}$ binds to pp90 and pp50 and that this complex has a half-life of approximately 15 minutes in SRA-RSV infect-ed cells.

This same type of analysis was performed on cells infected with different strains of nondefective RSV and mutant viruses (ts^{src}) contain-

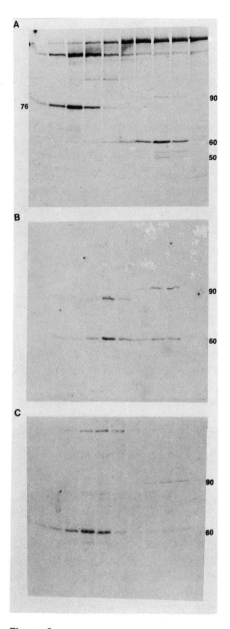

Figure 2

Analysis of the turnover of the pp50 : pp60^{src} : pp90 complex in RSV-transformed CEF. SRA-RSV-transformed CEF were labeled with 400 μCi/ml of [^{35}S]methionine and either harvested immediately (*A*) or chased for 30 minutes (*B*) or 300 minutes (*C*) with an excess of cold methionine. The cells were lysed in RIPA buffer and analyzed by velocity sedimentation as described in Fig. 1. Sedimentation is shown from left to right and fractions 4−26 are displayed.

ing temperature-sensitive defects in the *src* gene. It was found that the half-life of the complex varied from 5 to 15 minutes in different wild-type strains of RSV. In contrast, the complex was found to have a half-life of approximately 3 hours in cells infected with ts^{src} mutant viruses grown at the permissive temperature; little or no dissociation of the complex was detected in a 3-hour chase at the nonpermissive temperature. This suggests that mutations in the *src* gene which cause a *ts* transformed phenotype in mutant-infected cells results in an increased stability of the pp50 : pp60^{src} : pp90 complex.

It is also worth noting that at least 90% of the mutant form of the *src* protein is bound into a stable complex with pp90 and pp50 in *ts* virus-infected cells at the nonpermissive temperature. This feature may partially explain the transformation-defective phenotype of these mutant virus infected cells.

LOCALIZATION OF THE pp50 : pp60^{src} : pp90 COMPLEX

Other investigators have demonstrated that the newly synthesized pp60^{src} is synthesized on free polysomes (Lee et al. 1979; Purchio et al. 1980) and shortly after its synthesis associates with the plasma membrane (Krueger et al. 1980; Levinson et al. 1981). It was therefore of interest to determine if the interaction of pp60^{src} with pp50 and pp90 takes place within the soluble or particulate fraction of the cells. To answer this question, ^{32}P-labeled RSV-transformed CEF were fractionated by the two methods shown in Figure 3. In part A, lane 3, the cells were separated into soluble and particulate fractions by homogenization in buffers of low ionic strength by a modification of the procedure of Krueger and coworkers (1980). The fractions were immunoprecipitated with either normal rabbit, anti-pp90, or TBR serum. Essentially all of the pp60^{src} bound to pp90 and pp50 was found in the soluble fraction, while the majority of the pp60^{src} was found in the particulate fraction. Under the conditions of this fractionation, more than 99% of the pp90 was detectable in the soluble fraction. The second type of fractionation involved detergent solubilization of transformed cells using buffers designed to preserve the major cytoskeletal components of the cells. The procedure used was a modification of the protocol of Schliwa and Van Blerkom (1981). After washing the cells with STE and PHEM (Schliwa and Van Blerkom 1981) buffers, the cells were scraped into nonionic detergent-containing PHEM and treated in suspension. The detergent insoluble material was fractionated by centrifugation at 50,000*g* for 10 minutes. Both cellular fractions were then diluted in RIPA buffer (Brugge et al. 1981) and immunoprecipitated as described in Figure 3B. A majority of pp60^{src} fractionated with the detergent insoluble material as previously described by Burr and coworkers (1980). pp90 was found to

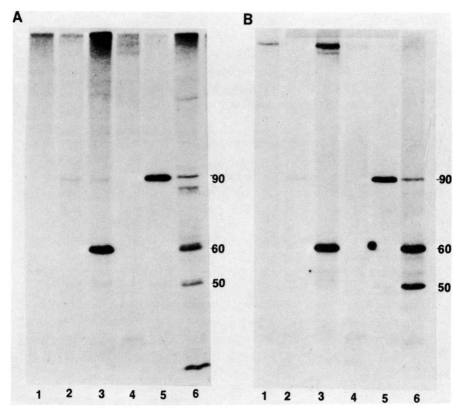

Figure 3

Fractionation of RSV-transformed CEF. SRA-RSV-transformed CEF were labeled with ^{32}P and either fractionated by homogenization in buffers of low ionic strength (*A*) or by solubilization in 0.15% Triton X-100 (*B*) as described in the text. The cell fractions were immunoprecipitated with either normal rabbit serum (lanes 1 and 4), monoclonal antibody against pp90 (lanes 2 and 5), or TBR serum (lanes 3 and 6). (*A*) Lanes 1–3, particulate fraction; lanes 4–6, soluble fraction. (*B*) Lanes 1–3, detergent-insoluble fraction; lanes 4–6, detergent-soluble fraction.

be very soluble in the detergent extraction, with little or no pp90 fractionating with the detergent-insoluble pellet. Similarly, nearly all of the pp60^{src} bound to pp50 and pp90 was found in the detergent-soluble fraction. Therefore by both methods of cellular fractionation employed, the pp50 : pp60^{src} : pp90 complex was associated with soluble cellular material, while the majority of the unbound pp60^{src} was found in the particulate detergent-insoluble cell fraction. According to the fractionation procedures used, we can also conclude that the heat-shock protein, pp90, is found in the soluble fraction of the cell.

SUMMARY

We have described the interaction of the transforming protein of RSV with two cellular proteins. One of the cellular proteins, designated here as pp90, has been identified as a heat-shock protein. A model of the interaction of pp60^{src} with pp90 and pp50 based upon the findings described in this report is presented in Figure 4.

1. pp60^{src} is synthesized on soluble polysomes and during or shortly after its synthesis, it is phosphorylated by the cAMP-dependent protein kinase and enters into a protein complex with pp90 and pp50, both of which are phosphorylated on serine.
2. During this complexation, pp60^{src} mediates the phosphorylation of pp50 on tyrosine.
3. The interaction is transient and leads to the release of pp60^{src}, pp50, which is now phosphorylated on tyrosine, and pp90, which is believed to be unmodified by this complexation. At this time, the event which triggers the release of the pp50 : pp60^{src} : pp90 complex is not known.
4. During the course of these events, or shortly after release, pp60^{src} is transported to the inner surface of the plasma membrane where it is anchored. The "mature" pp60^{src} is phosphorylated on tyrosine.

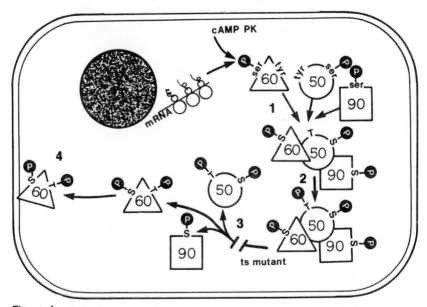

Figure 4
Model of the interaction of pp60^{src}, the transforming protein of RSV, with two cellular proteins, pp90 and pp50.

Since the function of the pp50 : pp60^{src} : pp90 complex is unknown, it is worthwhile to speculate on the possible roles of this interaction. The complex may provide a transport shuttle mechanism to bring the newly synthesized pp60^{src} to the plasma membrane. Little is known about the transport of proteins from soluble polysomes to specific membrane systems in the cell. The complex may also facilitate the solubility of pp60^{src} in the cytosol. Since pp60^{src} contains a small hydrophobic region, which presumably is necessary for attachment to the plasma membrane, the association of pp90 and pp50 may prevent aggregation of the *src* protein within the cytosol. Alternatively, the complex may regulate the functional activity of pp60^{src}. It has been shown that pp60^{src} molecules complexed with pp90 and pp50 are unable to phosphorylate IgG in the immune complex assay (Brugge et al. 1981). Further studies are in progress to investigate this complex in RSV transformation. The understanding of the pp50 : pp60^{src} : pp90 complex may help to elucidate the role of pp90 in the heat-shock response.

ACKNOWLEDGMENTS

This work was supported by grants CA-27951 from the National Cancer Institute.

REFERENCES

Bishop, J.M. and H. Varmus. 1982. Functions and origins of retroviral transforming genes. *RNA tumor viruses* (eds. R.A. Weiss, et. al.), p. 999–1108. Cold Spring Harbor Laboratory, Cold Spring Harbor, New York.

Brugge, J. and D. Darrow. 1982. Rous sarcoma virus-induced phosphorylation of a 50,000 molecular weight cellular protein. *Nature* **295:** 250–253.

Brugge, J., E. Erikson, and R.L. Erikson. 1981. The specific interaction of the Rous sarcoma virus transforming protein, pp60 ^{src}, with two cellular proteins. *Cell* **25:** 363–372.

Burr, J.G., G. Dreyfuss, S. Penman, and J.M. Buchanan. 1980. Association of the *src* gene product of ⏐Rous sarcoma virus with cytoskeletal structures of chicken embryo fibroblasts. *Proc. Natl. Acad. Sci.* **77:** 3484–3488.

Courtneidge, S.A., A.D. Levinson, and J.M. Bishop. 1980. The protein encoded by the transforming gene of avian sarcoma virus (pp60^{src} and a homologous protein in normal cells (pp60$^{proto-src}$ are associated with the plasms membrane. *Proc. Natl. Acad. Sci.* **77:** 3783–3787.

Gilmore, T., K. Radke, and G.S. Martin. 1982. Tyrosine phosphorylation of a 50K cellular polypeptide associated with the Rous sarcoma virus-transforming protein, pp60^{src}. *Mol. Cell. Biol.* **2:** 199–206.

Hunter, T. and B. Sefton. 1978. Transforming gene product of Rous sarcoma virus phosphorylates tyrosine. *Proc. Natl. Acad. Sci.* **77:** 1311–1315.

Kitamura, N., A. Kitamura, K. Toyoshima, Y. Hirayama, and M. Yoshida. 1982. Avian sarcoma virus Y73 genome sequence and structural similarity of its transforming gene product to that of Rous sarcoma virus. *Nature* **297:** 205–208.

Krueger, J.G., E. Wang, and A.R. Goldberg. 1980. Evidence that the *src* gene product of Rous sarcoma virus is membrane associated. *Virology* **101:** 25–40.

Lee, J.S., H.E. Varmus, and J.M. Bishop. 1979. Virus-specific messenger RNAs in permissive cells infected by avian sarcoma virus. *J. Biol. Chem.* **254:** 8015–8022.

Levinson, A.D., S.A. Courtneidge, and J.M. Bishop. 1981. Structural and functional domains of the Rous sarcoma virus-transforming protein (pp60^{src}). *Proc. Natl. Acad. Sci.* **78:** 1624–1628.

Lipsich, L.A., J. Cutt, and J.S. Brugge. 1982. Association of the transforming proteins of Rous, Fujinami and Y73 avian sarcoma viruses with the same two cellular proteins. *Mol. Cell. Biol.* **2:** 875–880.

Oppermann, H., W. Levinson, and J.M. Bishop. 1981a. A cellular protein that associates with a transforming protein of Rous sarcoma virus is also a heat-shock protein. *Proc. Natl. Acad. Sci.* **78:** 1067–1071.

Oppermann, H., A.D. Levison, L. Levintow, H.E. Varmus, J.M. Bishop, and S. Kawai. 1981b. Two cellular proteins that immunoprecipitate with the transforming protein of Rous sarcoma virus. *Virology* **113:** 736–751.

Purchio, A.F., S. Jonanovich, and R.L. Erikson, 1980. Sites of synthesis of viral proteins in avian sarcoma virus-infected chicken cells. *J. Virol.* **35:** 629–636.

Schliwa, M. and J. Van Blerkom. 1981. Structural interaction of cytoskeletal components. *J. Cell Biol.* **90:** 222–235.

Sefton, B.M., T. Hunter, K. Beemon, and W. Eckhart. 1980. Evidence that the phosphorylation of tyrosine is essential for cellular transformation by Rous sarcoma virus. *Cell* **20:** 807–816.

Stress-induced Changes in Nuclear Proteins of *Tetrahymena*

Martin A. Gorovsky, Claiborne V. C. Glover,
Susan D. Guttman, Karen J. Vavra,
Stuart Horowitz, and David S. Pederson
Department of Biology
University of Rochester
Rochester, New York 14627

In this report, we summarize our studies to date on the stress response in *Tetrahymena*.

EVOLUTIONARY CONSERVATION OF THE STRESS (HEAT-SHOCK) RESPONSE

Many features of the stress response in *Tetrahymena*[1] are similar to the heat shock response in *Drosophila* and other organisms. The following similarities have been noted (see Guttman et al. 1980 for discussion and references) (Fig. 1).

1. A variety of apparently unrelated agents (heat, anoxia, deciliation, amino acid analogs, etc.) induce the response.
2. Preexisting polysomes break down.
3. A small group of stress proteins (sp) rapidly comes to dominate the cell's protein synthesizing machinery.
4. The molecular weights of the induced proteins are similar to those of *Drosophila* heat-shock proteins (hsp).
5. Some induced proteins appear to be synthesized de novo while others are detectable in uninduced cells.
6. Induction is transient, but the induced proteins themselves are relatively stable.

---

[1] *Tetrahymena pyriformis* (formerly *T. pyriformis* strain GL) is an amicronucleate strain. *T. thermophila* (formerly *T. pyriformis*, syngen I) is a micronucleate strain.

299

NEPHGE ⟶

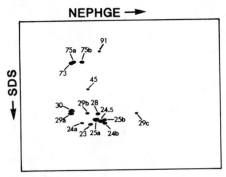

Figure 1
Diagramatic sketch indicating the electrophoretic mobilities and nomenclature for the major stress-induced proteins of starved *Tetrahymena pyriformis*.[2] The proteins diagramed represent those which are reproducibly labeled with radioactive amino acids during a 50-minute period of continuous heat shock or after deciliation or release from anoxia. The numbers indicate the approximate molecular weight of each protein in kilodaltons. One-letter suffixes, beginning with "a" for the most acidic, distinguish proteins having the same molecular weight.

7. Induction of at least some of the proteins requires new RNA synthesis.
8. Superinduction of some of the proteins occurs upon release of a block to protein synthesis (cycloheximide) administered for short periods after induction.

Although the above list represents entirely circumstantial evidence, the conclusion seems inescapable that the stress response in *Tetrahymena* can be equated with the classical heat-shock response of *Drosophila* and other organisms.

While the stress response itself appears to be highly conserved, some of the stress proteins themselves may be variable. Thus, the stress proteins of two different species of *Tetrahymena* have not remained invariant in sequence since (Fig. 2A,B) the divergence of *T. thermophila* and *T. pyriformis*.

[2]Starved cells were used in these studies because starvation greatly reduces the basal level of protein synthesis allowing easy detection of newly formed polysomes and proteins whose synthesis is induced by deciliation (Guttman and Gorovsky 1979). Since heat shock also induces a rapid breakdown of preexisting polysomes in *Tetrahymena* (Klemperer and Rose 1974; F. Calzone, pers. comm.), starvation is probably superfluous in studies of stress-induced proteins. A survey of the literature as well as unpublished experiments of G. Bannon (pers. comm.) indicate that in every case where comparisons can be made, the stress response in starved and growing cells is similar.

THERE MAY BE FUNCTIONAL GROUPS
OF HEAT-SHOCK PROTEINS

Cell fractionation studies with *Tetrahymena* originally suggested that there were at least two (functional?) groups of stress proteins (Fig. 2A).

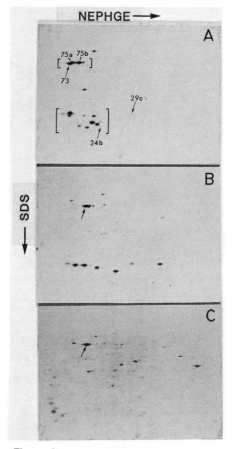

Figure 2
Fluorographs of two-dimensional gel electrophoretic analyses of whole-cell protein labeled with radioactive amino acids for 50 minutes following deciliation of *T. pyriformis* (*A*), heat shock of *T. thermophila* (*B*), or centrifugation of *T. pyriformis* (*C*). In *A*, the high- and low-molecular-weight groups identified by cell fractionation are indicated by brackets. sp29c is barely detectable in whole-cell protein, even though it is the major labeled protein in isolated nuclei (Guttman et al. 1980). The minor spots to the immediate right of sp73, sp75a, and sp75b are observed frequently but not invariably after stress. In *B*, *T. thermophila* were heat-shocked at 41°C, which is sublethal for this thermophilic strain. (→) In *B*, a protein whose electrophoretic mobility is very similar to that of sp73 in *T. pyriformis*; in *C*, sp73 in *T. pyriformis* whose synthesis is induced by mild stress (centrifugation and resuspension).

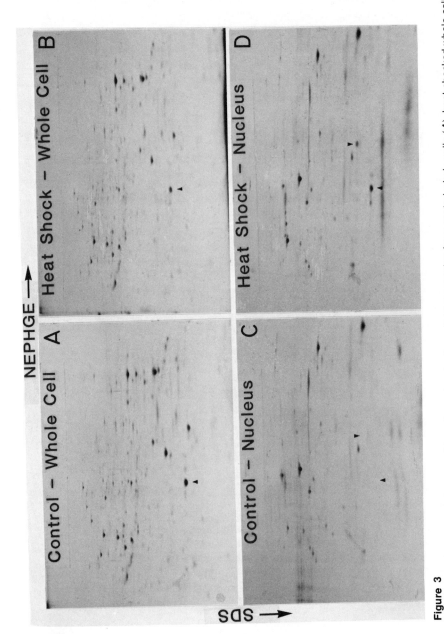

Figure 3
Coomassie Blue staining patterns of two-dimensional gels containing proteins from control whole cells (*A*), heat-shocked whole cells (*B*), control nuclei (*C*), and heat-shocked nuclei (*D*). (▲) The position of sp24b; (▼) the position of sp29c.

One group (sp73, sp75a, and sp75b) has a generalized distribution. They were found in similar relative amounts in all the cellular compartments examined and, as a group, their specific activities did not differ in the nucleus or cytoplasm. A low-molecular-weight group (sp23 to sp30, except for sp29c) appeared to be excluded from macronuclei but enriched in a 16,000g pellet (see below).

Recent evidence that, in *Drosophila,* the genes for hsp70 and hsp68 are part of a multigene family (Craig et al. 1982) and the genes for hsp28, 26, 23, and 22 are clustered (Corces et al. 1980; Craig and McCarthy 1980) also suggests that these groups may be evolutionarily conserved and may be functionally related.

NONCOORDINATE INDUCTION OF sp73 — A CONSERVED STRESS RESPONSE TRIGGER?

Although deciliation (a transient stimulus) and heat shock (a continuous stimulus) induce similar proteins, it is clear that at least one of these proteins, sp73, can be expressed noncoordinately. While this protein cofractionates with sp75a and sp75b, it differs from them in a number of ways. sp73 is present in unstressed starved cells in stainable amounts and is synthesized constitutively at low levels; sp75a and sp75b are not. Mild stress results in the strong induction of sp73 relative to the other stress proteins (Fig. 2C). Noncoordinate induction of heat-shock proteins in *Drosophila* has also been described (Lindquist 1980).

One or more proteins with a molecular weight of 73,000 is induced during stress in both *T. pyriformis* and *T. thermophila,* and a protein of similar molecular weight (~70K) is synthesized after heat shock in most organisms. In *Drosophila,* the gene for hsp70 (or a closely related gene) is being transcribed in unshocked cells (Findly and Pederson 1981). Sequences homologous to *Drosophila* hsp70 also can be found in yeast (Craig et al. 1982). This pattern of evolutionary conservatism, presence in uninduced cells, and ease of induction suggests an essential role for (h)sp70 in normal cells and in the stress response. One possibility is that it serves as a triggering agent for other features of the response.

STRESS-INDUCED NUCLEAR PROTEINS

While the high-molecular-weight group could have a nuclear function, it is not enriched in that compartment and more than 90% of sp73, 75a, and 75b is cytoplasmic. The low-molecular-weight group appears to be relatively excluded from isolated macronuclei (Guttman et al. 1980).

One induced protein, sp29c, is barely detectable in labeled whole-cell protein, but is the major radioactive spot in gels of nuclear proteins isolated from stressed *Tetrahymena* (Guttman et al. 1980). Stainable quantities of sp29c also are detectable in nuclei only after stress (see Fig. 3). These observations make it likely that sp29c has a nuclear function.

Another protein, sp24b, has surprising properties: it exists in large amounts in starved cells and appears to be translocated from cytoplasm to nucleus in response to stress (Fig. 3). While synthesis of this protein (or one comigrating with it) is induced to varying degrees by stress, little radioactively labeled sp24b appears in nuclei (Guttman et al. 1980).

Unfortunately, no consistent picture of stress-induced nuclear proteins in diverse organisms can be discerned at this time. None of the stress-induced proteins in *Tetrahymena* were histones (Guttman et al. 1980; Glover et al. 1981), although histone H2B is a stress-induced protein in *Drosophila* (Sanders 1981). A protein with properties similar to those of sp29c has been identified in *Chironomus tentans* (Vincent and Tanguay 1979), while Arrigo et al. (1980) and Loomis and Wheeler (1982) demonstrated a nuclear location for a low-molecular-weight group of heat-shock proteins in *Drosophila* and *Dictyostelium,* respectively. Thus, one or more newly synthesized, low-molecular-weight stress proteins appears to have a predominantly nuclear location. A number of investigators (Vincent and Tanguay 1979; Arrigo et al. 1980; Guttman et al. 1980; Loomis and Wheeler 1982) using a variety of organisms (*Tetrahymena, Drosophila, Chironomus, Dictyostelium*) have found that one or more proteins with a molecular weight of around 70K is found in isolated nuclei but is not necessarily enriched there. On the other hand, using electron microscope autoradiography, Velasquez et al. (1980) found that most of the newly synthesized hsp70 in *Drosophila* was nuclear. A number of factors could account for the lack of consistency in these observations. Many of these studies were done with a single stress treatment and the cellular distribution of specific proteins could vary with the time and severity of treatment. Proteins also could be lost or redistributed during isolation or fixation. Finally, the actual homologies (or lack thereof) between proteins of similar (or different) molecular weights in different organisms remain to be demonstrated.

INTRANUCLEAR LOCATION OF STRESS-INDUCED PROTEINS

In *Tetrahymena,* none of the proteins that appear in the nucleus after stress can be solubilized appreciably by treatment with staphylococcal nuclease under conditions that release most of the DNA and histones (Guttman et al. 1980). Similarly, Levinger and Varshavsky (1981) observed that *Drosophila* nuclear heat-shock proteins remain insoluble after nuclease digestion plus high-salt extraction. Taken together, these results argue that nuclear stress-induced proteins may be associated

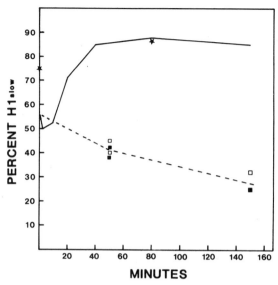

Figure 4

Effects of cycloheximide on the kinetics of stress-induced phosphorylation of H1. (-----) The pattern of phosphorylation in starved, heat-shocked or deciliated cells, and based on data published in Glover et al. (1981). (⋆) Data from growing, heat-shocked cells. Zero time in these cases is the initiation of stress. (-----) is based on PCA extracts obtained from whole starved cells (□) or heat-shocked cells (■) which had been treated with 10 μg cyclohexi-mide/ml for 10 minutes prior to zero time. Zero time in these experiments represents the time of resuspension at either 28°C or at 33.8°C in the presence of cycloheximide.

with structural proteins of the nuclear matrix. However, Arrigo et al. (1980) and Loomis and Wheeler (1982) found that, in *Drosophila* and *Dictyostelium,* respectively, the low-molecular-weight nuclear heat-shock proteins are chromatin associated. The basis for these differences is not clear.

STRESS-INDUCED PHOSPHORYLATION OF HISTONE H1

Both heat shock and deciliation of *T. pyriformis* resulted in a reversible increase in phosphorylation of macronuclear histone H1 (Glover et al. 1981). The final level of stress-induced H1 phosphorylation was the same in starved and growing cells, although starved cells had a much lower initial level. Stress-induced phosphorylation occurs at multiple sites on a large percentage of the H1 molecules. Thus, stress-induced phosphorylation of H1 is similar to cell-cycle-dependent H1 phosphoryla-tion (see Hohmann 1978 for review), even though stressed, starved *Tetrahymena* do not divide.

Stess-induced phosphorylation of H1 can be inhibited by cycloheximide (Glover et al. 1981). Coupled with the short lag that precedes phosphorylation, this raises the possibility that one of the newly synthesized nuclear stress proteins (possibly sp29c) is a histone kinase.

The function of the stress-induced phosphorylation of H1 in *Tetrahymena* macronuclei is unknown. A similar, extensive heat-shock-induced phosphorylation of H1 in growing *Drosophila* tissue culture cells has not been detected (see Figs. 1 and 2 in Glover 1982). However, it should be pointed out that the stress-induced increase in H1 phosphorylation in growing *Tetrahymena* is not nearly as great as in starved cells. In any event, an increase in H1 phosphorylation is the only chromatin-associated, stress-induced change detected so far in *Tetrahymena*. Recent observations that heat shock can effect the phosphorylation state of both nuclear (nonhistone) proteins (Caizergues-Ferrer et al. 1980) and of a ribosomal protein (Glover 1982) suggest that more extensive studies of stress-induced effects on protein phosphorylation may be warranted.

THE EFFECTS OF STRESS ON NUCLEOSOME PERIODICITY

H1 is associated with the linker region of chromatin (see Kornberg 1977 for review) and it has been suggested that the net charge on H1 might affect the spacing of nucleosomes (Noll 1976). Stress-induced phosphorylation not only lowers the net charge on H1 but also can induce a conformational change which alters the mobility of many (>30%) *Tetrahymena* H1 molecules on SDS gels (Glover et al. 1981). Therefore, it seemed reasonable to determine whether any change in nucleosome spacing accompanied stress-induced H1 phosphorylation. The periodicity of bulk chromatin was the same (~200 bp) in control and in heat-shocked cells at every stage of digestion by staphylococcal nuclease. The periodicity of the ribosomal genes was slightly shorter (~175 bp) than that of bulk chromatin and was indistinguishable in control and in stressed cells. We conclude that large differences in the phosphorylation of H1 have no effect on the average periodicity of bulk chromatin or of ribosomal gene-containing chromatin. It is possible, of course, that the positions of nucleosomes on specific DNA sequences (nucleosome phasing) changes without an accompanying change in the average nucleosome spacing. Whether stress induces alteration in the phasing of nucleosomes on specific genes and whether stress-induced phosphorylation of H1 plays any role in that process are subjects for future investigations.

ACKNOWLEDGMENTS

The work reported here was supported by research grants from the NIH.

REFERENCES

Arrigo, A.-P., S. Fakan, and A. Tissières. 1980. Localization of the heat shock-induced proteins in *Drosophila melanogaster* tissue culture cells. *Dev. Biol.* **78:** 86–103.

Caizergues-Ferrer, M., G. Bouche, F. Amalric, and J.-P. Zalta. 1980. Effects of heat shock on nuclear and nucleolar protein phosphorylation in Chinese hamster ovary cells. *Eur. J. Biochem.* **108:** 399–404.

Corces, V., R. Holmgren, R. Freund, R. Morimoto, and M. Meselson. 1980. Four heat shock proteins of *Drosophila melanogaster* coded within a 12-kilobase region in chromosome subdivision 67B. *Proc. Natl. Acad. Sci.* **77:** 5390–5393.

Craig, E. and B. McCarthy. 1980. Four *Drosophila* heat shock genes at 67B: Characterization of recombinant plasmids. *Nucleic Acids. Res.* **8:** 4441–4457.

Craig, E.A., T.D. Ingolia, M.R. Slater, and L.J. Mannseau. 1982. *Drosophila* and yeast have multigene families related to the major *Drosophila* heat shock inducible gene. *J. Cell. Biochem.* (Suppl.)**6:** 282.

Findly, R.C. and T. Pederson. 1981. Regulated transcription of the genes for actin and heat-shock proteins in cultured *Drosophila* cells. *J. Cell Biol.* **88:** 323–328.

Glover, C.V.C. 1982. Heat shock induces rapid dephosphorylation of a ribosomal protein in *Drosophila*. *Proc. Natl. Acad. Sci.* **79:** 1781–1785.

Glover, C.V.C., K.J. Vavra, S.D. Guttman, and M.A. Gorovsky. 1981. Heat shock and deciliation induce phosphorylation of histone H1 in *T. pyriformis*. *Cell* **23:** 73–77.

Gorovsky, M.A. and J. Keevert. 1975. Subunit structure of a naturally occurring chromatin lacking histones f1 and f3. *Proc. Natl. Acad. Sci.* **72:** 3536–3540.

Guttman, S.D. and M.A. Gorovsky. 1979. Cilia regeneration in starved *Tetrahymena:* An inducible system for studying gene expression and organelle biogenesis. *Cell* **17:** 307–317.

Guttman, S.D., C.V.C. Glover, C.D. Allis, and M.A. Gorovsky. 1980. Heat shock, deciliation and release from anoxa induce the synthesis of the same set of polypeptides in starved *T. pyriformis*. *Cell* **22:** 299–307.

Hohmann, P. 1978. The H1 class of histone and diversity in chromosomal structure. *Subcell. Biochem.* **5:** 87–127.

Klemperer, H.G. and V.A. Rose. 1974. Decrease of polysomes in *Tetrahymena* after synchronization shocks. *Nature* **248:** 443–446.

Kornberg, R.D. 1977. Structure of chromatin. *Annu. Rev. Biochem.* **46:** 931–954.

Levinger, L. and A. Varshavsky. 1981. Heat-shock proteins of *Drosophila* are associated with nuclease-resistant, high-salt-resistant nuclear structures. *J. Cell Biol.* **90:** 793–796.

Lindquist, S. 1980. Varying patterns of protein synthesis in *Drosophila* during heat shock: Implications for regulation. *Dev. Biol.* **77:** 463–479.

Loomis, W.F. and S.A. Wheeler. 1982. Chromatin-associated heat shock proteins of *Dictyostelium. Dev. Biol.* **90:** 412–418.

Noll, M. 1976. Differences and similarities in chromatin structure of *Neurospora crassa* and higher eukaryotes. *Cell* **8:** 349–355.

Sanders, M.M. 1981. Identification of histone H2b as a heat-shock protein in *Drosophila. J. Cell Biol.* **91:** 579–583.

Velazquez, J.M., B.J. DiDomenico, and S. Lindquist. 1980. Intracellular localization of heat shock proteins in *Drosophila. Cell* **20:** 679–689.

Vincent, M. and R.M. Tanguay. 1979. Heat-shock induced proteins present in the cell nucleus of *Chironomus tentans* salivary gland. *Nature* **281:** 501–503.

Heat-shock Proteins in the Protozoan *Tetrahymena:* Induction by Protein Synthesis Inhibition and Possible Role in Carbohydrate Metabolism

**James M. Wilhelm,
Peggy Spear, and Christina Sax**
*Department of Microbiology
University of Rochester Medical Center
Rochester, New York 14642*

The ciliated protozoan, *Tetrahymena thermophila,* is a unicellular eukaryote with a remarkably complex organization; it is of particular interest to cell biologists and geneticists because it carries, in a single cell, both a germ line and a somatic nucleus. We employ the organism for genetic and biochemical studies of ribosomal interactions with antibiotics, especially those that we showed stimulate errors in eukaryotic protein synthesis (Palmer and Wilhelm 1978). Our work on heat-shock induction stems from our observation that these antibiotics induce the heat-shock proteins in *Tetrahymena* (Wilhelm et al. 1979). We and others (Guttman et al. 1980) have found a variety of additional inducers, indicating that the induction of heat-shock proteins is indeed a "stress" response. We have formulated two hypotheses about heat-shock proteins in *Tetrahymena:* first, we propose that the stimulus is a block in the formation of aminoacyl-tRNA or a block to its utilization at the acceptor (A) site on the ribosome; second, we suggest that some of the induced proteins may be involved in gluconeogenesis or glycogen synthesis.

INDUCERS OF HEAT-SHOCK PROTEINS

When cultures of *T. thermophila* are transferred suddenly from 30°C to 39°C, the synthesis of over 25 polypeptides is induced (Fig. 1); most of these are barely detectable in growing cells at 30°C. The major species,

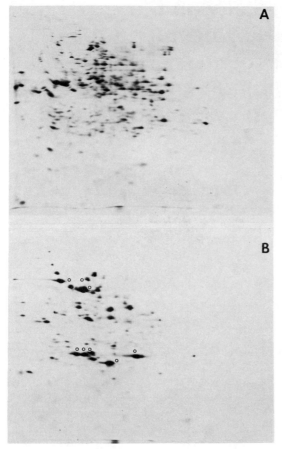

Figure 1
Patterns of polypeptide synthesis in growing *Tetrahymena* and during heat shock. Cells were washed into buffer and labeled immediately from [^{14}C]leucine for 30 minutes at 30°C (*A*) or after 20 minutes at 39°C (*B*). Cells were solubilized and polypeptides were separated by two-dimensional gel electrophoresis (nonequilibrium pH gradient/SDS). Radioactivity was detected by autoradiography. The most basic polypeptides are to the right; the largest are at the top. (○, *top to bottom*) Polypeptides 85K, 74K, 34K, and 30K. Labeling in buffer was employed because [^{14}C]leucine labeling is not feasible in growth medium.

prominent on one-dimensional gels, are polypeptides of $M_r = 85,000$, 74,000, and 30–34,000 (Fig. 2B). One interesting aspect of the induction is that synthesis of the heat-shock proteins reaches a maximum, then declines, and synthesis of normal proteins resumes even if the cells are maintained at high temperature. In additional experiments, we have found that actinomycin D will block induction; this result suggests that, as in other organisms, induction is dependent on new transcription.

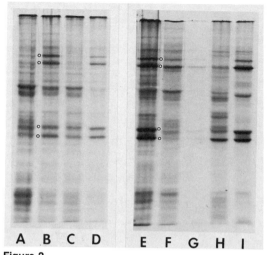

Figure 2
Patterns of polypeptide synthesis in *Tetrahymena* under various conditions of cell density and nutrition. Experimental details were similar to those of Fig. 1, but labeling was from [14C]lysine in growth medium and electrophoresis was one-dimensional (SDS). (○, *top to bottom*) The 85K, 74K, 34K, and 30K polypeptides. A, Growing cells (ca. 150,000/mL) at 30°C; B, heat shock (39°C, 20 min), with growing cells; C, growing cells at 30°C 3 hours after glucose addition (55 mM); D, heat shock after glucose addition with growing cells; E, same as B (for comparison with F); F, nongrowing cells (ca. 900,000/mL) at 30°C; G, heat shock with nongrowing cells; H, nongrowing cells supplemented with glucose 18 hours earlier; and I, heat shock with glucose-supplemented nongrowing cells.

A large, and seemingly disparate, array of metabolic inhibitors induce the heat-shock proteins (Table 1). Heat shock transiently inhibits total protein synthesis by 30–50%, as measured by amino acid incorporation; and strong induction occurs also with the inhibitors when used at concentrations that inhibit protein synthesis by about the same extent. We suggest that there is a property that all inducers share; namely all can act as protein synthesis inhibitors by (1) blocking the utilization of aminoacyl-tRNA at its binding site (the A site) on the ribosome or by (2) inhibiting the formation of aminoacyl-tRNA itself. Heat shock may also be explainable by this mechanism, in that aminoacyl-tRNA is bound to ribosomes from a complex of aminoacyl-tRNA, elongation factor I (EFI), and GTP. We have found that heat shock strongly reduces the labeling of GTP from inorganic phosphate. If this truly represents a reduction in the amount of cellular GTP, then aminoacyl-tRNA binding would be reduced. Further support for our hypothesis derives from the observation that cycloheximide, emetine, and hygromycin B, which block the translocation step of protein synthesis but do not interfere with aminoacyl-tRNA binding, do not induce heat-shock proteins; pactamycin, which inhibits initiation of polypeptide chains, is not an inducer

Table 1
Inducers of Heat-shock Proteins in *Tetrahymena*

Amino acid analogs	Ribosomal inhibitors
Canavanine[a]	Puromycin[a,b]
p-Fluorophenylalanine	
Ethionine	Tetracycline[b]
Alkylating agent	
TPCK[a,b]	Paromomycin[a,b,c]
	Heat shock[a]

[a] Strongest inducers.
[b] See Vazquez (1979) for reviews of the action of these inhibitors.
[c] See Wilhelm and Palmer (1978) for discussion of this antibiotic.

either. Our hypothesis relating inhibition of aminoacyl-tRNA function to regulation of gene expression has a certain analogy to the stringent response in prokaryotes, where the unusual nucleotide ppGpp couples translation and transcription. In our system, the mediator between protein synthesis inhibition and induction of gene expression remains unknown. We have not detected ppGpp in *Tetrahymena,* and there is no convincing evidence of its existence in any eukaryote.

POSSIBLE ROLE OF HEAT-SHOCK PROTEINS IN CARBOHYDRATE METABOLISM

A clue to a possible role for heat-shock proteins comes from the observation that at least three proteins (85K, a 74K, and a 30K) are prominently synthesized when *Tetrahymena* has ceased rapid growth at high cell densities (Fig. 2F). The identity of these polypeptides as heat-shock proteins was confirmed by two-dimensional gel electrophoresis. As cell growth slows, amino acid carbon is converted to glucose and stored as glycogen; the capacity for glycogen storage in this organism is remarkable — up to one-quarter of the dry weight. Because of the coincidence in synthesis of some heat-shock proteins and glycogen accumulation, we asked whether heat shock could stimulate glycogen production in cells where glycogen was low. With cells at low density, growing on a carbohydrate-free medium, heat shock does indeed stimulate a substantial accumulation of glycogen (Fig. 3). In addition, canavanine, puromycin, and paromomycin, when tested at concentrations that induce heat-shock proteins, stimulate *Tetrahymena* to accumulate glycogen (Fig. 4). We hypothesize that some of the

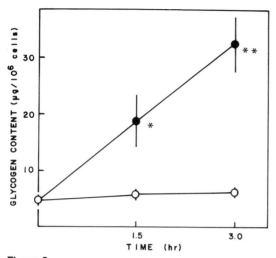

Figure 3

Accumulation of glycogen by *Tetrahymena* during heat shock. Replicate flasks ($n = 5-7$) of growing cells were maintained at 30°C (o) or transferred to 39°C (●). At the indicated times, glycogen was extracted and determined colorimetrically. Values are means ± S.E. (*) Significantly different from corresponding control at $P < 0.05$; (**) $P < 0.001$.

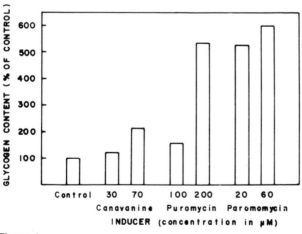

Figure 4

Accumulation of glycogen by *Tetrahymena* during treatment with metabolic inhibitors that induce heat-shock proteins. Replicate flasks ($n = 2-3$) were maintained at 30°C (control) or treated with the indicated inhibitors at 30°C. After 1.5 hours, glycogen was determined as in Fig. 3. Mean values (as glycogen/10^6 cells) have been normalized to mean control values, set at 100%.

heat-shock proteins may be involved in gluconeogenesis and glycogen synthesis; their role would be, of course, to elaborate an energy storage molecule during "stress." Two observations are consistent with our hypothesis. First, the addition of glucose at 30°C induces synthesis of several proteins; one is the 30K heat-shock protein (Fig. 2C), which is abundantly synthesized also in high-density cells supplemented with glucose (Fig. 2H). This protein could be involved in the conversion of glucose to glycogen. Second, glucose addition represses the synthesis of the 85K and 74K proteins, either in nongrowing cells or in growing cells after heat shock (Fig. 2, H and D). Thus, these proteins could be involved in the conversion of amino acid precursors to glucose itself, and exogenous glucose could repress their synthesis.

SUMMARY/CONCLUSION

Our studies with the lower eukaryote, *Tetrahymena,* have led to hypotheses concerning the mechanism of induction of heat-shock proteins and their role in the metabolic balance of the organism. There is a certain attractive simplicity in the notion that disparate kinds of "stress" might be unified by an inhibition of protein synthesis and that "stress" might result in the redirection of metabolism toward production of an energy storage molecule. Whether these ideas have application to other organisms remains to be seen. Our future work will be directed toward study of the fate of aminoacyl-tRNA during stress, and toward investigation of key enzymes of amino acid degradation, gluconeogenesis, and glycogen synthesis as possible heat-shock proteins.

ACKNOWLEDGMENTS

This work was supported by National Institutes of Health research grant GM-25168 to J.M.W.

REFERENCES

Guttman, S.D., C.V.C. Glover, C.D. Allis, and M.A. Gorovsky. 1980. Heat shock, deciliation, and release from anoxia induce the synthesis of the same set of polypeptides in starved *T. pyriformis. Cell* **22:** 299–307.

Palmer, E. and J.M. Wilhelm. 1978. Mistranslation in a eukaryotic organism. *Cell* **13:** 329–334.

Vazquez, D. 1979. *Inhibitors of protein biosynthesis.* Springer-Verlag, Berlin/Heidelberg/New York.

Wilhelm, J.M., M. Pacilio, and G. Weisburg. 1979. A common response of *Tetrahymena* to heat shock, amino acid analogs, and antibiotics that generate abnormal protein. *J. Cell Biol.* **83:** 402a (Abst.)

A Preliminary Comparison of Maize Anaerobic and Heat-shock Proteins

Philip M. Kelley
and Michael Freeling
Department of Genetics
University of California
Berkeley, California 94720

At least two stress-response systems have been characterized in plants. Anaerobic stress induces a set of proteins including the alcohol dehydrogenases (Sachs et al. 1980). Heat shock induces a set of proteins very similar in molecular weight to those observed in other eukaryotes (Key et al., this volume). The genetics and developmental expression of *Adh1* and its products have been extensively studied by Dr. Drew Schwartz and his students; Freeling and Birchler (1981) have recently reviewed much of this work. This report focuses primarily on the similarities and differences between the anaerobic and heat-shock response in maize.

COMPARISON OF ANAEROBIC AND HEAT-SHOCK PROTEINS

Both anaerobiosis and heat shock in plants are characterized by the exclusive expression during stress of a specific set of polypeptides. Both types of stress result in the translational shut-off of normal protein synthesis. Both responses are reversible: restoration of normal conditions results in a return to synthesis of normal proteins.

Comparison of the stress proteins by native-SDS two-dimensional gel electrophoresis shows that none of the major anaerobic proteins comigrate with major heat-shock proteins of maize (Fig. 1). A minor heat-shock protein of 33 kD does comigrate with a major anaerobic transitional protein. The transition proteins, a set of at least four dimeric proteins

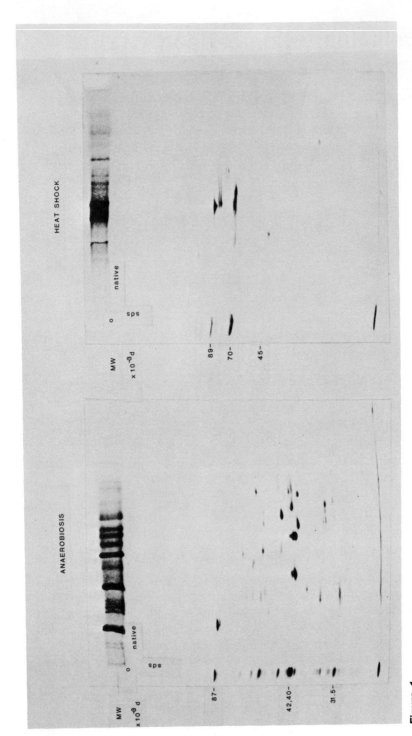

Figure 1
Autoradiogram of native-SDS two-dimensional polyacrylamide gels of maize primary root [^{35}S]methionine proteins. Maize Berkeley Fast seeds were germinated, treated, labeled, and harvested as described by Sachs et al. (1980). *Anaerobiosis:* anaerobic treatment for 6 hours then labeled for 12 hours. *Heat shock:* heat-shock treatment at 43°C for 30 minutes then labeled at the same temperature for 3 hours.

of the same polypeptide molecular weight, are synthesized immediately following the initiation of anaerobiosis.

While the heat-shock response has been observed in all tissues examined, including both leaves and roots (B.G. Atkinson et al., pers. comm.), the anaerobic response, while evident in the root, cannot be detected in the mature leaf (Okimoto et al. 1980). Response kinetics for these two stress conditions also differ. The heat-shock response is detectable in minutes. The anaerobic response is much slower; only transition protein synthesis is rapid. After 6 hours of an argon environment, anaerobic protein synthesis has reached maximal rate. mRNA-ADH1 levels reach steady-state anaerobic levels 3−6 hours after the onset of anaerobiosis (Gerlach et al. 1982; S. Hake, unpubl.).

INTERMEDIATE OXYGEN CONCENTRATIONS AFFECT MAIZE ROOT GENE EXPRESSION

Intermediate temperatures between 25°C and 40°C differentially effect the type and translational rate of the heat-shock proteins in *Drosophila*. Different concentrations of oxygen have similar effects on the expression of the maize anaerobic proteins. We compared the pattern of proteins synthesized following treatment with intermediate concentrations of oxygen (Fig. 2) and found such patterns to be distinct from either untreated or anaerobic roots. Pyruvate decarboxylase and alcohol dehydrogenase are known to be induced at intermediate oxygen concentrations (Wignarajah and Greenwood 1976) as well as under anaerobic conditions. The pattern of proteins synthesized after 2% oxygen treatment identify a set of genes not normally expressed under aerobic conditions, but also not expressed under complete anaerobiosis.

WHAT ARE THE FUNCTIONS OF THE ANAEROBIC PROTEINS?

The anaerobic genes are fully accessible as nucleotide sequences. The Peacock laboratory, CSIRO, reported the first DNA recombinant to mRNA-ADH1 and other anaerobic messages as well (Gerlach et al. 1982). This laboratory has similar probes. One of the alleles for *Adh1* has been cloned (J. Bennetzen, unpubl.), and the dozens of interesting mutant and variant *Adh1* alleles will soon be reduced to nucleotide sequence arrangement. Unfortunately, gene structure can not tell us much about the function served by the anaerobic proteins.

It seems reasonable to guess that ADH activity serves to regenerate oxidizing power as NAD and thus drive glycolysis. Schwartz (1969) discovered that mutants that lacked ADH1 enzyme subunit activity were easily drowned under conditions where *Adh1*+ siblings survived; this

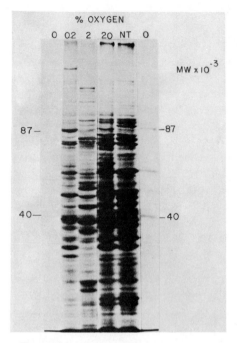

Figure 2
Autoradiogram of a SDS-polyacrylamide gel of maize primary root [35S]methionine proteins. Berkeley Fast seedlings were immersed in 200 ml of solution through which the following gas was bubbled: lane 1 and 6, argon gas; lane 2, 0.2% oxygen-balance argon; lane 3, 2% oxygen-balance argon; lane 4, air (20% oxygen); and lane 5, NT = no pretreatment. Roots were then labeled with [35S]methionine for 3 hours (Sachs et al. 1980) with the appropriate oxygen concentration in the gas phase.

renders *Adh1* expression as necessary in order to survive anaerobiosis. However, this does not imply that lines of maize with greater ADH activity would be more anaerobiosis resistant. Indeed, a popular hypothesis to explain differences in flood tolerance among plants predicts just the opposite (Crawford 1966). We suggest that a reasonable way to discover the exact relationships between anaerobic proteins and anaerobiosis survival is first to obtain mutants that are tolerant of anaerobiosis.

SUMMARY

Maize roots respond to anoxia by synthesizing a specific set of proteins, largely distinct from the heat-shock proteins. A possible exception is a 33-kD transition polypeptide that may be a common response to both forms of stress. The pattern of polypeptides translated at concentrations of oxygen intermediate (2%) between anaerobiosis and air (20%) was unique and not a simple blending of aerobic and anaerobic proteins.

ACKNOWLEDGMENTS

This research was supported by NIH grant GM21734.

REFERENCES

Crawford, R.M.M. 1966. The control of anaerobic respiration as a determining factor in the distribution of the genus *Senecio*. *J. Ecol.* **54:** 403–413.

Freeling, M. and J.A. Birchler. 1981. Mutants and variants of the alcohol dehydrogenase-1 gene in maize. In *Genetic engineering, principles and methods* (ed. J.K. Setlow and A. Hollaender), vol. 3, pp. 223–264.

Gerlach, W.L., A.J. Pryor, E.S. Dennis, R.J. Ferl, M.M. Sachs, and W.J. Peacock. 1982. cDNA cloning and induction of the alcohol dehydrogenase gene (*Adh*-1) maize. *Proc. Natl. Acad. Sci.* **79:** 2981–2985.

Okimoto, R., M.M. Sachs, E.K. Porter, and M. Freeling. 1980. Patterns of polypeptide synthesis in various maize organs under anaerobiosis. *Planta* **150:** 89–94.

Sachs, M.M., M. Freeling, and R. Okimoto. 1980. The anaerobic proteins of maize. *Cell* **20:** 761–767.

Schwartz, D. 1969. An example of gene fixation resulting from selective advantage in suboptimal conditions. *Am. Nat.* **103:** 479–481.

Wignarajah, K. and H. Greenway. 1976. Effect of anaerobiosis on activities of alcohol dehydrogenase and pyruvate decarboxylase in roots of *Zea mays*. *New Phytol.* **77:** 575–584.

The Synthesis of Heat-shock and Normal Proteins at High Temperatures in Plants and Their Possible Roles in Survival under Heat Stress

**Mitchell Altschuler
and Joseph P. Mascarenhas**
*Department of Biological Sciences
State University of New York at Albany
Albany, New York 12222*

Soybean cells or seedlings when rapidly transferred from 25°C to 40°C respond by the inhibition of synthesis of most normal proteins and the new synthesis of several heat shock proteins (hsp) (Barnett et al. 1980; Altschuler and Mascarenhas 1982). A sudden temperature shock is not a condition that organisms are likely to encounter in nature where they are exposed to gradually increasing temperatures during the day rather than a sudden increase of 12–15°C. When the temperature is gradually increased at the rate of 3° per hour, soybean seedlings respond differently to the temperature increase than do seedlings given a sudden heat shock. The temperature at which maximum protein synthesis occurs is shifted several degrees higher (from 37°C to 43°C) and in addition there appears to be a protection of normal protein synthesis from heat-shock inhibition. In addition, a prior mild heat shock or a gradual increase in temperature into the heat-shock range (40–43°C) enables soybean cells to survive when subsequently exposed to a potentially lethal temperature (52°C) (Altschuler and Mascarenhas 1982).

The aim of this work was to characterize further the response of plants to a gradual temperature increase and to determine whether heat-shock proteins play a direct role in the survival of cells under extreme heat stress when they are given a prior exposure to gradually increasing temperatures.

EFFECT OF A GRADUAL INCREASE IN TEMPERATURE ON PROTEIN SYNTHESIS

The incubation temperature of soybean seedlings was gradually raised 3° per hour from 25°C to 52°C and the seedlings labeled for 1 hour at various intermediate temperatures prior to homogenization and analysis of the extracts by SDS-polyacrylamide gel electrophoresis (Fig. 1). When exposed to gradually increasing temperatures, most normal proteins continue to be synthesized up to 43°C and heat-shock proteins are synthesized in appreciable amounts even at 49°C. A small amount of label is incorporated into heat-shock proteins, even at 52°C. This is quite different from the situation where seedlings are rapidly shifted from 25°C to the heat-shock temperature when most normal protein synthesis is inhibited by 40°C and neither heat-shock proteins nor normal proteins are synthesized by 45°C (Altschuler and Mascarenhas 1982). Normal protein synthesis is thus protected even at extremely high temperatures

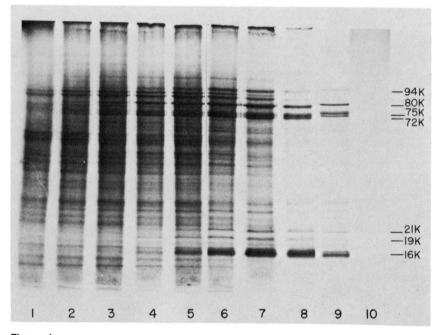

Figure 1
Effect of a gradual increase in temperature of 3°C per hour on protein synthesis in soybean seedlings. The seedlings were labeled during the terminal hour of incubation at the temperature shown below with [³⁵S]methionine as described by Altschuler and Mascarenhas (1982). Lane 1, 25°C; lane 2, 28°C; lane 3, 31°C; lane 4, 34°C; lane 5, 37°C; lane 6, 40°C; lane 7, 43°C; lane 8, 46°C; lane 9, 49°C; lane 10, 52°C. Extracts from equivalent amounts of tissue were loaded into the lanes.

by a gradual temperature increase, conditions more representative of those that plants are exposed to in the field. Similar results have been described for *Drosophila* cells (Lindquist 1980).

IS HEAT-SHOCK PROTEIN SYNTHESIS REQUIRED FOR THE PROTECTION OF CELLS FROM DEATH WHEN EXPOSED TO HIGH TEMPERATURES?

A prior heat-shock treatment or a gradual increase in temperature protects cells from death that might be caused by subsequent extreme heat stress (Mitchell et al. 1979; McAlister and Finkelstein 1980; Altschuler and Mascarenhas 1982).

In order to be able to determine unambiguously whether heat-shock proteins are required for the protection from cell death, it would be ideal to work with a system in which heat-shock proteins are not present and are not synthesized. It might then be possible to separate the effects of heat-shock proteins from those of other factors that might be involved in the protection. Pollen tubes might fulfill the requirements of such a system. When pollen grains, which are haploid male organisms constituting the male haploid life cycle of plants, are either transferred to growth medium at 37°C (Fig. 2) or 41°C (data not presented) or initially cultured at 25°C and rapidly transferred to 37°C (data not presented), no heat-shock proteins are synthesized. Heat-shock proteins are synthesized by nonpollen tissues of *Tradescantia paludosa* and the heat-shock

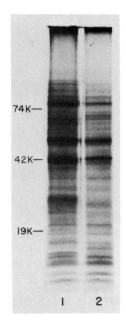

Figure 2
Protein synthesis in *Tradescantia paludosa* pollen tubes grown at 25°C (lane 1) and 37°C (lane 2). Autoradiogram of an SDS-polyacrylamide gel of proteins from pollen tubes labeled for 1 hour with [^{35}S]methionine. Extracts from equivalent amounts of tissue were loaded in the lanes.

proteins are very similar in their sizes to those synthesized by soybean cells and tissues (Altschuler and Mascarenhas, unpubl.). At 37°C, there is a reduction of total incorporation of label into protein as compared with

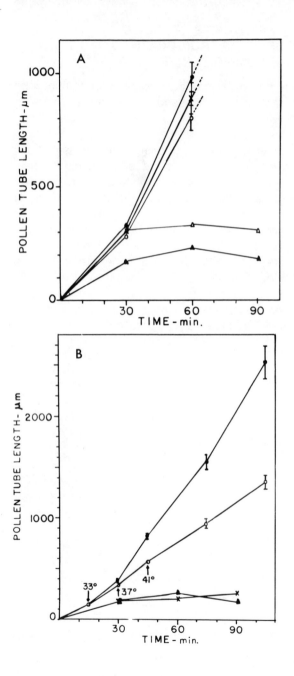

25°C and there might be a reduction in the synthesis of a few protein bands, but no new bands of sizes similar to the heat-shock proteins are synthesized by pollen (Fig. 2).

Pollen grains were grown at different temperatures to determine the effects of temperature on growth. The data in Figure 3A show that pollen tube growth is optimal at 25°C and is very similar although slightly poorer at 29°C and 33°C. Growth at these temperatures was linear over 90 minutes. At 37°C and at 41°C, growth of pollen tubes occurred during the initial 30 minutes up to lengths of about 300 and 200 μm, respectively, and thereafter further pollen tube growth was inhibited. It is of interest that initial pollen tube growth in *Tradescantia* up to a length of about 250 μm is insensitive to inhibition by protein synthesis inhibitors (Mascarenhas 1975). When pollen grains are initially germinated and grown at 41°C for 30 minutes and subsequently shifted to 25°C, there is no recovery of growth during a subsequent period of 60 minutes (Fig. 3B). If, however, the temperature of incubation is raised at 4°C per 15 minutes from 29°C to 41°C, pollen tube growth continues for at least 60 minutes at 41°C, the maximum time studied. There is a substantial protection of pollen tube growth from inhibition because of the gradual temperature increase. If, however, pollen tubes are grown at 29°C for 30 minutes and then suddenly transferred to 41°C, further growth of the tubes is immediately inhibited (data not shown). No heat-shock proteins are synthesized in pollen tubes given a gradual temperature increase to 41°C and then labeled with [35S]methionine at 41°C (data not shown). These results seem to indicate that the new synthesis of heat-shock proteins is not required for the protection observed. Whether the ungerminated pollen grain already contains presynthesized heat-shock proteins has not yet been determined. Even if one were to detect hsp spots in two-dimensional gels of proteins from ungerminated pollen, an interpretation of these results would not be straightforward because of the presence of possible cognate hsp genes and the translation of the mRNAs from these genes in nonheat-shocked cells. If pollen contained heat-shock proteins and these were involved in protection, one might expect pollen tubes grown at, or suddenly transferred to, 37°C or 41°C to show protection of their growth at these temperatures. No such protec-

Figure 3
Growth of pollen tubes as a function of temperature treatment. (*A*) Pollen grown in media continuously at different temperatures. (•) 25°C; (x) 29°C; (o) 33°C; (Δ) 37°C; (▲) 41°C. (*B*) (•) pollen grown continuously at 29°C; (o) pollen grown 15 minutes at 29°C, followed by a temperature increase of 4°C per 15 minutes up to 41°C and further incubation at 41°C for 60 minutes; (▲) pollen grown continuously at 41°C; (x) pollen grown 30 minutes at 41°C, followed by transfer to 25°C for 60 minutes. Pollen was grown in drops of medium (Lafleur et al. 1981) on slides in moist petri plates, and tube lengths were measured after fixing and staining the preparation with Methyl Green.

tion is observed. It might be argued that the heat-shock proteins might be present in the ungerminated pollen grain in some inactive form and are converted into the active form during a gradual temperature increase or incubation at a mild heat-shock temperature. Soybean cells are partially protected from cell death at 52°C if incubated at an intermediate temperature of 34°C, 37°C, or 41°C for 1 hour prior to exposure to 52°C (Altschuler and Mascarenhas 1982 and unpubl.). If inactive forms of the heat-shock proteins were present in the ungerminated pollen grain, one might thus expect the conversion of inactive to active forms by incubation directly at 37°C or 40°C, and accordingly pollen should grow at these temperatures. Pollen tube growth is, however, inhibited at 37°C and at 41°C unless tubes are given a prior exposure to gradually increasing temperatures. These results seem to imply that normal heat-shock proteins or their postulated inactive counterparts are not present in the ungerminated pollen grain, and are thus not likely to be directly involved in thermoprotection. This does not include cognate heat-shock proteins, whose role is at present not clear.

CONCLUSIONS

The fact that normal protein synthesis is protected at extremely high temperatures when the temperature increase is gradual suggests that the response of organisms to a sudden temperature increase, i.e., heat shock, might not be truly representative of the reactions of organisms in nature. In pollen tubes—a system in which heat-shock proteins are not synthesized and, based on physiological evidence, in which heat-shock proteins do not seem to be present—a gradual increase in the incubation temperature protects growth and viability of the pollen, indicating that heat-shock proteins may not be involved in the mechanism of this protection. Pollen tubes do not recover from a sudden exposure to a temperature of 37°C or 41°C, on subsequent return to 25°C, unlike vegetative plant tissues in which heat-shock protein synthesis occurs (Barnett et al. 1980). Heat-shock protein synthesis might thus be required for the process of recovery from a sudden heat shock.

ACKNOWLEDGMENTS

The work reported here was supported by NSF Grant PCM-7806810.

REFERENCES

Altschuler, M. and J.P Mascarenhas. 1982. Heat shock proteins and effects of heat shock in plants. *Plant Mol. Biol.* (in press).

Barnett, T., M. Altschuler, C.N. McDaniel, and J.P. Mascarenhas. 1980. Heat shock proteins in plant cells. *Dev. Genet.* **1:** 331–340.

Lafleur, G.J., R.E. Gross, and J.P. Mascarenhas. 1981. Optimization of culture conditions for the formation of sperm cells in pollen tubes of *Tradescantia. Gamete Res.* **4:** 35–40.

Lindquist, S. 1980. Varying patterns of protein synthesis in *Drosophila* during heat shock. Implications for regulation. *Dev. Biol.* **77:** 463–479.

Mascarenhas, J. P. 1975. Biochemistry of angiosperm pollen development. *Bot. Rev.* **41:** 259–314.

McAlister, L. and D.B. Finkelstein. 1980. Heat shock proteins and thermal resistance in yeast. *Biochem. Biophys. Res. Comm.* **93:** 819–824.

Mitchell, H.K., G. Moller, N.S. Petersen, and L. Lipps-Sarmiento. 1979. Specific protection from phenocopy induction by heat shock. *Dev. Genet.* **1:** 181–192.

The Heat-shock Response in Plants: Physiological Considerations

Joe L. Key, Chu-Yung Lin,
Ewa Ceglarz, and Fritz Schöffl
Department of Botany
University of Georgia
Athens, Georgia 30602

When soybean seedlings (Key et al. 1981) or cultured cells of soybean and tobacco (Barnett et al. 1980) are shifted from a normal growing temperature of 28°C to one of 40°C, the synthesis of a new set of proteins (heat-shock proteins) is induced while the synthesis of most proteins made at 28°C is reduced (see Fig. 3, lanes F and G). A number of other plants respond similarly to soybean (e.g., pea, millet, corn, sunflower, cotton, wheat) in that a large number of new proteins of similar molecular weight are induced by a heat-shock treatment (E. Ceglarz, C.Y. Lin, and J.L. Key, unpubl.). The major differences among species are the optimum temperature of induction of heat-shock proteins, the break-point temperature, the distribution of the 15–20-kD heat-shock proteins on two-dimensional gels, and the relative level of normal protein synthesis that occurs during heat shock. This response in plant systems seems to share many properties in common with the heat-shock response of *Drosophila* (Ashburner and Bonner 1979) and other eukaryotic systems which have been studied. In etiolated soybean seedlings or tissue excised from them, synthesis of heat-shock proteins is detected during the initial 15 minutes after the temperature shift and persists at least for several hours. Heat-shock proteins are detected at 35°C and increase as the temperature is increased up to 40–41°C (Key et al. 1981); when seedlings are returned to 28°C after 4 hours at 40°C, normal protein synthesis resumes and heat-shock protein synthesis essentially ceases after 3 or 4 hours. This pattern of heat-shock protein synthesis is

329

closely paralleled by the accumulation at 40°C and subsequent loss at 28°C of heat-shock-specific mRNAs as detected by Northern blot hybridization to cloned heat-shock cDNAs; some heat-shock mRNAs are detected within 3 to 4 minutes after transfer to 42°C and accumulate up to 20,000 copies per cell (Schöffl and Key 1982). While there is some loss in total poly(A) RNA complexity during 40–42°C heat-shock treatment along with some shift in relative abundance of 28°C sequences, a high proportion of those sequences persists even though they are poorly translated during heat-shock (Schöffl and Key 1982 and unpubl.); these data coupled with the rapid polyribosome-to-monoribosome transition which occurs at 40°C (Key et al. 1981) implies a strong translational regulatory response to heat-shock in the soybean system.

In addition to studies designed to understand the heat-shock system of plants at the molecular level, we also are investigating the possible physiological significance of the heat-shock response in plants. Here we report on some of the physiological aspects of the heat-shock response in soybean.

PROPERTIES OF HEAT-SHOCK cDNA CLONES

Several heat-shock cDNA clones have been characterized in some detail; Table 1 summarizes some properties of 15 of the heat-shock clones. These data are suggestive of several families of highly conserved heat-shock sequences (hs domains) for the 15–27-kD heat-shock proteins of plants. Of special note are the group-I clones that hybrid select mRNAs which translate into 13 different 15–18-kD proteins (see also Schöffl and Key 1982). To date we have not isolated cDNA clones to the 68–70, 84, and 92-kD heat-shock proteins of the soybean system.

PROTECTIVE INFLUENCE OF THE HEAT-SHOCK SYSTEM

Only limited progress has been realized in gaining an understanding of the physiological significance of the heat-shock response. There is evidence in some systems, however, that intermediate heat-shock temperatures provide protection to higher otherwise lethal temperatures (e.g., Loomis and Wheeler 1980). In the soybean seedling, the incorporation of externally supplied [³H]leucine into protein occurs at a high level as the incubation temperature is increased up to about 40°C (Fig. 1); above 40–41°C (the break-point temperature), there is a precipitous drop in [³H]leucine incorporation into protein. If, however, the seedlings are preincubated for about 2 hours at 40°C prior to assessing [³H]leucine incorporation at temperatures above the break point, the level of incorporation is much higher at these elevated temperatures with substantial [³H]leucine incorporation persisting out to 47.5°C (Fig. 1). This protective influence of (or acclimation

Table 1
Classification of Soybean Heat-shock cDNA Clones

Class	Clone	Insert (kb)	RNA (kb)	Protein (kD) one-dimensional	two-dimensional
I	1920	.37	.8−.9	15−18	13 proteins
	2059	.5	.8−.9	15−18	13 proteins
	1968	.43	.8−.9	15−18	13 proteins
	2005	.4	.8−.9	15−18	13 proteins
II	2019	.4	.8−.9	18	one protein
	2026	.4	.8−.9	18	one protein
III					
A	2004	.5	.8−.9; HMW	ND[a]	ND[a]
B	1991	.5	.8−.9	ND[a]	ND[a]
IV	2033	.5	.8−.9	21−23	ND
V	2036	.14	4; LMW	ND[a]	ND[a]
	2014	.32	4; LMW	ND[a]	ND[a]
VI	0075	ND	.8−.9	15	ND
VII	0053	ND	.8−.9	15	ND
VIII	0054	ND	ND	28	four proteins
	0055	ND	ND	28	four proteins

[a]No translatable RNA's hybrid selected by the cloned inserts; the clones of group V represent sequences homologous with rRNA.

to) a 40°C heat shock persists even if the seedlings are incubated for 4 hours at 28°C prior to the shift to 45°C or 47.5°C (J. L. Key, unpubl.).

The pattern of heat-shock proteins synthesized at 45°C following a 40°C heat-shock period is similar to the normal 40°C pattern. The data presented in Figure 2 show that the ability to accumulate heat-shock-specific mRNAs at the elevated (45−47.5°C) heat-shock temperatures is one response of the soybean seedling to the protective action of the 40°C heat shock. Thus, these data show that some event(s) or response (e.g., possibly synthesis and functioning of heat-shock proteins) which occurs at 40°C protects the transcription/translation systems resulting in synthesis of heat-shock mRNAs and heat-shock proteins at otherwise nonpermissive temperatures.

LOCALIZATION OF HEAT-SHOCK PROTEINS

As noted above and reported previously (Key et al. 1981), soybean seedlings rapidly resume normal protein synthesis when shifted from 40°C heat-shock conditions to 28°C; there is no detectable synthesis of heat-shock proteins after 3−4 hours at the normal temperature. However, those heat-shock proteins ([³H]leucine labeled) that were synthesized during the 40°C

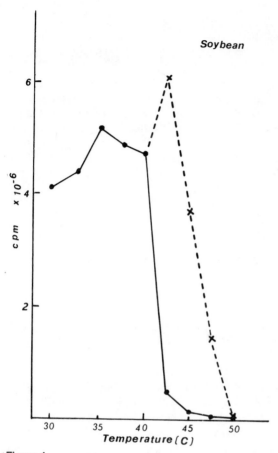

Figure 1
The influence of temperature on amino acid incorporation in soybean seedlings. Two-day germinated seedlings 2–2.5 cm in length were incubated in shake culture in a buffer-sucrose medium (6 ml) containing 50 μg/ml chloramphenicol. After 1 hour at the indicated temperature (●), 50 μCi of [³H]leucine were added to each flask and incubation was continued for 2 hours. Samples incubated similarly at 40°C for 2 hours were subsequently incubated for 2 hours at the temperatures as noted (x) in buffer medium containing 50 μCi [³H]leucine. After the 2-hour labeling periods, the seedlings were rinsed and total protein in the embryonic axis extracted by homogenization in an SDS buffer. Data are reported as protein cpm/5 seedlings.

heat shock are very stable during a subsequent chase in nonradioactive leucine, independent of whether the chase is accomplished at 28°C or 40°C; some 80% of the label is retained in the heat-shock proteins during a 20-hour chase (C.Y. Lin, Y.M. Chen, and J.L. Key, unpubl.). There is, however, a marked shift in cellular distribution of some of the heat-shock proteins, depending upon whether the chase was accomplished at 28°C or

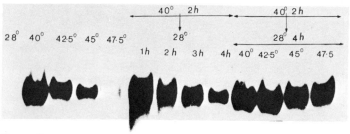

Figure 2
Northern blot analysis of heat-shock-specific mRNA production under different temperature regimes. The five left lanes display the relative level of accumulated poly(A) RNA which hybridizes to p53 (class-VII clone of Table 1) after seedlings are incubated for 2 hours at the indicated temperature. Note the low level of heat-shock-specific mRNA accumulation at 45°C and 47.5°C. Seedlings incubated for 2 hours at 40°C were shifted to 28°C and incubated for 1, 2, 3, or 4 hours prior to RNA extraction (lanes 6–9); note the typical decline in heat-shock mRNA during 28°C incubation (Schöffl and Key 1982). Additional samples were incubated for 2 hours at 40°C followed by 4 hours at 28°C; these samples (lanes 10–13) were then shifted to the indicated temperature for 2 hours before RNA extraction. Note the significant accumulation of heat-shock-specific poly(A) RNA at 45°C and 47.5°C compared with the initial heat shock at those temperatures (lanes 3, 4). One μg of poly(A) RNA was added to each lane.

40°C. Localization of heat-shock proteins in nuclei of *Drosophila* cells has been reported (e.g., Arrigo et al. 1980).

The SDS gel presented in Figure 3 shows the profile of heat-shock proteins associated with Percoll gradient-purified nuclei (lanes B–E) compared with a display of total ^{3}H-labeled proteins of 41°C (lane F) and 28°C (lane G) tissue. An intense double band of the 15–18-kD heat-shock proteins represents a high proportion of ^{3}H-labeled protein associated with the nuclear fraction at the end of the 3-hour heat-shock treatment (lane B); there are lesser amounts of the 70-kD and 92-kD heat-shock protein bands in the nuclear fraction along with trace levels of other protein bands. Following a 4-hour chase at 28°C, the 15–18-kD heat-shock proteins are reduced in the nuclear fraction to barely detectable levels (lane C); lesser amounts of the other bands chase during the 4-hour period at 28°C. The 15–18-kD heat-shock proteins remain associated with the nuclei during a chase at 40°C. Significantly, when the seedlings which were chased for 4 hours at 28°C are returned to 41°C, the 15–18-kD heat-shock proteins again become associated with the nuclear fraction (lanes D and E). It is not possible to compare directly the label intensity of lane B with lanes D and E because of dilution of the ^{3}H-labeled heat-shock protein's specific activity resulting from the synthesis of nonradioactive heat-shock proteins during early hours of the chase and subsequent heat-shock period. However,

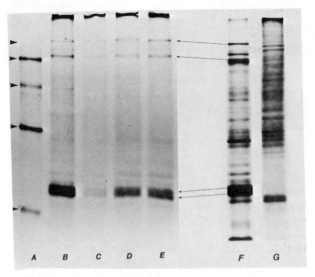

Figure 3

SDS gel pattern of heat-shock proteins associated with isolated nuclei. Seedlings were incubated at 41°C for 3 hours and 500 μCi of [³H]leucine were added for 2 hours prior to isolating nuclei (lane B). Similarly, treated seedlings were incubated for an additional 4 hours at 28°C in 1 mM nonradioactive leucine prior to nuclei isolation (lane C). Seedlings treated as in C were subjected to an additional 30 minutes (lane D) or 2 hours (lanes E) heat shock at 41°C before isolating the nuclei. Total 41°C (lane F) and 28°C (lane G) ³H-labeled proteins run on a separate gel are displayed for comparison with the nuclear-associated proteins. Arrowheads mark the molecular-weight standards (lane A), which are from top to bottom: 92.5 kD, 69 kD, 46 kD, 32 kD, and 12.3 kD.

significant levels of the 15–18-kD heat-shock proteins do become nuclear-associated during the subsequent 30-minute (lane D) or 2-hour (lane E) incubation at 41°C.

SUMMARY

Attention is focused in this report on preliminary studies designed to assess the physiological role(s)/significance of the heat-shock system in plants. The data certainly support the view that heat-shock at or near the break point temperature (40–41°C for soybean) provides some sort of protection that permits the plant to function at otherwise nonpermissive temperatures. Presumably the accumulation of heat-shock proteins in a time- and temperature-dependent manner (about 2 hr at 40°C is required to provide maximum protection in the soybean for heat-shock mRNA and heat-shock protein synthesis at 45–47.5°C) and selective localization within the various subcellular fractions in some way provides the protection or acclimation to high-temperature heat shock. While subcellular fractions that have been purified on Percoll (nuclei) or sucrose (mitochondria and ribosomes) gradients have identifying features relative to the localized heat-shock proteins, there is a considerable overlap especially as relates to the complex group of 15−18-kD heat-shock proteins (C.Y. Lin and J.L. Key, unpubl.). Some of the heat-shock proteins (e.g., the 27 kD, 68 kD, and 84 kD) seem not to localize significantly if at all in the organelle fractions and to concentrate preferentially in the 140,000g supernatant fraction.

While more studies are needed to provide definitive insights, it appears that the association during heat shock of some heat-shock proteins with the organelle fractions provides the basis of acclimation, or thermal protection, which permits, for example, nuclei to synthesize heat-shock mRNAs and ribosomes to synthesize heat-shock proteins at otherwise nonpermissive temperatures. This protection of the transcription apparatus permitting heat-shock mRNA synthesis at the very high temperature and normal mRNA synthesis to resume upon return to 28°C could be afforded by the heat-shock proteins becoming "chromatin" proteins or possibly a part of the matrix structure; both suggestions have been offered from localization studies in the *Drosophila* system. Any functional mechanism of "protection" must accommodate the fact that a similar set of heat-shock proteins (primarily the 15−18-kD group in soybean) associates in some way with the different organelle fractions during heat-shock and affords "protection" for very different processes (e.g., transcription and translation). It may be, however, that a much lower level of the heat-shock proteins more or less specific to each fraction relates to the functional differences. While it seems unlikely, the possibility must be considered also that heat-shock protein synthesis and selective localization are not the essential responses elicited by heat shock that permit tolerance to the otherwise nonpermissive temperatures.

ACKNOWLEDGMENTS

This research was supported by DOE contract DE-AS09-80ER10678 to JLK and a DFG fellowship to FS. The excellent assistance of C. Mothershed and Y.M. Chen is appreciated.

REFERENCES

Arrigo, A.P., S. Fakan, and A. Tissières. 1980. Localization of the heat shock-induced proteins in *Drosophila melanogaster* tissue culture cells. *Dev. Biol.* **78:** 86−103.

Ashburner, M. and J. Bonner. 1979. The induction of gene activity in *Drosophila* by heat shock. *Cell* **17:** 241−254.

Barnett, T., M. Altschuler, C.N. McDaniel, and J.P. Mascarenhas. 1980. Heat shock induced proteins in plant cells. *Dev. Genet.* **1:** 331–340.

Key, J.L., C.Y. Lin, and Y.M. Chen. 1981. Heat shock proteins of higher plants. *Proc. Natl. Acad. Sci.* **78:** 3526–3530.

Loomis, W.F. and S. Wheeler. 1980. Heat shock response of *Dictyostelium. Dev. Biol.* **79:** 399–408.

Schöffl, F. and J.L. Key. 1982. An analysis of mRNAs for a group of heat shock proteins of soybean using cloned cDNAs. *J. Mol. Appl. Gen.* (in press).

Heat-shock Induction of Abnormal Morphogenesis in *Drosophila*

**Herschel K. Mitchell
and Nancy S. Petersen**
*California Institute of Technology
Division of Biology
Pasadena, California 91125*

The theme for this presentation was established long ago, particularly by the intriguing observation of Hans Gloor (1947) that a heat shock (or ether) applied to very early embryos of *Drosophila* yielded adult flies with four wings instead of two (a bithorax phenocopy). The four-winged character was not inherited and there was no mutation involved; there was only a diversion, at a critical moment, in the normal programs of expression of normal genes. Such diversions are easily created by heat shock or other environmental influences, and abnormalities are common if the time is right. Clearly, the temporal coordination of developmental events and the maintenance of molecular activities in a normal sequence are essential to the production of normal adult organisms. The fragility of the temporal control mechanism is indicated by the fact that heat shock and a great variety of common chemicals can induce phenocopies in *Drosophila* and birth defects in mammals. The heat-shock effects involving changes of gene expression, production of certain proteins, and repression of production of others are all somehow involved in temporal control. The work described here represents a brief summary of our efforts to make use of *Drosophila* to study both heat-shock effects on a higher eukaryote and to find a system for evaluation of regulatory mechanisms in the sequential program.

337

RESULTS AND DISCUSSION

We have for several years explored the use of phenocopies as temporal markers for gene expression and as subjects for extended studies on the regulation of morphogenesis. For the most part, we have confined our attention to pupal stages of the life cycle of Drosophila as opposed to embryonic stages, since morphogenesis without cell division is common in pupal stages. Furthermore, normal metamorphosis provides synchronous activities in cell groups large enough for studies in molecular biology. Some examples are given here and described in more detail elsewhere (Mitchell and Petersen 1982).

Phenocopies

In Figure 1, we show 12 abnormal phenotypes in adult flies that result from heat shocks of Drosophila pupae at specific sensitive periods. In all of these cases, the specific abnormalities can be induced to at least the 90% level among the animals treated. In each of these cases, the abnormality is induced within a specific sensitive period of less than 4 hours. Of the examples given (Fig. 1), the four in the left column all involve scutellar bristles and all four have similarities to certain mutants. These and the specific sensitive periods in hours from puparium formation are as follows from top to bottom: hook, 36; javelin, 41; forked or singed, 34; hook (sternopleural), 44. In the second column, hooklike bristles and no bristles between the eye facets do not have mutant counterparts nor does the third phenocopy which shows knobs at the tips of the bristles on the anterior edge of the wing. Sensitive periods for these three are 44, 28, and 22 hours, respectively. The bottom picture in the center row shows a phenocopy (22 hr) similar to the mutant crossveinless. This particular picture shows a clumping of hairs in part of the area where a crossvein should be. In the last column on the right, two hair phenocopies are shown. The first (38 hr) resembles the mutant multiple wing hair in some respects (Mitchell and Petersen 1981) while the second (45 hr) is similar to hairs present in the mutants forked and singed. The third phenocopy (81 hr) resembles the mutant curly and others that have turned up wings, while the last example (72 hr) is a double phenocopy with similarities to the mutants taxi and blistered. In addition to these examples, we have presented elsewhere (Mitchell and Petersen 1982) a temporal map that includes 34 phenocopies which are induced by heat shock, each at its given sensitive period within the range of 20 to 92 hours.

Clearly these phenocopies are useful in a variety of ways. They may be useful in the present context—that of evaluating functions of the heat-shock proteins. In all cases examined so far, the phenomenon of protection against a lethal heat shock by a lower temperature pretreatment

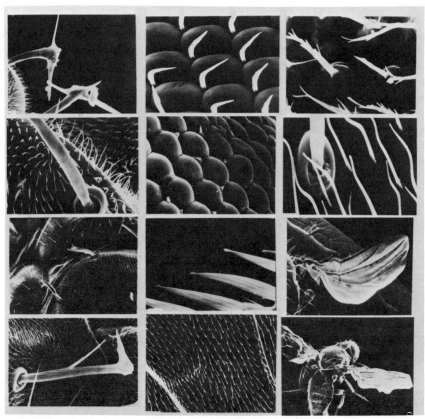

Figure 1
Examples of phenocopies resulting from heat shock. Each character appears in at least 90% of the animals treated and each has its own sensitive period(s) in pupal stages. Induction conditions in these examples were 40.8°C (in a water bath) for 35 minutes. The photographs are from the scanning electron microsope.

also protects against phenocopy induction. Conditions that prevent phenocopies are those which induce heat-shock protein synthesis without inhibiting ongoing cellular protein synthesis. The pretreatment must also be within the sensitive period for phenocopy induction. Heat-shock proteins may be involved in the prevention of phenocopies, probably via effects on regulation of gene expression.

Multihair Phenocopy

Of the various phenocopies we have described, the one most suitable for detailed investigations of heat shock and its specific effects is the one in the upper right in Figure 1, that by our nomenclature is called *multihair* (MtH). As we have noted earlier (Mitchell and Petersen 1981), epithelial cells of *Drosophila* differentiate to produce hairs on many areas on the fly and they do so after cell division has been completed. The largest area is that on the wing where there are about 30,000 cells (including both cell layers), each of which produces a single hair. These cells produce hairs with remarkable synchrony as we have observed by light microscopy and thin-section electron microscopy, as well as by sensitivity to phenocopy induction. In the last situation, a heat shock (40.8°C) of 35 minutes duration at 38 hours resulted in abnormal hairs in virtually all of the distal 90% of the wing cells. That is we found less than four normal hairs among the approximately 13,000 on one wing surface. There were not more than this of abnormal hairs in the proximal cells (about 2000) of the wing. Obviously this kind of synchrony is unusual (for in vivo) systems and highly advantageous for studies of heat shock effects at the molecular level (Petersen and Mitchell 1981).

As mentioned earlier, cell hairs appear on many areas of the adult fly but as shown in Figure 2, the sensitive periods for induction of the multihair phenocopy follow a general anterior-to-posterior gradient. There is a pattern of protein synthesis that follows the same time relation as phenocopy induction in wings versus dorsal thorax (Mitchell

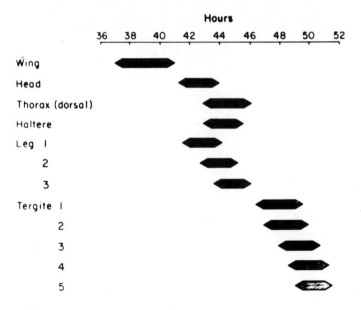

Figure 2
Sensitive periods for induction of the multihair phenocopy on different areas of *Drosophila*. The synchrony of induction is high within each area.

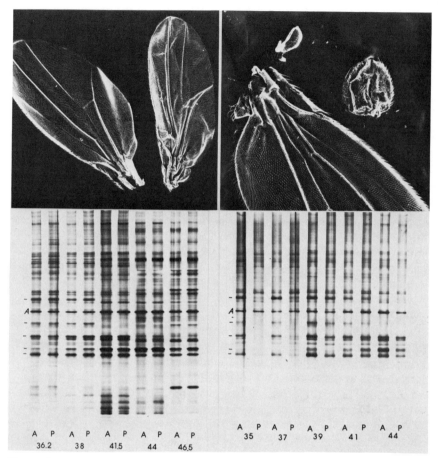

Figure 3
Protein synthesis patterns in two different *bithorax* mutant combinations. Pictures at the top show the relative wing sizes for the anterior (A) and posterior (P) wings. A normal haltere is included in the picture at the right. Corresponding patterns of protein synthesis (autoradiographs of acrylamide gels from [³⁵S]methionine label) are shown below. The five markers at the left of each autoradiograph series indicate positions of the five most abundant proteins that appear just prior to the sensitive period for the multihair phenocopy.

and Petersen 1981). That is, phenocopy induction comes 4 hours later on the dorsal thorax than it does on wings (Fig. 2), and a pattern of production of the same five abundant proteins also comes 4 hours later on the dorsal thorax. The same relation is true for the haltere (see also Fig. 3) and from limited data it appears to be true also for the tergites on the abdomen. A major conclusion then from these observations is that a heat shock on a whole animal, or any nonhomogeneous system for that matter, depends on the state of the cell at the time of heat shock. It is of course of interest to know if the apparent gradient in the whole animal is based on humoral effects or predetermination.

Bithorax Mutants

We have shown by means of the multihair phenocopy that epithelial cells in different areas on the fly do not react to heat shock in the same way at the same time. They do react in a similar fashion on their own schedules and this takes roughly the form of an anterior-to-posterior gradient. In order to distinguish between a heat-shock reaction due to position of a cell group in the animal and one due to predetermination, we have examined the situation in two types of bithorax mutants. These were generously supplied by Dr. E. B. Lewis. One of these (abx bx^3 pbx/Df(3)Ubx109) yields adult animals with four full wings (Fig. 3). The posterior wings and thorax are derived from the haltere discs rather than wing. The second mutant combination (abx bx^3 pbx/Ubx^{61d}) yields adult flies with a normal anterior wing and a very small winglike structure in the place of the haltere. This is shown at the top right in Figure 3 along with the anterior wing and a normal haltere to show the relative sizes. With the first mutant, a heat shock at 38 hours (the normal sensitive period for production of abnormal wing hairs) causes both anterior and posterior wings to give the multihair phenocopy. Furthermore, both wing pairs show the same changing patterns of protein synthesis (Fig. 3, lower left). In contrast to these results, the mutant complex that gives small wings reacts to a heat shock at 38 hours to give all abnormal hairs on the normal anterior wing but only small patches on the small wings. The situation is reversed at 42 hours (the sensitive period for induction of abnormal hairs on halteres, Fig. 2). At this time, the treatment results in normal hairs on the anterior wing but the small wing has mostly abnormal hairs with small patches of normal hairs. These results are confirmed completely by the protein synthesis patterns in anterior (large) wings and posterior (small) wings as they change with time. As shown in Figure 3 (lower right), the five protein bands that are associated with hair differentiation (Fig. 3, lower left, and Mitchell and Petersen 1981) appear in the small wing samples on haltere time and not on wing time. Since these small wings have less than 5% of their cells that react on wing time, this is quite a reasonable result. We conclude from these results that the gradients observed with respect to sensitivity to heat shock and to protein synthesis patterns are the result of predetermination rather than position in the animal. Thus, each cell must have its own temporal program but groups of cells in given areas can be in synchrony with respect to sensitive periods and the predetermined programs of gene expression that are reflected directly in the changing patterns of protein synthesis.

SUMMARY

Heat-shock proteins with their ubiquitous occurrence and apparent conservation in structure among species are most interesting objects for investigation. At the same time, one should not lose sight of the strong evidence that each of the heat-shock genes has its own place in normal expression (without a heat shock) in some tissue at some time. What is in fact most unique in the system is the simultaneous induction and repression of the heat-shock genes and other genes, respectively. If there is something special in common among the heat-shock genes then there must be something special in common among the genes that are turned off at elevated temperatures. However, even the heat-shock genes become repressed when the temperature is high enough to induce phenocopies. From this vantage point, apparent special categories of genes are less meaningful from the standpoint of regulatory mechanisms than from that of functions of the gene products. It seems most likely that each of these products has its own function and the only thing coordinant about the system is the reaction to heat shock.

In any case, regardless of mechanisms of shutdown of gene expression, the phenocopies studied so far are produced only under conditions that repress virtually all transcription and translation. The system yields phenocopies only when the recovery following a heat shock is slow. They do not appear when the recovery is relatively fast, as it is under conditions that yield thermotolerance. Thus, it still seems most likely that the critical matter in determining whether a differentiating cell is to become abnormal or normal rests entirely on temporal coordination of recovery from an arrested state. Specific responses depend on predetermined programs of gene expression in progress at the time of treatment.

With respect to prospects for the future, the events of differentiating wing tissue provide the basis for a variety of investigations. One direction in particular rests on the hypothesis that programs of sequential differentiation may depend on recycling of gene products or portions thereof through the nucleus where a given component can act as an inducer of the next gene in one sequence of a program. It is a fact that proteins at least do recycle and could easily provide a great variety of information in passage (Mitchell and Lipps 1975). We are thinking especially about the various situations in wing differentiation where sequential events in hair construction can be related to specific gene products (H.K. Mitchell, J. Roach, and N.S. Petersen, in prep.).

ACKNOWLEDGMENTS

This work was supported in part by grants from the Public Health Service Numbers GM 25966A and AG 01931A and the Biomedical Research Support Grant.

REFERENCES

Gloor, H. 1947. Versuche mit Aether an Drosophila. *Rev. Suisse Zool.* **54:** 637–712.

Mitchell, H.K. and L.S. Lipps. 1975. Rapidly labeled proteins of the salivary gland chromosomes of *Drosophila melanogaster. Biochem. Genet.* **15:** 585–602.

Mitchell, H.K. and N.S. Petersen. 1981. Rapid changes in gene expression in differentiating tissues of *Drosophila. Dev. Biol.* **85:** 233–242.

————. 1982. Developmental abnormalities in *Drosophila* induced by heat shock. *Dev. Genet.* (in press).

Petersen, N.S. and H.K. Mitchell. 1981. Recovery of protein synthesis after heat shock: Prior heat treatment affects the ability of cells to translate in RNA. *Proc. Natl. Acad. Sci.* **78:** 1708–1711.

Effects of Heat Shock on Gene Expression during Development: Induction and Prevention of the Multihair Phenocopy in *Drosophila*

**Nancy S. Petersen
and Herschel K. Mitchell**
*Division of Biology
California Institute of Technology
Pasadena, California 91125*

The induction of developmental defects by heat can be used to study regulation of gene expression during differentiation and to look at the effects of heat-shock gene expression on specific developmental processes. Heating can cause developmental abnormalities in a wide variety of organisms including mammals. In *Drosophila,* the defects are called phenocopies because of their resemblance to mutant defects. They are most likely the result of interruption of gene expression at crucial periods in development when there are rapid changes in gene expression (Mitchell and Lipps 1978; Mitchell and Petersen 1981). In agreement with this hypothesis, a heat shock that induces phenocopies stops RNA and protein synthesis; also the sensitive period for induction of a particular phenocopy is short, usually less than 2 hours. A mild heat treatment (30 min at 35°C) during the sensitive period will prevent phenocopy induction by a subsequent high-temperature shock in each case we have looked at so far. These include the blond phenocopy, spread wing and blister phenocopies, and the hook phenocopy, as well as the multihair phenocopy on wings. The conditions that best prevent phenocopies are those which maximally induce heat-shock protein synthesis without inhibiting ongoing protein synthesis. The phenocopies that can be prevented involve a wide range of biochemical processes from enzyme activation to structural changes. What they have in common may involve regulation of gene expression. By looking in detail at the way specific phenocopies are induced and prevented we hope to discover information about how gene expression is regulated during development and about the effects of heat-shock proteins on gene expression. In this paper, we will discuss the induction and prevention of the multihair phenocopy on *Drosophila* wings and the effects of heat shock on RNA and protein synthesis during hair development.

NORMAL WING HAIR DEVELOPMENT

Drosophila wings develop from the imaginal wing disc. This is a group of cells set aside in the embryo that undergoes morphological differentiation to produce wing tissue and part of the thorax during the pupal period. Wings are particularly convenient for studying cell differentiation because the developing wing consists of a double layer of cells which differentiate synchronously. Almost all of the cells are the same type; each cell produces one hair, lays down cuticle, and undergoes shape changes which are probably related to these processes. During the time when there are extensive morphological changes, there are also rapid changes in the proteins being synthesized. Figure 1 shows wing cells at two different stages in development. The first picture shows hairs immediately after they are formed lying on the surface of the cells. This is the stage at which the multihair phenocopy can be induced. The next process in hair formation is to lay down a coating called cuticulin around the outside of the hair. The hair is then moved into an upright position as shown in Figure 1B and fibers are deposited on the inside (H.K. Mitchell et al., in prep.). The proteins that are being synthesized in these cells at the different stages of development are shown in Figure 1C. The rapid changes in the major proteins being made are evident. These changes reflect changes in mRNA concentration (Mitchell and Petersen 1981). We are looking for direct connections between gene expression and visible cellular differentiation. It can be seen that there are, for example, several proteins that are made at the right time to be components of cuticulin. We have identified one protein that is probably involved in the extensive cell movement which takes place between 42–48 hours and is associated with the movement of the cell hair into an upright position. This protein is actin and there is a peak of actin synthesis during the time when cell movement occurs. Fortunately, actin is a gene that has been cloned (Fyrberg et al. 1980). In collaboration with Beverley Bond and Norman Davidson (in prep.), we have been able to show that the changes in actin synthesis reflect changes in the concentration of actin message and that three of the six actin genes in *Drosophila* are expressed in 46–48-hour wings. The details of this connection have yet to be worked out; however, the cell movement at this time in development is probably important in the induction of phenocopies discussed below.

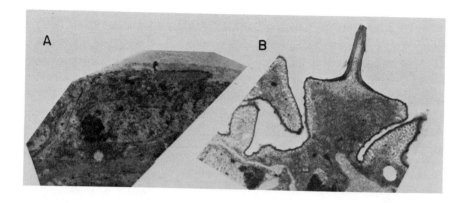

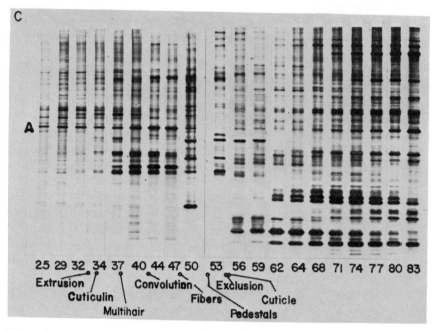

Figure 1

Protein synthesis and developmental changes during wing differentiation. (*A*) An electron micrograph section showing one wing cell at 37 hours after puparium formation. The cell is rectangular in shape and a newly formed hair is lying on the surface. (*B*) An electron micrograph picture of a 53-hour wing cell. At this point the hair is upright in the center of the cell and it stands in the center of a pedestal. Note that the hair has acquired an internal structure of fibers which continue to form the top of the pedestal. Also cytoplasm organelles are excluded from an area just under the hair. Cuticle deposition has started at the convoluted cell surface as indicated by the numerous microvilli. (*C*) Protein synthesis in pupal wings of different ages. For each lane on the SDS gel, three wing pairs of the age indicated were dissected and labeled for 25 minutes with [^{35}S]methionine (25−50 hr) or [^{3}H]leucine (53−83 hr). Samples were dissolved in SDS sample buffer at 100°C and layered directly on 7−20% linear gradient acrylamide gels. The position of actin is indicated as A.

MULTIHAIR PHENOCOPY INDUCTION AND PREVENTION

When pupae are heated to 40.8°C for 35 minutes at 38 hours after puparium formation, the flies which emerge have branched hairs on the wings. Figure 2A is a scanning electron micrograph picture of normal wing hairs; Figure 2B shows the multihair phenocopy, and Figure 2C shows wing hairs that have been heated 30 minutes at 35°C before the 40.8°C treatment. The 35°C treatment allows the hairs to develop in a nearly normal fashion in spite of the 40.8°C treatment. The phenocopy is induced after wing hairs have already been formed and are lying on the surface of the wing. The light microscope pictures in Figure 2D show that wing hairs immediately after the heat shock are not branched. Only 15–20 hours after the heat shock does hair branching occur. Figure 2E shows hair branching occurring. It seems that cytoplasm is pushed into the hair as the hair is pushed into its upright position. Figure 2F is an electron micrograph picture of a hair caught in the process of branching. The cytoplasm is pushing out in places where the cuticulin has not been completed. Our hypothesis is that the multihair phenocopy results from cell movement to right the hair before completion of the cuticulin. The relationship of these events to changes in gene expression induced by heat shock is shown in Figure 3. The most obvious effect of 35°C pretreatment on protein synthesis is to allow much more rapid recovery after the 40.8°C shock. The recovery of protein synthesis would allow the cuticulin to be completed before the cell movement necessary to push the hair upright. Since we have not yet identified the proteins involved in cuticulin synthesis, nor have we shown that actin synthesis is related to cell movement, we cannot prove this hypothesis at present. However, it reinforces our earlier conclusion (Petersen and Mitchell 1981) that effects of the 35°C pretreatment on recovery of protein synthesis whether direct or indirect are important to cells.

Another very interesting observation that may also be related to the timing of cellular events and phenocopy is the effect of a pretreatment on the recovery of the program of mRNA synthesis and decay. Heat shock stabilizes messages that would normally decay as part of the developmental program. Figure 3, C and D, shows translation products of wing mRNA during recovery from heat shock. It can be seen that the normal messages for 38-hour wings are present for the entire recovery period of 15–25 hours even though some of these messages (for example, those encoding translation products a and c) would have decayed within 4 hours during normal development (Mitchell and Petersen 1981). The resumption of mRNA decay is coincident with the synthesis of new mRNAs in the developmental program. The decay of specific messages and the appearance of new messages in the developmental program (as indicated by translation products b and d) occurs

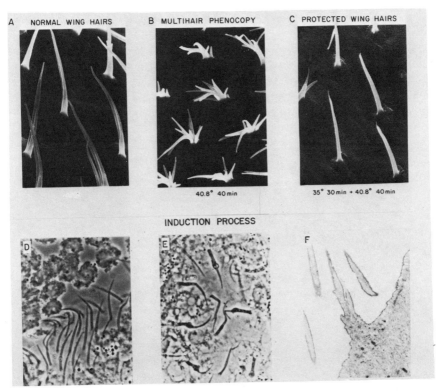

Figure 2
Induction of the multihair phenocopy. (*A*) Normal wing hairs: scanning electron micrograph picture of adult wing hairs made by pupae that were heat-shocked before the sensitive period for phenocopy induction (34 hr, 40 min at 40.8°C). (*B*) Multihair phenocopy: scanning electron micrograph picture of wing hairs resulting from a 40-minute heat shock of 38-hour pupae at 40.8°C. (*C*) Protected wing hairs: scanning electron micrograph picture of hairs made when the 40.8°C, 40-minute heat shock as in (*B*) is preceded by a 30-minute shock at 35°C. (*D*) Wing hairs following heat shock: phase-contrast picture of wing cells with hairs taken 10 hours after a 35-minute heat shock at 40.8°C shows unbranched hairs. (*E*) Hair bulbs and branching: phase-contrast picture taken 17 hours after the heat shock shows bulbs of cytoplasm inside hairs and some hair branching. (*F*) Hair branching: electron microscope picture of a hair caught in the process of branching. Cytoplasm appears to be pushing through breaks in the cuticulin causing the branching.

sooner in wings which received a 35°C pretreatment. Heat-shock messages decay before the resumption of the program of synthesis and decay of normal messages. This indicates that there are very specific controls for the synthesis and decay of messages during development and that these two processes are coordinated in some way.

SUMMARY

Heating at 40–41°C turns off RNA and protein synthesis and induces stage-specific developmental defects in *Drosophila* pupae. These defects can be prevented by a lower temperature (35°C) pretreatment which induces heat-shock gene expression but does not shut off normal protein synthesis. We are making use of this system to study regulation

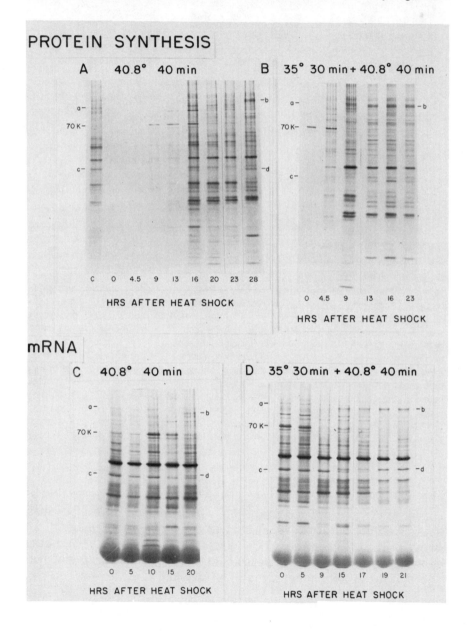

of gene expression during development and the effects of heat-shock protein synthesis on differentiation. Initially we have chosen to look at the multiple hair phenocopy on wings. Biochemical analysis is simplified by the facts that wing tissue is almost entirely one cell type and the cells differentiate synchronously. Either mutation or heat shock can alter the normal course of hair formation to produce multiple hairs or branched hairs. The multihair phenocopy is induced by heating during the process of hair formation at 38 hours after puparium formation. During this time there is a rapidly changing program of protein synthesis. The changes in protein synthesis reflect in changes in the concentrations of major mRNAs. A heat shock sufficient to induce the multihair defect shuts off both RNA and protein synthesis and stabilizes messages that would normally decay as part of the developmental program. The resumption of development following heat shock involves resumption of message decay as well as new message synthesis.

The multihair phenocopy seems to be the result of the noncoordinate recovery of two processes involved in hair construction: the synthesis of cuticulin on the outside of the hair and the cell movement involved in moving the hair from a supine to an upright position. When the cell is pushed into its upright position before the completion of the cuticulin, hair branches are formed. The prevention of this phenocopy may be due to effects of the pretreatment on recovery of protein synthesis. In wings that receive a 35°C treatment before the 40.8°C shock, protein synthesis recovers much faster so that the cuticulin can be completed before the hair is moved upright. In order to determine whether this hypothesis is correct and to follow in detail the molecular events in hair morphogene-

Figure 3
Protein synthesis and mRNA translations following heat shock in 38-hour wings. (*A*, *B*) Each lane represents three wing pairs from 38-hour pupae (c) or 38-hour pupae which had been heated and allowed to recover at 25°C for the time indicated. Wings were dissected, labeled for 30 minutes with [^{35}S]methionine, dissolved in sample buffer, and run on 7–20% SDS acrylamide gels. (*A*) Recovery of protein synthesis following a 40-minute heat shock at 40.8°C. (*B*) Recovery of protein synthesis following a double heat shock, 30 minutes at 35°C followed by 40 minutes at 40.8°C. The position of the 70K heat-shock protein is indicated as well as the positions of two proteins normally made in 38-hour wings (a and c) and two proteins which are normally synthesized later than a and c (at 44–48 hr) in the developmental program (b and d). (*C*, *D*) In vitro translation of mRNA present in pupal wings recovering from heat shock. For each time indicated, RNA was extracted from wings from 10 animals and translated in vitro in the rabbit reticulocyte system. It can be seen that except for appearance of heat-shock gene products the translation products are essentially the same until 15–20 hours after the heat shock (40 min at 40.8°C) in *C*, and until between 9 and 15 hours after the double heat shock in *D*. In both cases, the disappearance of translation products a and c is coincident with the appearance of products b and d.

sis, it will be necessary to identify the gene products involved and follow their synthesis and incorporation into cellular structures.

ACKNOWLEDGMENTS

This work was supported in part by Public Health Service Grants GM 25966A and AGO1931A and the Caltech Biomedical Research Support Grant.

REFERENCES

Fyrberg, E.A., K.L. Kindle, N. Davidson, and A. Sodja. 1980. The actin genes of *Drosophila:* A disperse multigene family. *Cell* **19:** 365–378.

Mitchell, H.K. and L.S. Lipps. 1978. Heat shock and phenocopy induction in *Drosophila. Cell* **15:** 907–919.

Mitchell, H.K. and N.S. Petersen. 1981. Rapid changes in gene expression in differentiating tissues of *Drosophila. Dev. Biol.* **85:**233–242.

Petersen, N.S. and H.K. Mitchell. 1981. Recovery of protein synthesis following heat shock; pretreatment affects mRNA translation. *Proc. Natl. Acad. Sci.* **78:** 1708–1711.

The Physiological Role of Heat-shock Proteins in *Dictyostelium*

**William F. Loomis
and Steven A. Wheeler**
*Department of Biology
University of California, San Diego
La Jolla, California 92093*

The universal response of eukaryotic cells to a brief sublethal heat shock implies a conservation of the functions induced. Thus, analysis of the physiological roles of the heat-shock proteins in any organism is likely to have bearing on cellular processes in all higher organisms. Several years ago we showed that a heat shock at 30°C induced the synthesis of a small group of proteins in amoebae of *Dictyostelium* that were similar to the heat-shock proteins of *Drosophila* (Loomis and Wheeler 1980). Recently, Kelley and Schlesinger (1982) prepared anti-serum against a major heat-shock protein, hsp70, of chick cells and showed that the antibodies cross-reacted with hsp70 of yeast, *Dictyostelium*, *Drosophila*, and human cells. Another heat-shock-induced protein of *Dictyostelium*, hsp82, has been found to comigrate on two-dimensional electrophoresis with its analog in *Drosophila*. Moreover, just as in *Drosophila*, heat shock in *Dictyostelium* induces a group of proteins of about 30,000 daltons (low-molecular-weight proteins) which are preferentially localized in the nucleus (Loomis and Wheeler 1982). Thus, the proteins which are induced by a heat shock appear to have been conserved to a considerable extent during evolution.

Despite broad efforts to recognize discrete enzymatic or structural roles for the heat-shock proteins, no discrete functions have yet been assigned to the individual proteins. However, in yeast, *Dictyostelium*, and *Drosophila*, it has been shown that a brief heat shock results in protection of the cells from the lethal effects of higher temperatures. In

353

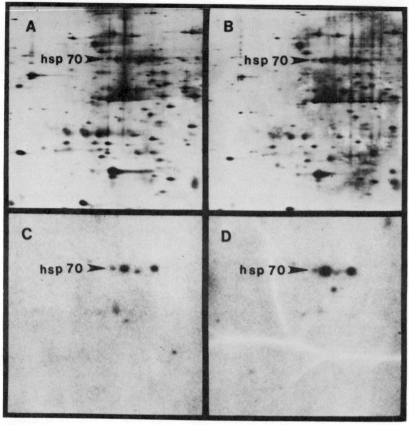

Figure 1

Heat-shock proteins. Extracts of *Dictyostelium discoideum* Ax3 were analyzed before (*A* and *C*) and 3 hours after (*B* and *D*) a heat shock at 30°C. Proteins were separated on two-dimensional gels as described by Loomis and Wheeler (1980) with the acidic end shown to the left. Proteins were silver-stained (*A* and *B*) using the technique of Morrissey (1981) or immunostained (Towbin et al. 1979) with antibodies prepared against hsp70 of chick fibroblasts (*C* and *D*). The anti-serum was described by Kelley and Schlesinger (1982) and was kindly provided by them.

Dictyostelium, we were able to show that this thermal protection requires de novo synthesis of the heat-shock proteins following a heat shock (Loomis and Wheeler 1980). Thus, analysis of the mechanism of thermal protection may indicate the physiological roles of some of the heat-shock proteins.

hsp 70

The most prevalent protein synthesized following a heat shock, hsp70, is a major protein in unheat-shocked cells, which can be recognized by silver staining of two-dimensional gels (Fig. 1). hsp70 separates into four spots on isoelectric focusing and consists of two related but distinct proteins of the same molecular weight but different isoelectric points; each protein is present in a phosphorylated and an unphosphorylated form (Loomis et al. 1982). Threonine residues of hsp70 were found to be in rapid equilibrium between phosphorylation and dephosphorylation. Since protection is maximal within 30 minutes at 30°C, it is not clear how a brief period of synthesis of an already abundant protein, hsp70, could effect thermal protection unless the hsp70 induced by heat shock were qualitatively different. However, analysis of the fragments produced by partial proteolysis of purified hsp70 synthesized before and after heat shock indicated no qualitative change in the protein or its phosphorylation pattern. Immunostaining of the heat-shock proteins separated on two-dimensional gels with the antiserum prepared against chicken hsp70 (kindly provided by Milton Schlesinger) revealed no newly reactive proteins (Fig. 1). It is possible that hsp70 is not directly involved in thermal protection and may simply play a highly conserved physiological role in normal cellular metabolism.

LOW-MOLECULAR-WEIGHT HEAT-SHOCK PROTEINS

On the other hand, synthesis of the low-molecular-weight heat-shock proteins cannot be observed before heat-shock induction and rapidly increases following that shock. Within 10 minutes, the low-molecular-weight heat-shock proteins enter the nucleus and become associated with the chromatin (Loomis and Wheeler 1982). Therefore they could account for the rapid increase in thermal protection. Upon returning the cells to growth conditions at 22°C, the low-molecular-weight heat-shock proteins exit the nucleus within 3 to 6 hours, at which time the cells have lost all thermal protection.

HEAT-SHOCK-IMPAIRED MUTANT

We reasoned that if we could isolate mutant strains which failed to become thermally resistant following a heat shock, an altered pattern of synthesis of heat-shock proteins in such strains might indicate which ones were essential for protection. Therefore 5000 clones of a mutagenized population were screened for protection from the lethal effects of 4 hours at 34°C. A strain, HL122, was found that grew well at 22°C and synthesized proteins at the normal rate at 30°C, but was seriously impaired in thermal protection elicited at 27°C, 30°C, or 32°C (Loomis and Wheeler 1982). Diploids selected between HL122 and X2, a strain wild-type for thermal protection, showed normal thermal protection, indicating that the mutation was recessive. Haploids induced to segregate

Table 1

Mapping of *hsi*A500

Phenotypes of haploid segregants of DL11[a]		Linkage group	Genotype of DL11 haploid parents	
Tsg	Tsg[+]		HL122	X2
HsiA	0	7	II	*axe*A1	*axe*A1
HsiA[+]	6	0	III	*axe*B1 *hsi*A500	*axe*B1 *tsg*A1
			IV		*bwn*A1

[a]Haploidization of the diploid DL11 was induced by benlate, and clones that formed small haploid spores were tested. One haploid carried both *hsi*A500 and *bwn*A1. The Hsi phenotype was scored as mutant if the half-life of cells at 34°C was less than 1 hour after a 30-minute heat shock at 30°C. Segregants carrying *tsg*A1 failed to grow at 27°C (Tsg phenotype).

from one such diploid, DL11, showed linkage of the mutation to *tsg*A1[+], indicating that strain HL122 carried the heat-shock-impaired mutation *hsi*A500 on linkage group III (Table 1).

The patterns of protein synthesis following heat shock in strain HL122 and two of the segregants carrying *hsi*A were compared with those in a wild-type strain on two-dimensional gels. Synthesis of hsp70 occurred at a reduced rate but the proteins were unaltered in the strains carrying *hsi*A. Strikingly, there were no observable low-molecular-weight heat-shock proteins in nuclei of the *hsi*A strains following a heat shock (Fig. 2). There are about eight of these proteins with molecular weights between 26,000 and 32,000 and none of them are made in the mutant strains (Loomis and Wheeler 1982). Thus, it is likely that the mutation has affected a component necessary for the common induction of all of them. The cosegregation of the impairment in thermal protection and loss of heat-shock induction of the low-molecular-weight heat-shock proteins implicates them in the mechanism of thermal protection. Their preferential localization in nuclei, tightly associated with the chromatin, suggests that they may protect it from thermal damage or limit transcription so as not to subject critical genes to error-prone transcription or translation at the lethal temperature.

Transcription of poly(A)[+] RNA is restricted to less than a dozen species following heat shock in wild-type cells (R. Firtel, S. Rosen, and W. Loomis, unpubl.). Most likely these are the products of the heat-shock genes. Since the *hsi*A mutation results in no observable synthesis of the low-molecular-weight heat-shock proteins and the residual synthesis of hsp70 in the mutant cells could be due to translation of mRNA made prior to heat-shock induction, it is conceivable that the *hsi*A mutation gives rise to thermolabile RNA polymerase. However, neither

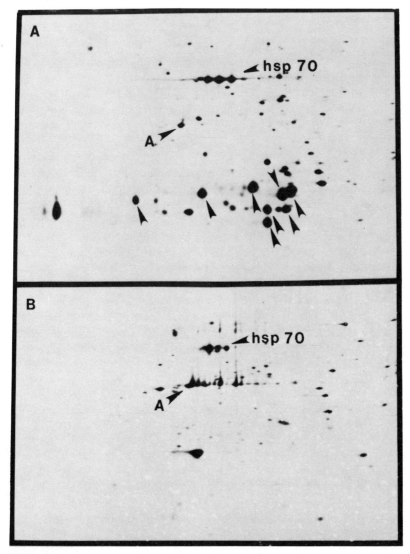

Figure 2
Synthesis of nuclear proteins in wild-type and mutant heat-shocked cells. 10^6 cells of wild-type strain Ax3 (*A*) and mutant strain HL122 (*B*) were labeled with 200 μCi [^{35}S]methionine 60 minutes after being shifted to 30°C. Nuclei were prepared and the extracts analyzed by autoradiography as described previously (Loomis and Wheeler 1982). Actin is indicated by A and the positions of the heat-shock proteins are indicated by arrows.

RNA polymerase I nor II activity was found to be thermolabile at 30°C in extracts of mutant or wild-type strains. Of course, factors not measured in the in vitro assay used (Pong and Loomis 1973) could be affected.

LOW-TEMPERATURE HEAT-SHOCK MUTANT

To strengthen the implication of the low-molecular-weight heat-shock proteins in thermal protection, we sought to isolate a strain which became protected at 22°C, a temperature at which wild-type cells are completely sensitive to the lethal effects of 34°C. If the low-molecular-weight heat-shock proteins are essential for thermal protection, they should be induced in such a mutant at 22°C. We have a strong selection for such mutants since wild-type populations are 99.9% killed after 4 hours at 34°C if induction of proteins is inhibited by cycloheximide at 22°C before shifting up. This selection was applied three times sequentially to a mutagenized population. Between each round of selection, the survivors were grown at 18°C in case induction of the heat-shock response resulted in stasis of the cells at 22°C. Following the third round of selection, the survivors were cloned. One strain, HL123, was found which reproducibly had 10-fold more viable cells than the wild-type after 4 hours at 34°C. However, the difference in apparent half-lives was only about 50%. This strain will be used in an attempt to select a double-mutant strain in which the low-temperature induction of thermal resistance is more pronounced. At present we can only conclude that low-temperature induction of the heat-shock response is not a highly mutable characteristic and may require mutations in multiple alleles in *Dictyostelium.*

SUMMARY

The studies to date have shown that the heat-shock proteins of *Dictyostelium* share several characteristics with those in other organisms. The observation that newly made proteins induced by heat shock are necessary for thermal protection and the finding that the low-molecular-weight heat-shock proteins are not synthesized in mutants carrying *hsi*A500 directs our attention to the physiological role of this family of chromatin-associated, heat-shock proteins in thermal protection. The surprising conservation of the heat-shock response and the thermal protection it affords suggests that this process has been selectively advantageous under a variety of conditions and may underlie the beneficial effects of fever and hyperthermic treatment in humans. Further analyses in *Dictyostelium* should benefit from the molecular analyses of cloned heat-shock genes that are now in progress.

ACKNOWLEDGMENTS

This work was supported by a grant from the NIH (GM23822).

REFERENCES

Kelley, P. and M. Schlesinger. 1982. Antibodies to two major chicken heat shock proteins cross-react with similar proteins in widely divergent species. *Mol. Cell Biol.* **2:** 267–274.

Loomis, W.F. and S.A. Wheeler. 1980. Heat shock response of *Dictyostelium*. *Dev. Biol.* **79:** 399–408.

————. 1982. Chromatin associated heat shock proteins of *Dictyostelium*. *Dev. Biol.* **90:** 412–418.

Loomis, W.F., S.A. Wheeler, and J. Schmidt. 1982. Phosphorylation of the major heat shock protein of *Dictyostelium discoideum*. *Mol. Cell Biol.* **2:** 484–489.

Morrissey, J. 1981. Silver stain for proteins in polyacrylamide gels: A modified procedure with enhanced uniform sensitivity. *Anal. Biochem.* **117:** 307–310.

Pong, S. and W.F. Loomis. 1973. Multiple nuclear RNA polymerases during development of *Dictyostelium discoideum*. *J. Biol. Chem.* **248:** 3933–3939.

Towbin, H., T. Staehelin, and J. Gordon. 1979. Electrophoretic transfer of proteins from polyacrylamide gels to nitrocellulose sheets. *Proc. Natl. Acad. Sci.* **76:** 4350–4354.

Physiologically Relevant Increases in Body Temperature Induce the Synthesis of a Heat-shock Protein in Mammalian Brain and Other Organs

Ian R. Brown, James W. Cosgrove, and Bruce D. Clark

Department of Zoology
Scarborough College
University of Toronto
West Hill, Ontario, Canada M1C 1A4

While many studies have examined the effect of elevation of ambient temperature on the induction of heat-shock proteins in a wide range of tissue culture systems, unicellular organisms, insects, and plants, comparatively few attempts have been made to examine the possible induction of heat-shock proteins within the intact body tissues of thermoregulating animals. Our studies suggest that an increase in body temperature of 3°C, which is similar to that attained during fever reactions, can induce the synthesis of a protein of molecular weight 74,000 in the brain, heart, and kidney of the rabbit.

Our laboratory is engaged in studies on the regulation of protein synthesis in the mammalian brain. The approach which we have adopted is to determine how translational processes in the brain are affected by the introduction of physiologically relevant treatments. We have demonstrated that the powerful psychotropic drug LSD is a useful tool with which to probe macromolecular synthesis in the mammalian brain (for recent review see Brown et al. 1982). This drug binds to neurotransmitter receptors in the brain, altering patterns of nerve firing and profoundly influencing sensory perception. LSD rapidly induces specific changes in protein synthesis in the rabbit brain, i.e., activation of a translational control mechanism which causes a global inhibition of translation in the brain. LSD also results in the induction of synthesis of a protein similar in molecular weight to one of the major heat-shock proteins reported in other systems. The induction of this 74K protein appears to be due to drug-induced hyperthermia which increases body temperature from 39.7°C to 42.5°C. The 74K protein can also be induced in various organs of the adult and fetal rabbit by hyperthermia generated by other means, i.e., elevation of ambient temperature.

EFFECT OF LSD ON PROTEIN SYNTHESIS IN THE RABBIT BRAIN

The intravenous injection of LSD to rabbits at 10−100 μg/kg body weight induces a transient inhibition of protein synthesis in all regions of the brain as reflected by a rapid disaggregation of polysomes to monosomes (Brown et al. 1982). This effect appears to be mediated by the binding of LSD to neurotransmitter receptors since the disaggregation of brain polysomes is inhibited by the prior injection of receptor blockers. Mild stress when applied in combination with low doses of LSD accentuates the disaggregation of brain polysomes such that a massive effect is observed even at 1 μg/kg. Administration of LSD to pregnant female rabbits induces a disaggregation of polysomes in the maternal brain and in fetal brain, kidney, and liver. Our analysis of the mechanism of the LSD-induced disaggregation of brain polysomes has ruled out RNase degradation of mRNA and premature termination as possible mechanisms. The evidence suggests that LSD induces a lesion that affects reinitiation of protein synthesis (Brown et al. 1982).

EFFECT OF HYPERTHERMIA ON BRAIN PROTEIN SYNTHESIS

LSD induces an elevation of body temperature in rabbits (and also in humans following drug overdoses). The degree of LSD-induced brain polysome disaggregation in rabbits is correlated with the extent of LSD-induced hyperthermia (Heikkila and Brown 1979a; Brown et al. 1982). Treatments that accentuate the drug-induced polysome shift (i.e., increasing drug dose or the application of mild stress) also accentuate LSD-induced hyperthermia. Conversely, pretreatment with neurotransmitter blocking agents that prevent LSD-induced polysome disaggregation also inhibit LSD-induced hyperthermia.

Given the correlation between LSD-induced hyperthermia and brain polysome disaggregation, it was of interest to determine the effect on brain protein synthesis of increasing body temperature by other means. Elevation of body temperature by 2−3°C by either placing animals at a higher ambient temperature or by generation of fever by injection of a bacterial pyrogen, induces a disaggregation of brain polysomes (Heikkila and Brown 1979b; Brown et al. 1982). These results suggest that protein synthesis in the rabbit brain is sensitive to physiologically relevant increases in body temperature that are induced by three different procedures, i.e., LSD, elevation of ambient temperature, and injection of a bacterial pyrogen.

EFFECT OF HYPERTHERMIA ON SUBSEQUENT CELL-FREE PROTEIN SYNTHESIS IN BRAIN

In order to facilitate a more detailed analysis of the translational steps that are affected by hyperthermia, an initiating cell-free translation system derived from brain was developed. Induction of hyperthermia in the intact animal results in a subsequent decrease in protein synthesis capacity in the cell-free system. Subfractionation of the cell-free system suggests that hyperthermia induces the appearance of an inhibitory factor in the postribosomal supernatant (Cosgrove et al. 1981).

INDUCTION OF A 74K PROTEIN FOLLOWING HYPERTHERMIA

To determine whether the inhibitory effect of hyperthermia on protein synthesis in the brain was general or selective, proteins labeled in vivo were analyzed by gel electrophoresis and fluorography (Freedman et al. 1981; Brown et al. 1982). The relative labeling of a 74K protein was specifically increased during the period of general reduction in incorporation of labeled amino acid. Some of the newly synthesized 74K protein was found associated with brain nuclei. Hyperthermia also induces the synthesis of the 74K protein in the kidney.

The 74K protein is similar in molecular weight to one of the major heat-shock proteins induced in vertebrate tissue culture cells following elevation of ambient temperature at a time when overall protein synthesis is greatly reduced (Kelley and Schlesinger 1978). While elevation of ambient temperature induces the synthesis of a set of heat-shock proteins in tissue culture systems, in the intact mammalian organs a marked increase in synthesis of only a 74K protein is observed.

CELL-FREE TRANSLATION OF POLYSOMES FOLLOWING HYPERTHERMIA

To investigate whether the increased labeling of the 74K protein following hyperthermia was due to an increase in the level of $mRNA_{74K}$ associated with polysomes, purified polysomes were translated in a reticulocyte cell-free system (Heikkila et al. 1981). Analysis of the labeled translation products of polysomes isolated from the cerebral hemispheres demonstrated a marked increase in synthesis of the 74K protein relative to controls whose body temperature did not change (Fig. 1).

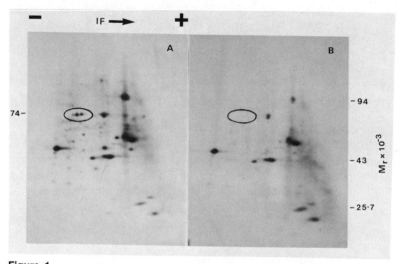

Figure 1
Induction of a 74K heat-shock protein in the rabbit brain following elevation of body temperature. An mRNA-dependent reticulocyte lysate was programmed with free polysomes isolated from the cerebral hemispheres of the rabbit brain following elevation of body temperature from 39.7°C to 42.5°C. Hyperthermia was induced by placement of animals in an incubator at 37°C for 2 hours. The body temperature of control animals, kept at room temperature, remained at 39.7°C. Equal amounts of acid-precipitable radioactivity from the [^{35}S]methionine-labeled translation products of brain polysomes from heat-treated and control animals were analyzed by two-dimensional gel electrophoresis and fluorography. The position of the 74K heat-shock protein is encircled. (A) Heat treatment; (B) control.

Selective increases in labeling of a 74K protein following elevation of body temperature from 39.7°C to 42.5°C were also observed in the translation products of heart and kidney polysomes. Since an increase in labeling of the 74K protein is apparent when polysomes are translated in the heterologous cell-free translation system, it is likely that the phenomenon is due to an increase in level of mRNA$_{74K}$ associated with polysomes rather than a change in protein modification.

TEMPORAL CHANGES IN SYNTHESIS OF THE 74K PROTEIN

LSD-induced hyperthermia is transient, reaching a maximum body temperature of 42.5°C by 1 hour after intravenous injection of the drug at 100 μg/kg. By 4 hours, body temperature decreases to 41°C and by 8 hours recovery to a normal body temperature of 39.7°C occurs. Translation in a reticulocyte cell-free system of brain polysomes isolated at 1, 4, and 8 hours after drug administration reveals a temporal decrease in the synthesis of the 74K protein as drug-induced hyperthermia subsides.

INDUCTION OF THE 74K PROTEIN IN FETAL ORGANS

To determine the effect of physiologically relevant increases in body temperature on protein synthesis in fetal organs, the body temperature of pregnant female rabbits was elevated by 3°C by increasing ambient temperature. Two hours following commencement of the heat treatment, polysomes were isolated from various fetal organs and translated in a reticulocyte cell-free system. Analysis of the translation products on two-dimensional gels indicates a marked induction of synthesis of a 74K heat-shock protein in fetal brain, liver, kidney, and heart.

EFFECT OF HYPERTHERMIA ON PROTEIN SYNTHESIS IN THE RETINA AND BRAIN MICROVASCULAR SYSTEM

Recently we have begun to analyze the effect of hyperthermia on more specific populations of cells in the brain. The microvascular (i.e., capillary) system of the brain, which is composed primarily of endothelial cells, is the site of the blood-brain barrier. This system controls the entry and exit of molecules between the brain and the blood system. At times of stress, homeostatic mechanisms might be particularly important at the blood-brain barrier. Hyperthermia induced by either LSD or by elevation of body temperature induces the synthesis of a 74K protein in the brain microvascular system as analyzed by in vivo labeling techniques (Inasi and Brown 1982).

The retinal system has allowed us to examine the effect of hyperthermia on the synthesis and axonal transport of brain proteins. Hyperthermia generated by LSD induces a transient disaggregation of retinal polysomes and a marked induction of synthesis of a 74K protein (Clark and Brown 1982). The 74K protein is transported into nerve axons, since it appears in the optic nerve 3 hours after the induction of hyperthermia.

CONCLUSIONS

A physiologically relevant increase in body temperature of 3°C, similar to that attained during fever reactions, induces a marked increase in synthesis of a 74K protein in the brain, heart, liver, and kidney of the rabbit. Induction of the 74K protein is also observed in fetal organs following elevation of maternal body temperature by 3°C. Synthesis of the 74K protein decreases as hyperthermia subsides.

In many systems, elevation of ambient temperature induces the synthesis of a set of heat-shock proteins, however in intact mammalian organs only a 74K protein is induced. Since mammalian cell lines have genes coding for a set of heat-shock proteins, it is of interest that in intact organs only one of the major heat-shock proteins is induced by hyperthermia. The pattern and extent of induction of heat-shock proteins may be related to the magnitude and rate of increase in ambient temperature. In the intact organ experiments, the rate of increase of body temperature is gradual in comparison with conditions in many tissue culture experiments. It would appear that constraints exist on the induction of the full set of heat-shock proteins in intact organs during physiologically relevant increases in body temperature. The functional role of the 74K protein in the cellular reaction of intact mammalian organs to thermal stress remains to be determined. The present results indicate that the heat-shock phenomenon, which in mammalian cells is frequently studied under somewhat artificial laboratory conditions, is relevant to natural cellular responses to physiological stress in the intact animal.

ACKNOWLEDGMENTS

The technical assistance of Sheila Rush is gratefully acknowledged. These studies were supported by grants to I.R.B. from the Medical Research Council of Canada.

REFERENCES

Brown, I.R., J.J. Heikkila, and J.W. Cosgrove. 1982. Analysis of protein synthesis in the mammalian brain using LSD and hyperthermia as experimental probes. In *Molecular approaches to neurobiology* (ed. I.R. Brown), pp. 221–253. Academic Press, New York.

Clark, B.D. and I.R. Brown. 1982. Protein synthesis in the mammalian retina following the intravenous injection of LSD. *Brain Res.* **247:** 97–104.

Cosgrove, J.W., B.D. Clark, and I.R. Brown. 1981. Effect of intravenous administration of d-lysergic acid diethylamide on subsequent protein synthesis in a cell-free system derived from brain. *J. Neurochem.* **36:** 1037–1045.

Freedman, M.S., B.D. Clark, T.F. Cruz, J.W. Gurd, and I.R. Brown. 1981. Selective effects of LSD and hyperthermia on the synthesis of synaptic proteins and glycoproteins. *Brain Res.* **207:** 129–145.

Heikkila, J.J. and I.R. Brown. 1979a. Disaggregation of brain polysomes after LSD *in vivo:* Involvement of LSD-induced hyperthermia. *Neurochem. Res.* **4:** 763–776.

———. 1979b. Hyperthermia and disaggregation of brain polysomes induced by bacterial pyrogen. *Life Sciences* **25:** 347–352.

Heikkila, J.J., J.W. Cosgrove, and I.R. Brown. 1981. Cell-free translation of free and membrane-bound polysomes and polyadenylated mRNA from rabbit brain following administration of d-lysergic acid diethylamide *in vivo. J. Neurochem.* **36:** 1229–1238.

Inasi, B.S. and I.R. Brown. 1982. Synthesis of a heat-shock protein in the microvascular system of the rabbit and brain following elevation of body temperature. *Biochem. Biophys. Res. Commun.* **106:** 881–887.

Kelley, P.M. and M.J. Schlesinger. 1978. The effect of amino acid analogues and heat shock on gene expression in chick embryo fibroblasts. *Cell* **15:** 1277–1286.

Preferential Synthesis of Rat Heat-shock and Glucose-regulated Proteins in Stressed Cardiovascular Cells

Lawrence E. Hightower*
and Fredric P. White†
*Microbiology Section
Biological Sciences Group
The University of Connecticut
Storrs, Connecticut 06268

†Faculty of Medicine
Memorial University of Newfoundland
St. John's, Newfoundland, Canada A1B 3V6

Our laboratories seek functions for the avian and mammalian heat-shock proteins. Our perception of the heat-shock response and heat-shock proteins has been strongly influenced by our first encounters with these proteins. They were observed in the Hightower laboratory as proteins that rapidly accumulated in cultured chicken embryo cells treated with the arginine analog, canavanine (Hightower and Smith 1978; Hightower 1980) and in the White laboratory as proteins rapidly synthesized in explants of rat brain tissue (White 1980) which had not been stressed purposefully. We subsequently showed that the major 71-kD stress protein (SP71) synthesized by brain explants had an apparent size and peptide map identical to SP71 synthesized by canavanine-treated rat embryo cells in culture (Hightower and White 1981). In brain explants, SP71 accumulated to high levels in cells associated with cerebral microvessels (White 1981). However, SP71 synthesis is not unique to brain explants. Incubated tissue slices from a variety of rat organs all rapidly accumulate SP71 and several other stress proteins. Heat-shocked rats accumulate these same stress proteins in most tissues in vivo (Currie and White 1981). As exemplified by cardiac tissue, the pattern of protein synthesis in stressed tissue in vivo is virtually identical to that of incubated explants. Therefore, at the level of protein synthesis, explants are models for stressed tissue and not for normal tissue in vivo. SP71 is not an abundant protein in either normal tissues in vivo or in

369

nonstressed cells in culture; however, it is closely related to an abundant 73 kD protein (P73) in nonstressed tissues in vivo and in cultured cells.

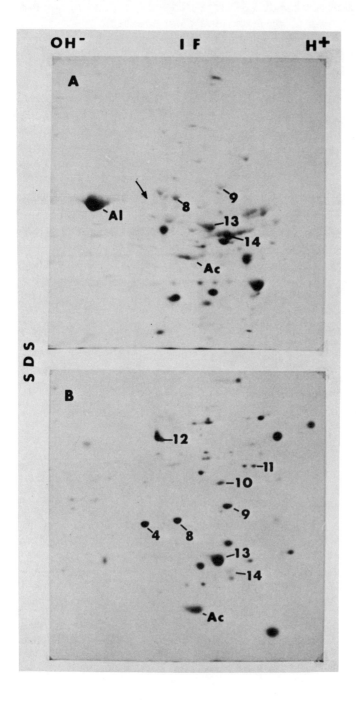

These experiments encouraged our working hypothesis that stress proteins may function in traumatized animals and may be part of generalized tissue responses to wounding and other stresses.

A second major working hypothesis evolved from attempts to find a unifying thread among the variety of stressors found by us and others to induce stress proteins in avian and mammalian cells in culture. The stressors that we have either found or confirmed as inducers of stress proteins within 3 hours in cultured rat embryo cells include: L-canavanine (0.6 mM), fluorophenylalanine (0.2 mM), puromycin (10 μg/ml), diamide (0.3 mM), sodium arsenite (25 μM), Cd^{++} (10 μM), Zn^{++} (100 μM), Cu^{++} (500 μM), and heat shock (42°C for 30 min). Based on the observation that several different amino acid analogs were stressors, and having ruled out direct effects of the analogs, we reasoned that the cellular accumulation of abnormal proteins (analog-substituted in this case) triggered the production of stress proteins. Stressors would then be agents that either cause cells to accumulate abnormal proteins biosynthetically or are protein denaturants. This hypothesis correctly predicted that intermediate doses of puromycin, which causes cells to accumulate randomly chain-terminated puromycylpolypeptides, and diamide, which attacks protein sulfhydryls, both cause cells to produce stress proteins. The other stressors listed above are all known protein denaturants. The nature of the stressor in tissue explants is unknown. However, one possible link to the abnormal protein hypothesis would be membrane permeability changes during excision or initial incubation leading to protein denaturation in the explant.

Mammalian stress proteins have been studied mainly in cultured embryonic cells and cell lines. Here, we present some of our initial comparisons of stress responses in differentiated cells and tissues. Rat cardiovascular cells were chosen because incubated heart slices and blood microvessels isolated from incubated brain tissue mounted extraordinarily strong stress responses, and because rat heart cells can be easily separated into myocardial and mesenchymal populations based on differential rates of attachment to cell culture plates. Sections of rat

Figure 1
(A) Fluorograms of two-dimensional gels containing [^{35}S]methionine-labeled proteins from rat thoracic aorta radiolabeled in vivo (Currie and White 1981) and (B) [^{3}H]leucine-labeled proteins from segments of longitudinally slit aortas incubated in vitro for 2 hours. Incubation conditions and sample preparation for gel analysis are described in Hightower and White (1981). Proteins were separated by isoelectric focusing (IF) in the first dimension and SDS-PAGE in the second dimension after the methods of O'Farrell (Hightower and White 1981). (→) The expected position of SP71; Al, albumin; Ac, actin; spot 4, SP71; spot 8, P73; spot 9, GR78; spot 10, SP88; spot 11, GR99; spot 12, SP110; spot 13, vimentin and α-tubulin; and spot 14, β-tubulin.

thoracic aorta provided another opportunity to study separate populations of differentiated cells because the vascular endothelium can be

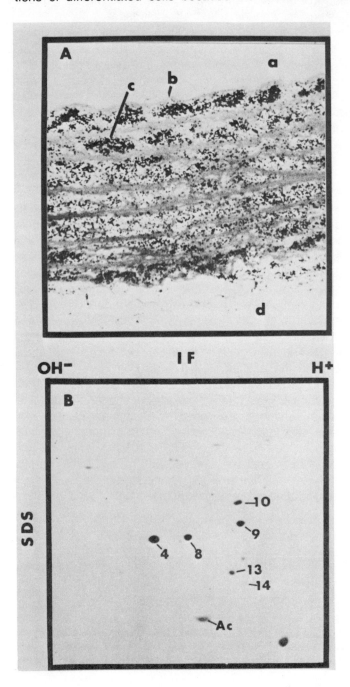

stripped away leaving an almost pure population of vascular smooth muscle cells for separate analysis.

PREFERENTIAL SYNTHESIS OF STRESS AND GLUCOSE-REGULATED PROTEINS BY INCUBATED AORTAS

The pattern of radioactively labeled proteins synthesized by the descending thoracic aorta in vivo is shown in Figure 1A. The aortic cells synthesized large amounts of a 73-kD protein (P73, spot 8) related to SP71 and a 78-kD glucose-regulated protein (GR78, spot 9) but no detectable SP71 (position marked by arrow). Other prominent landmark proteins, identifiable by their relative positions in the two-dimensional gels, included actin (Ac), albumin (Al), β-tubulin (spot 14), and a region which probably contained both vimentin and α-tubulin (spot 13). Glucose-regulated proteins have been defined by others as proteins that accumulate to higher levels in glucose-deficient medium, in cells treated with various inhibitors of protein glycosylation, or in paramyxovirus-infected cells. Their functions are unknown. We have identified glucose-regulated proteins here by comigration with authentic glucose-regulated proteins and by peptide mapping for GR78.

The pattern of radioactive proteins synthesized by sections of aorta which were slit longitudinally and incubated in oxygenated Krebs Ringer phosphate buffer supplemented with 1 mM glucose for 2 hours (Fig. 1B) was strikingly different than the in vivo patterns. P73, GR78, and proteins in the vimentin region, all of which were abundant in vivo, accumulated to even higher relative levels in vitro. SP71 along with two other known stress proteins (SP88, spot 10; SP110, spot 12) and a second glucose-regulated protein (GR99, spot 11), all of which were scarce in vivo, accumulated to relatively higher levels in vitro. Finally, the accumulation of some proteins such as β-tubulin (spot 14) decreased dramatically in the explant.

Figure 2
Protein synthesis in incubated aorta scraped on its lumenal side to remove endothelium prior to incubation and radiolabeling. (*A*) Autoradiogram of a 0.5-μm transverse section of scraped aorta incubated with [³H]leucine for 2 hours and prepared as described in White (1981). a, lumen; b, internal elastic lamina; c, smooth muscle cell; and d, adventitia. (*B*) Fluorogram of [³H]leucine-labeled proteins synthesized by scraped aorta during 2 hours of incubation in vitro. Two-dimensional gel analysis and numbering of proteins as in Fig. 1.

SYNTHESIS OF STRESS AND GLUCOSE-REGULATED PROTEINS BY AORTIC SMOOTH MUSCLE CELLS

The intimal sides of longitudinally cut segments of aorta were scraped to remove endothelial cells and the remaining tissue incubated with [^{3}H]leucine as described above. An autoradiogram of a transverse section of scraped, incubated aorta is shown in Figure 2A. The internal elastic lamina (b) was free of endothelial cells. Smooth muscle cells (c) active in protein synthesis were distributed throughout the media, and only a small fraction of the total grains were over adventitia (d). Especially high grain densities were distributed over the layer of smooth muscle cells proximal to the scraped internal lamina, possibly reflecting a stronger response to damage.

The pattern of radioactive proteins synthesized by scraped aorta in vitro (Fig. 2B) was dominated by SP71, P73, GR78, and a highly acidic protein ($M_r = 40,000$) which was abundant in vivo as well (Fig. 1A). The relative synthesis of other cellular proteins was suppressed. Interestingly, the pattern of proteins synthesized by incubated aorta scraped on its ablumenal side more closely resembled that in Figure 1B than that in Figure 2B. It is possible that the disruption of vascular endothelium, which occurred to some extent even in unscraped aorta (Fib. 1B) as determined by light microscopy, may have been a crucial event in eliciting or at least enhancing the stress response in incubated aortic smooth muscle tissue.

MYOCARDIAL AND CARDIAC MESENCHYMAL CELLS IN CULTURE RETAIN THE POTENTIAL TO SYNTHESIZE STRESS PROTEINS

The patterns of radioactive polypeptides synthesized by cardiac mesenchymal (Fig. 3A) and myocardial cells (Fig. 3C) in culture were compared. Both types of cultures synthesized P73 and GR78, however myocardial cells were relatively deficient in SP88. Other distinguishing features were the synthesis in myocardial cells but not mesenchymal cells of a protein (marked *my* in Fig. 3C), which is probably desmin, and the synthesis of several proteins (one of which is marked *me* in Fig. 3A) by mesenchymal cultures that were not detectable in myocardial cultures. Thus, in addition to morphologic differences, our cultures showed the expected differences in intermediate filament proteins between cardiac muscle (desmin) and nonmuscle (vimentin) cells.

To analyze the capacity of each type of culture to produce stress proteins, the patterns of radioactive proteins synthesized by heat-shocked mesenchymal (Fig. 3B) and myocardial (Fig. 3D) cells were compared. Both cultures rapidly accumulated SP71. However, the synthesis of several proteins at the expected positions for major cytoskeletal

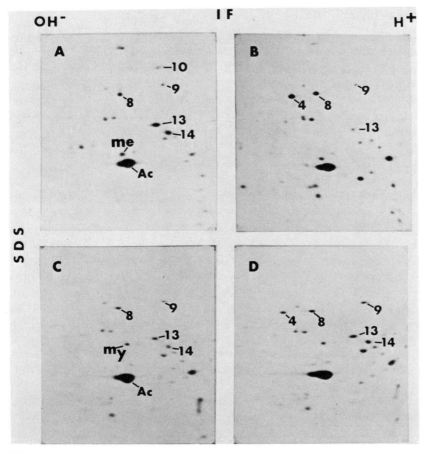

Figure 3
Fluorograms of [³⁵S]methionine-labeled proteins synthesized by control (*A*) or heat-shocked (*B*) cultured cardiac mesenchymal cells and control (*C*) or heat-shocked (*D*) myocardial cells in culture. Hearts were removed from newborn rats, minced, trypsinized, and plated on plastic cell culture plates. Mesenchymal cells attached within 2 hours. The unattached cells were transferred to another plate and allowed to attach over the next 24 hours. These cultures contained mostly myocardial cells which exhibited characteristic morphologies and contractile activity. Confluent cultures were exposed to 45°C for 10 minutes, allowed to recover at 37°C for 2 hours, and then labeled for 30 minutes with [³⁵S]methionine. Sample preparation and two-dimensional gel analyses have been described previously (Hightower and White 1981). me, an abundant protein in mesenchymal but not myocardial cells; my, putative desmin, which was abundant in myocardial but not mesenchymal cells.

proteins responded differently in the two cultures. Heat-shocked mesen-chymal cells did not synthesize detectable proteins corresponding to β-tubulin and the synthesis of spot-13 proteins (probably a mixture of α-tubulin and vimentin) was much reduced. In contrast, the relative levels of synthesis of putative desmin, vimentin, and β-tubulin were not diminished in stressed myocardial cells. A comparison of relative amounts of intermediate filament protein synthesis (intermediate filament proteins were isolated by Triton-KCl extraction) in control and stressed cardiac cell cultures showed that the synthesis of these proteins had not changed. Therefore, the decreases in radioactive spots 13 and 14 appeared to be due to decreased accumulation of α- and β-tubulins, respectively.

CONCLUSION

Several differentiated cardiovascular cell types, including aortic smooth muscle cells, myocardial cells, and cardiac mesenchymal cells can all be stimulated to synthesize heat-shock proteins, and in some cases, glucose-regulated proteins. In experiments not presented here, radioactive proteins from vascular endothelial cells scraped from incubated aortas and analyzed on two-dimensional gels also included SP71. Therefore a variety of differentiated mammalian cells can mount stress responses, reinforcing the idea that this is a general cellular and tissue response. In this respect, the stress response described here is like inflammatory and phagocytic responses. All of these may be general protective mechanisms at the tissue or cellular level, and could be interrelated.

The effects of hyperthermia on the synthesis of tubulins was cell-type-specific. Newly synthesized tubulin accumulated to much lower levels in heat-shocked mesenchymal cells than in myocardial cells. This may indicate differences in the control of tubulin synthesis in the two cell types. Alternatively, microtubules may be disassembled in stressed mesenchymal but not myocardial cells. Hypotheses that levels of tubulin subunits may regulate tubulin synthesis have been entertained by others.

The stimulated synthesis of glucose-regulated proteins along with stress proteins is tissue explants, even in the presence of exogenous glucose, and in canavanine-treated rat embryo cells (Hightower and White 1981) is intriguing. We favor the hypothesis that glucose-regulated proteins function in glycoprotein biosynthesis. It is possible that the accumulation of conformationally aberrant nascent glycoproteins on internal membranes may trigger synthesis of glucose-regulated proteins. This may be the link between certain stressors, such as canavanine, which can be incorporated into nascent glycopolypeptides and inducers of glucose-regulated proteins, such as 2-deoxyglucose or tunicamycin, that could generate partially or nonglycosylated polypeptides which then fold abnormally. We are currently testing the hypothesis that animal cells sense abnormal proteins and respond by synthesizing stress proteins.

ACKNOWLEDGMENTS

Research support to F.P. White was from The Medical Research Council of Canada (MA-5405) and to L.E. Hightower from the Public Health Service (HL 23588 and CA 14733) and from the National Science Foundation (PCM 78-08088 and PCM 81-18285).

REFERENCES

Currie, R.W. and F.P. White. 1981. Trauma-induced protein in rat tissues: A physiological role for a "heat shock" protein? *Science* **214:** 72–73.

Hightower, L.E. 1980. Cultured animal cells exposed to amino acid analogues or puromycin rapidly synthesize several polypeptides. *J. Cell. Physiol.* **102:** 407–427.

Hightower, L.E. and M.D. Smith. 1978. Effects of canavanine on protein metabolism in Newcastle disease virus-infected and uninfected chicken embryo cells. In *Negative strand viruses and the host cell* (ed. B.W.J. Mahy and R.D. Barry), pp. 395–405. Academic Press, London.

Hightower, L.E. and F.P. White. 1981. Cellular responses to stress: Comparison of a family of 71–73 kilodalton proteins rapidly synthesized in rat tissue slices and canavanine-treated cells in culture. *J. Cell. Physiol.* **108:** 261–275.

White, F.P. 1980. Differences in protein synthesized *in vivo* and *in vitro* by cells associated with the cerebral microvasculature. A protein synthesized in response to trauma? *Neuroscience* **5:** 1793–1799.

———. 1981. Protein and RNA synthesis in cerebral microvessels: A radioautographic study. *Brain Res.* **229:** 43–52.

A Mammalian Response to Trauma: The Synthesis of a 71-kD Protein

**Fredric P. White and
R. William Currie**
*Faculty of Medicine
Memorial University of Newfoundland
St. John's, Newfoundland A1B 3V6 Canada*

We have been studying a stress response in mammalian tissues that was first observed in brain slices. We found that the major protein synthesized by rat brain slices was not normally synthesized by the brain in vivo (White 1980a). This protein, which we will refer to as stress protein 71 (SP71), has a molecular weight of 71 kD, as determined by SDS-PAGE, and an isoelectric point of approximately 5.8. The synthesis of SP71 starts within 30 minutes of slice preparation and can be inhibited during this time by actinomycin D. It appears to be synthesized by the cells of the cerebral microvasculature, most likely the endothelial cell or pericyte, in response to the trauma of slice preparation (White 1981a). This response to trauma is not brain specific (Hightower and White 1981). Lung, liver, kidney, thymus, and heart all respond to slicing by synthesizing SP71.

SP71 can also be induced in rat embryo cultures by heat, canavanine, and Cd^{++} (Hightower and White 1981). SP71 appears to be related to one of the heat-shock proteins, hsp70, of *Drosophila* (Tissières et al. 1974). The molecular weights and isoelectric points of SP71 and hsp70 are similar and they are induced in vitro by similar treatments (Johnston et al. 1980; Hightower and White 1981).

As well as being synthesized in vitro, SP71 is synthesized by all rat organs following stresses such as hyperthermia and ligation of large vessels (Currie and White 1981), and a similar protein is induced in

379

rabbit brain by hyperthermia and LSD (Heikkila and Brown 1981). We have continued to characterize this stress response using both mammalian tissue slices and in vivo hyperthermia and describe in this report the postnatal development of the response, the species differences between mouse and rat SP71, and the accumulation of SP71 in vivo after hyperthermia.

SYNTHESIS OF SP71 BY ORGAN SLICES

The basal level of in vivo SP71 synthesis was undetectable in most organs from unstressed rats. Of the organs studied, only the adrenals and bladder synthesized SP71 and even in these organs SP71 levels were low. After exposing tissues to the trauma of slice preparation, hyperthermia, or the ligation of major arteries, SP71 became the major protein synthesized by all organs. The synthesis of SP71 in slices is shown in Figure 1. Both mouse and rat heart slices are shown. Although the SP71 synthesized by all organs of the same species had identical molecular weights and isoelectric points, there was a difference between the SP71 synthesized by rat and mouse. The mouse SP71 had an isoelectric point which was slightly more acidic than the rat SP71. The molecular weights of both proteins appeared to be identical.

SP71 has been shown by peptide mapping to be closely related to

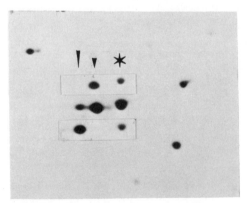

Figure 1
SP71 synthesized by heart slices from rat and mouse. Heart slices were incubated with [³H]leucine for 2 hours after preparation. Proteins were solubilized in O'Farrell lysis buffer and separated by two-dimensional gel electrophoresis. Acid proteins are on the right-hand side of the fluorogram. (*Top insert*) SP71 and P73 synthesized by mouse heart slices. (*Bottom insert*) SP71 and P73 synthesized by rat heart slices. The middle set shows SP71 and P73, when proteins from mouse and rat heart slices were separated on the same gel. (*) P73; (▮) rat SP71; (▼) mouse SP71.

another protein P73 (Hightower and White 1981). P73 has a molecular weight of 73 kD and an isoelectric point of 5.6. Unlike SP71, P73 is a major protein in most cells. It was synthesized at high rates in vivo and was usually synthesized by slices in vitro (Fig. 1). The P73s synthesized by both mouse and rat slices appear to be identical.

The response of mammalian cells to trauma not only entails the induction of SP71, but also a general inhibition of protein synthesis. This can be seen by comparing the proteins synthesized by rat lung in vivo (see Fig. 4, a1) with those synthesized by either rat lung slices in vitro (see Fig. 4, a4) or rat lung in vivo after hyperthermia (see Fig. 4, a5). In some organs, i.e., brain and heart, the induction of SP71 and the inhibition of normal protein synthesis is so drastic that only SP71, P73, and sometimes actin can be detected.

ACCUMULATION OF SP71 IN VIVO

We have examined the accumulation of SP71 in various rat organs after raising the temperature of rats to 42°C for 15 minutes. SP71 was synthesized by all organs 30 minutes, 1 hour, and 2 hours after hyperthermia but not 1 or 2 days after the initial trauma. SP71 accumulated quickly and became a major protein constituent in all organs. Figure 2 shows the accumulation of SP71 in the adrenals. By 6 hours after hyperthermia, the concentration of SP71 was maximum. The concentration remains high for 1 or 2 days and declines only slowly to a basal level in about 8 days. The decline in the concentration of SP71 to basal levels in rat organs varied between 4 days in the brain and 8 to 16 days in most other organs. The turnover rate of SP71, therefore, appears to be relatively slow.

SUBCELLULAR LOCALIZATION OF SP71 IN LIVER

We have tried to determine the subcellular localization of SP71 using liver, an organ which is easily fractionated by standard techniques and which accumulates large quantities of SP71 in response to hyperthermia. Earlier localization studies with brain slices were useful in determining the cellular localization of SP71 synthesis but not its subcellular localization (White 1980b). Livers were removed 1 day after the rats were subjected to hyperthermia, and homogenized in 0.32 M sucrose. Standard subcellular fractionation techniques were used (Ehrenreich et al. 1973). Many subcellular fractions contained SP71 but only the soluble/cytosol fraction was enriched with SP71. Thus, possible adsorption artifacts complicated the localization of SP71 to specific subcellular organelles. A similar situation was found for P73 in studies on subcellu-

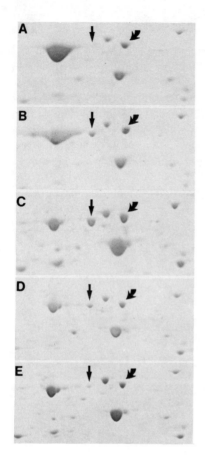

Figure 2
Protein accumulation in rat adrenal gland after hyperthermia. Proteins from rat adrenal glands were separated by two-dimensional electrophoresis and stained with Coomassie Blue. The location of SP71 is marked by straight arrow and location of P73 by curved arrow. (A) Proteins in rat adrenal gland of a normal rat; serum albumin is large spot on left. (B) Proteins in rat adrenal gland 150 minutes after the rats' temperature was raised to 42°C for 15 minutes. (C) 6 hours; (D) 2 days; and (E) 8 days after hyperthermic shock.

lar fractions of brain homogenates (Strocchi et al. 1981; F.P. White, unpubl.).

SP71 and P73 are both easily solubilized by homogenization, yet appear to bind under certain circumstances to almost any membrane. We, therefore, feel that extreme caution must be used before assigning a subcellular localization to these proteins and are currently trying to quantify possible adsorption phenomena with reconstruction experiments.

ISOELECTRIC POINT VARIANTS OF SP71 AND P73

All rat organs synthesized identical species of SP71 as determined by two-dimensional gel electrophoresis. This was the case whether the organs were sliced or the rat heated to induce SP71 and was independent of whether the tissue was added directly to O'Farrell lysis buffer or homogenized in 0.32 M sucrose prior to solubilization in lysis buffer. However, isoelectric point variants of both SP71 and P73 were produced with certain extraction procedures and these variants appeared to be organ specific (Hightower and White 1981). Figure 3, a and b, show SP71 and P73 synthesized by rat lung slices. The slices were homogenized in 10 mM phosphate buffer (pH 7.4) prior to solubilization in lysis buffer. An identical pattern for SP71 and P73 would have resulted if the slices were added directly to lysis buffer after incubation or homogenized first in 0.32 M sucrose, fractionated, and then solubilized in lysis buffer. Figure 3c shows SP71 and P73 as a group of eight charge variants that were obtained when the lung homogenate was cleared of particulate matter by centrifugation. Variants one through four appear to be derived from SP71 while five through eight appear to be derived from P73, since the latter comigrated with stainable protein (Fig. 3d). Other proteins did not form multiple spots with this procedure. Altering the protein to ampholyte concentration in the isoelectric focusing gel had no effect on the production of the variants. The variants might be produced by oxidation or proteolysis. However, it is possible that the variants are either gene products or have undergone some specific posttranslational modification and that the single SP71 and P73 species usually obtained by two-dimensional gel electrophoresis is an artifact.

SYNTHESIS OF SP71 AS A
FUNCTION OF POSTNATAL DEVELOPMENT

The synthesis of SP71 by heart, lung, and brain was examined as a function of postnatal development in rat. SP71 was induced either by preparing tissue slices or in separate experiments by heating animals to 42°C for 15 minutes. Newly synthesized proteins were labeled with either [³H]leucine or [³⁵S]methionine for 2 hours, 30 minutes after slicing or hyperthermia. Proteins were separated by two-dimensional gel electrophoresis, fluorograms prepared, and SP71 identified by its position in the autofluorogram. The proteins synthesized in vivo with and without hyperthermia, and in slices of brain, heart, and lung are shown in Figure 4. The heart and lung always synthesized SP71 in response to trauma. Even newborn rats synthesized large quantities of SP71 in heart, lung, kidney, and liver (data not shown). However, prior to 3 weeks neither slicing nor heat shock induced the synthesis of SP71 in brain. After 3

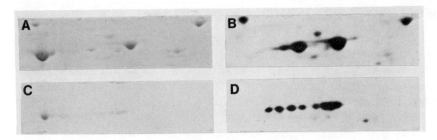

Figure 3
Proteins isolated from lung slices. Lung slices were labeled with [³⁵S]methionine for 2 hours after preparation. The slices were homogenized in 10 mM phosphate buffer after incubation. (A) Stained protein and (B) radioactive protein, coincident with SP71 and P73 in the homogenate; albumin is toward the left. (C) Stained protein and (D) radioactive protein from the same homogenate after particulate matter was removed by centrifugation. SP71 appears now as four proteins as does P73.

weeks, SP71 was the major protein synthesized by brain slices and brains in heat-shocked rats. The cells synthesizing SP71 in brain slices were enriched in microvasculature fractions (White 1981b), and the increase in SP71 synthesis by these cells coincided with the final maturation of the blood brain barrier.

SUMMARY/CONCLUSIONS

The synthesis of SP71 appears to be an important response of mammalian cells to trauma. In this report we have shown that SP71 synthesized by mouse has a different isoelectric point from the SP71 synthesized by rat; that although SP71 is synthesized in vivo for less than 8 hours following a hyperthermic episode, its concentration in tissues can be substantial even 4 to 8 days after hyperthermia; that SP71 appears to be easily solubilized from cells by simple homogenization; that multiple isoelectric point variants of SP71 either exist in some organs or are produced under specific conditions of extraction; and that the brain cannot synthesize SP71 until 3 weeks postpartum while all other organs tested show little or no change in SP71 synthesis with postnatal development.

We suggest that the synthesis of SP71 is an important consequence of vascular injury, and that the function of SP71 might be revealed by studying the differences between the repair mechanisms of mature and immature cerebral microvasculature since SP71 is synthesized only in the former.

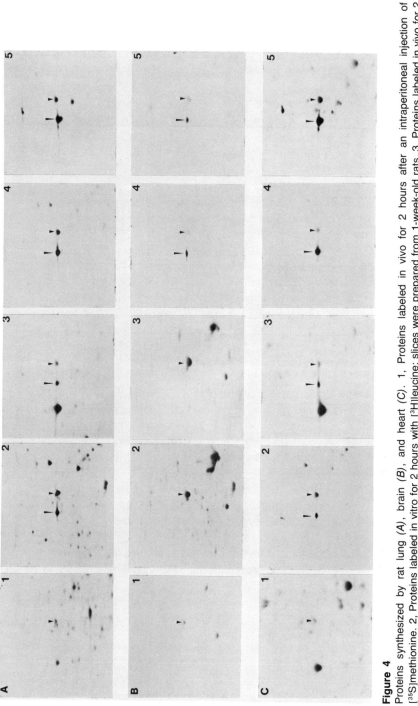

Figure 4

Proteins synthesized by rat lung (*A*), brain (*B*), and heart (*C*). 1, Proteins labeled in vivo for 2 hours after an intraperitoneal injection of [³⁵S]methionine. 2, Proteins labeled in vitro for 2 hours with [³H]leucine; slices were prepared from 1-week-old rats. 3, Proteins labeled in vivo for 2 hours after 15 minutes at 42°C with an intraperitoneal injection of [³⁵S]methionine; rats were 2 weeks old at time of experiment. 4, Proteins labeled in vitro for 2 hours with [³H]leucine; slices were prepared from 3-week-old rat. 5, Proteins labeled in vivo for 2 hours after 15 minutes at 42°C with an intraperitoneal injection of [³⁵S]methionine; rats were 4 weeks old.

ACKNOWLEDGMENTS

This work was supported by MRC (Canada), grant MT 5405. We thank Ms. Norma Churchill for her technical assistance.

REFERENCES

Currie, R.W. and F.P. White. 1981. Trauma-induced protein in rat tissues: A physiological role for a "heat shock" protein? *Science* **214:** 72–73.

Ehrenreich, J.H., J.J.M. Bergeron, P. Siekevitz, and G.E. Palade. 1973. Golgi fractions prepared from rat liver homogenates. 1. Isolation procedure and morphological characterization. *J. Cell Biol.* **59:** 45–72.

Heikkila, J.J. and I.R. Brown. 1981. Comparison of the effects of intravenous administration of *d*-lysergic acid diethylamide on free and membrane bound polysomes in the rabbit brain. *J. Neurochem.* **36:** 1219–1228.

Hightower, L.E. and F.P. White. 1981. Cellular responses to stress: Comparison of a family of 71–73 kilodalton proteins rapidly synthesized in rat tissue slices and canavanine-treated cells in culture. *J. Cell. Physiol.* **108:** 261–275.

Johnston, D., H. Oppermann, J. Jackson, and W. Levinson. 1980. Induction of four proteins in chick embryo cells by sodium arsenite. *J. Biol. Chem.* **255:** 6975–6980.

Strocchi, P., B.A. Brown, J.D. Young, J.A. Bonventre, and J.M. Gilbert. 1981. The characterization of tubulin in CNS membrane fractions. *J. Neurochem.* **37:** 1295–1307.

Tissières, A., H.K. Mitchell, and U.M. Tracy. 1974. Protein synthesis in salivary glands of *Drosophilia melanogaster:* Relation to chromosome puffs. *J. Mol. Biol.* **84:** 389–398.

White, F.P. 1980a. Differences in protein synthesized *in vivo* and *in vitro* by cells associated with the cerebral microvasculature. A protein synthesized in response to trauma? *Neuroscience* **5:** 1793–1799.

————. 1980b. The synthesis and possible transport of specific proteins by cells associated with brain capillaries. *J. Neurochem.* **35:** 88–94.

————. 1981a. Protein and RNA synthesis in cerebral microvessels: A radioautographic study. *Brain Res.* **229:** 43–52.

————. 1981b. The induction of "stress" proteins in organ slices from brain, heart, and lung as a function of postnatal development. *J. Neurosci.* **1:** 1312–1319.

The Induction of a Subset of Heat-shock Proteins by Drugs That Inhibit Differentiation in *Drosophila* Embryonic Cell Cultures

Carolyn H. Buzin and
Nicole Bournias-Vardiabasis
Division of Cytogenetics and Cytology
City of Hope Medical Center
Duarte, California 91010

Cells taken from *Drosophila* embryos and put into culture just after gastrulation undergo differentiation in vitro over the next 24 hours. The temporal and morphogenetic sequence of development for muscles and neurons in these cultures follows closely that which occurs in vivo (Seecof et al. 1973). Myoblasts give rise to myocytes, which fuse to form multinucleated myotubes; neuroblasts give rise to neurons, which cluster to form miniature ganglia, send out axons, and form functional neuromuscular junctions.

A number of drugs that are known to be teratogens in mammalian systems significantly inhibit muscle and/or neuron differentiation in these developing cultures (Bournias-Vardiabasis and Teplitz 1982). We have examined, by two-dimensional gel electrophoresis, the proteins synthesized by *Drosophila* embryonic cells in the presence of several of these drugs. Drug-treated cells exhibit a significant increase in the synthesis of three small proteins that have the same electrophoretic mobilities as heat-shock proteins (hsp) 22 and 23. We show a correlation between drugs that inhibit embryonic differentiation and those that induce the synthesis of the three proteins. In addition, we describe conditions that protect differentiation in these cells from inhibition by drug or heat treatment.

HEAT-SHOCK PROTEINS IN PRIMARY EMBRYONIC CELL CULTURES

We have examined, by two-dimensional gel electrophoresis, the [^{35}S]methionine-labeled proteins synthesized by *Drosophila* primary embryonic cell cultures exposed to a heat shock of 30°C or 37°C for 1 hour (Fig. 1, a−c). The pattern of heat-shock proteins, particularly the multiple components in the 70 kD region, corresponds closely to those described recently in salivary glands and in two *Drosophila* cell lines (Buzin and Petersen 1982). Several points should be noted:

1. hsp83 is synthesized to an appreciable extent at 25°C, as well as under heat-shocked conditions. This is also true in several *Drosophila* cell lines.

2. A few proteins in the 70 kD region are synthesized at low levels at 25°C, but protein synthesis in this region is vastly increased with heat shock.

3. hsp26−28, 23, 22a, and 22b are synthesized at moderate levels at 30°C and in increased amounts at 37°C. Of the hsp22 variants, both 22a and 22b (designated numbers 72 and 73, respectively, in Buzin and Petersen [1982]) are synthesized in heat-shocked salivary glands; only one is synthesized in Schneider line 2 (hsp22a) and Oregon (hsp22b) cell lines.

EFFECT OF DRUGS ON PRIMARY EMBRYONIC CELL CULTURES

Muscle and/or neuron differentiation in *Drosophila* primary embryonic cell cultures is significantly inhibited by a number of drugs that are known to be teratogens in mammalian systems (Bournias-Vardiabasis and Teplitz 1982). We have examined protein synthesis in primary cultures allowed to differentiate for 18−20 hours in the presence of coumarin or diphenylhydantoin (Fig. 1, d and e), both drugs that inhibit muscle and neuron differentiation in these cells. The proteins synthesized are essentially identical to those of untreated cultures (Fig. 1a), except that the synthesis of three proteins (one at 23 kD, two at 22 kD) was induced in the drug-treated cells. These proteins migrate with identical electrophoretic mobilities on two-dimensional gels as the small hsp23, 22a, and 22b. Studies are in progress to compare the peptide digests of the three small heat-shock proteins with the drug-induced proteins; preliminary evidence indicates that they are the same set of proteins. If this is true, we have found conditions in which a subset of heat-shock proteins, rather than the entire set, is synthesized.

In order to determine whether synthesis of proteins 23, 22a, and 22b is correlated with the effects of a drug on embryonic differentiation, we tested a number of drugs, both teratogens and nonteratogens, on the *Drosophila* primary cultures (Table 1). A group of seven drugs that do not inhibit differentiation to a significant extent also do not induce proteins 23, 22a and 22b to any appreciable extent (for example, see cortisone, Fig. 1f). In contrast, coumarin, diphenylhydantoin, and pentobarbital all inhibit muscle and/or neuron differentiation; they also induce the synthesis of the three proteins 20- to 65-fold over that of untreated cells. Dexamethasone, methyltestosterone, diethylstilbestrol, ecdysterone, and tolbutamide cause a modest increase in synthesis of proteins 23, 22a, and 22b. Two exceptions were found: amaranth and amethopterin, drugs that inhibit differentiation in embryonic cultures, do not increase the synthesis of proteins 23, 22a and 22b. Perhaps the action of these drugs on the cells does not proceed via some step common to that of the other teratogens.

Ireland and Berger (1982) have recently described an ecdysterone-stimulated synthesis of hsp22, 23, 26, and 27 in Schneider line 3, a *Drosophila* cell line. We have found that ecdysterone inhibits muscle and neuron differentiation in primary embryonic cells at concentrations of 10^{-5} M to 10^{-7} M (Table 1) and increases the synthesis of proteins 23, 22a, and 22b about sevenfold over that observed in untreated cells. Although the genes for hsp22, 23, 26, and 27 are all located at the 67B locus on the *Drosophila* chromosome (Wadsworth et al. 1980), we find no evidence of the synthesis of proteins 26 and 27 in ecdysterone-treated primary cells, using either a pH $5-7$ or pH $6-8$ gradient for the isofocusing (first) dimension. In addition, cells treated with each of the drugs given in Table 1 also show no synthesis of proteins 26 and 27 on either type of gel (data not shown).

For pentobarbital, a correlation has been made between the concentration required for inhibition of differentiation and that required for increased synthesis of proteins 22a, 22b, and 23. At 10^{-4} M, no significant inhibition occurs and the level of the three proteins is only marginally above that of control cells (Table 1). However, at 10^{-3} M, muscle differentiation is inhibited by about 50% and the proteins are induced about 25-fold over control levels.

PROTECTION EXPERIMENTS

A high heat shock (40.2°C for 40 min), given at specific periods during *Drosophila* pupal development, will induce specific abnormalities in the resulting adults, termed phenocopies. Mitchell et al. (1979) have shown that a mild heat pretreatment (which induces the synthesis of heat-shock proteins, but does not turn off normal protein synthesis) results in protection against phenocopy induction by a subsequent high heat shock. Similarly, a mild heat pretreatment prior to an otherwise lethal heat shock increases the survival of *Drosophila* larvae, pupae, adults, and continuous cell lines.

In *Drosophila* primary embryonic cell cultures, hyperthermia acts as a teratogen; cells exposed to 40.2°C for 2 hours (6.5–8.5 hr after oviposition) and then shifted down to 25°C show a significant reduction in the

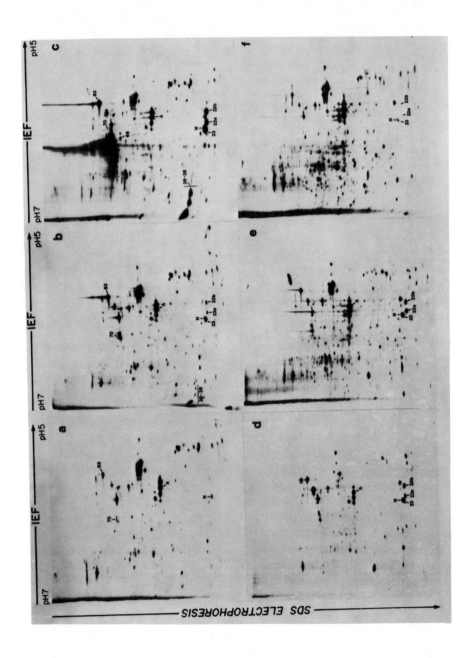

number of differentiated myotubes and neurons at 24 hours compared with untreated control cells. Short treatments with diphenylhydantoin or coumarin at 6.5−8.5 hours also result in an inhibition of differentiation. Table 2 describes three experiments designed to show whether a mild heat pretreatment can protect cell differentiation otherwise inhibited by hyperthermia or drugs. Embryonic cultures given a 0.5-hour pretreatment at 35°C just prior to a 2-hour heat treatment at 40.2°C are partially protected from the effects of the hyperthermia; a similar pretreatment protects the cells from a subsequent 2-hour drug treatment. For diphenylhydantoin, differentiation in pretreated cultures is close to that of untreated control cells. For coumarin, pretreatments given immediately before or 1 hour before drug treatment give about the same partial protection.

Currently we are carrying out the converse experiment to those described above. Embryonic cell cultures are pretreated with a drug that induces proteins 23, 22a, and 22b and then heated at 40.2°C for 2 hours. Preliminary results indicate that pretreated cultures show at least a twofold increase in differentiation compared with those heated without a pretreatment. If so, this may indicate that proteins 22 and 23 play some role in the protection effect.

CONCLUSION

We have examined, by two-dimensional gel electrophoresis, the proteins synthesized by *Drosophila* primary embryonic cells in the presence of a number of drugs that inhibit differentiation in these cells. Drug-treated cells show a significant increase in the synthesis of three proteins that have the same electrophoretic mobilities as hsp23, 22a, and 22b synthesized by these cells in response to a 37°C heat shock. Heat shock of the embryonic cells produces the full complement of heat-shock proteins, whereas drug treatment stimulates synthesis of only the subset 23, 22a,

Figure 1

Two-dimensional gel electrophoresis patterns from heat-shocked or drug-treated primary embryonic cells. (*a−c*) *Drosophila* primary embryonic cell cultures were prepared as described previously (Buzin and Seecof 1981), allowed to differentiate at 25°C for 24 hours, and then labeled for 1 hour with [^{35}S]methionine (200 μCi/ml) at (*a*) 25°C, (*b*) 30°C, or (*c*) 37°C. (*d−f*) Primary embryonic cell cultures were allowed to differentiate at 25°C in the presence of the indicated drug for 18−20 hours, beginning shortly after plating (about 3.5−4 hr after oviposition). Cells were labeled for 1 hour with [^{35}S]methionine at 25°C in the presence of the drugs (*d*) 10^{-3} M coumarin, (*e*) 10^{-4} M diphenylhydantoin, and (*f*) 10^{-3} M cortisone. Solubilization of cells, two-dimensional gel electrophoresis, and fluorography were carried out as in Buzin and Seecof (1981). Exposures (cpm × days) were as follows: (*a*) 223,000; (*b*) 307,000; (*c*) 254,000; (*d*) 362,000; (*e*) 424,000; (*f*) 594,000. A = actins; N = a nonheat-shock protein that migrates very close to hsp23.

Table 1

Comparison of Drug Effects on Embryonic Development and the Synthesis of Proteins 22 and 23

Treatment	Concentration (M) [a]	class	Drosophila assay [b] percent of control N	M	22K protein, [c] (relative increase over control)
Control	—	—	100	100	1
Coumarin	10^{-3}	T	19*	7*	64
Diphenylhydantoin	10^{-4}	T	45*	49*	21
Pentobarbital	10^{-4}		79	97	2
Pentobarbital	10^{-3}	T	78	48*	25
Dexamethasone	10^{-4}	T	54*	38*	10
Methyltestosterone	10^{-5}	T	42*	54*	8
Diethylstilbestrol	10^{-5}	T	82	48*	7
β-Ecdysterone	10^{-5}	T	32*	51*	7
Tolbutamide	10^{-3}	T	45*	30*	6
Amaranth	10^{-3}	T	114	11*	tr
Amethopterin	10^{-3}	T	53*	45*	0.4
Caffeine	10^{-3}	NT	99	62	2
Antipyrine	10^{-3}	NT	77	102	1
Dimethylsulfoxide	10^{-3}	NT	129	95	0.6
Cortisone	10^{-3}	NT	88	93	tr
Saccharin	10^{-3}	NT	68	112	0.8
Sulfanilamide	10^{-3}	NT	94	95	0.2
Progesterone	10^{-5}	NT	122	64	0.1

[a]Teratogens (T) were tested at concentrations that were not cytotoxic but inhibited muscle and/or neuron differentiation by 50% or more; nonteratogens (NT) were tested at 10^{-3} M or the highest concentration that was not cytotoxic.

[b]Drosophila primary embryonic cell cultures were prepared as described (Bournias-Vardiabasis and Teplitz 1982) and allowed to differentiate at 25°C in the presence of the drug for 24 hours. Cultures were stained and numbers of differentiated myotubes (M) and neuron clusters (N) were counted using a Bausch and Lomb automated image analysis system. Results were expressed as a percentage of the number of myotubes and neuron clusters in control cultures prepared on the same day. A total of four dishes per trial were scored and each drug was tested in three or more separate trials. A drug was classified as a teratogen in this system if, on three separate trials, the average number of myotubes and/or neuron clusters was significantly lower than the controls (Wilcoxon's signed rank test $\alpha < 0.05$, indicated by *). The average number of neuron clusters in control cultures ranged from 200 to 400; the number of myotubes ranged from 100 to 500 on any given day.

[c]From fluorograms of two-dimensional gels (see Fig. 1, d–f), the integrated optical densities of proteins 22a and 22b were determined, using a Bausch and Lomb automated image analysis system. Densities were determined from several exposures of a gel, as well as from different experiments using the same drug, and those in the linear range of the film were averaged. Values were normalized to the integrated optical density of actin II on the same fluorogram. The normalized values for proteins 22a and 22b (approximately the same) were averaged and compared with normalized densities of these proteins in control cells (very low). Because protein 23 is not well separated from the nonheat-shock protein N, these densities are not reported. tr, Trace amount, below sensitivity of image system.

Table 2
Mild Heat Pretreatment Protects *Drosophila* Embryonic Cells from Teratogenic Effects

			Percent of Control[a]	
Experiment	Pretreatment	Treatment	N	M
1. Control	—	—	100	100
2. High heat	—	40.2°C; 2 hr (6.5–8.5 hr AO)	25*[b]	23*
3. Low heat/high heat	35°C; 0.5 hr (6–6.5 hr AO)	40.2°C; 2 hr (6.5–8.5 hr AO)	50*	73
4. DPH	—	DPH, 26°C; 2 hr (6.5–8.5 hr AO)	69	48*
5. Heat/DPH	35°C; 0.5 hr (6.5–8.5 hr AO)	DPH, 26°C; 2 hr (6.5–8.5 hr AO)	83	98
6. COU	—	COU, 26°C; 2 hr (6.5–8.5 hr AO)	62	38*
7. Heat/COU	35°C; 0.5 hr (5–5.5 hr AO), then 26°C (5.5–6.5 hr AO)	COU, 26°C; 2 hr (6.5–8.5 hr AO)	83	73
8. Heat/COU	35°C; 0.5 hr (5–5.5 hr AO)	COU, 26°C; 2 hr (5.5–7.5 hr AO)	83	66

Drosophila primary embryonic cell cultures were prepared as described (Bournias-Vardiabasis and Teplitz 1982). The temperature, drug used, and duration of treatments are indicated; the age of the embryonic cells (hour after oviposition) at time of treatment is given below in parentheses. Medium containing drugs was removed after treatment and replaced with drug-free medium. Drugs used were diphenylhydantoin (DPH) (10^{-4} M) and coumarin (COU) (10^{-3} M).

[a]After treatment, embryonic cell cultures were allowed to differentiate at 25°C in the absence of the drug until 24 hours after oviposition. Scoring of cultures for myotubes (M) and neuron clusters (N) was carried out as described in Table 1.

[b](*) Values are significantly different from those of control (Wilcoxon's signed rank test $\alpha < 0.05$).

and 22b, but not hsp26, 27, or any other heat-shock proteins. Eight out of ten drugs tested that inhibit differentiation in primary embryonic cultures stimulate the synthesis of the three proteins; seven drugs that do not inhibit differentiation also do not induce these proteins. For pentobarbital, a correlation between the dose required to inhibit differentiation in the cells and that required to induce the proteins has been made.

Protection experiments have shown that a mild heat pretreatment (that will induce heat-shock proteins) partially protects cells from inhibition of differentiation caused by a subsequent 2-hour period of hyperthermia or drug treatment.

Since the synthesis of a subset of heat-shock proteins can be separated from that of the entire complement of heat-shock proteins, studies on the function and control of hsp22 and hsp23 can be carried out in the absence of the other heat-shock proteins. We plan to follow up preliminary experiments suggesting that hsp23, 22a, and 22b may be involved in the protection effect.

ACKNOWLEDGMENTS

We thank Erica Gollub and Cheryl Clark for their excellent technical assistance. This work was supported in part by a Biomedical Research Support Grant and grants from the Muscular Dystrophy Association and March of Dimes.

REFERENCES

Bournias-Vardiabasis, N. and R.L. Teplitz. 1982. The use of *Drosophila* embryonic cell cultures as an *in vitro* teratogen assay. In *Teratogenesis, Carcinogenesis, and Mutagenesis* (in press).

Buzin, C.H. and N.S. Petersen. 1982. A comparison of the multiple *Drosophila* heat shock proteins in cell lines and larval salivary glands by two-dimensional gel electrophoresis. *J. Mol. Biol.* **157** (in press).

Buzin, C.H. and R.L. Seecof. 1981. Developmental modulation of protein synthesis in *Drosophila* primary embryonic cell cultures. *Dev. Genet.* **2**: 237–252.

Ireland, R.C. and E.M. Berger. 1982. Synthesis of low molecular weight heat shock peptides stimulated by ecdysterone in a cultured *Drosophila* cell line. *Proc. Natl. Acad. Sci.* **79**: 855–859.

Mitchell, H.K., G. Moller, N.S. Petersen, and L. Lipps-Sarmiento. 1979. Specific protection from phenocopy induction by heat shock. *Dev. Genet.* **1**: 181–192.

Seecof, R.L., I. Gerson, J.J. Donady, and R.L. Teplitz. 1973. *Drosophila* myogenesis *in vitro:* The genesis of 'small' myocytes and myotubes. *Dev. Biol.* **35**: 250–261.

Wadsworth, S.C., E.A. Craig, and B.J. McCarthy. 1980. Genes for three *Drosophila* heat-shock-induced proteins at a single locus. *Proc. Natl. Acad. Sci.* **77**: 2134–2137.

Correlations between Synthesis of Heat-shock Proteins and Development of Tolerance to Heat and to Adriamycin in Chinese Hamster Fibroblasts: Heat Shock and Other Inducers

Gloria C. Li*, Dennis C. Shrieve*,
and Zena Werb†
*Department of Radiation Oncology
†Department of Anatomy and
Laboratory of Radiobiology and Environmental Health
University of California
San Francisco, California 94143

Mammalian cells exposed to a nonlethal heat treatment have been shown to acquire a transient resistance to subsequent heat challenge as determined by an increase in cell survival. This phenomenon has been termed thermotolerance. The biophysical and/or biochemical bases of thermotolerance are not known, but experimental evidence suggests that protein synthesis may be required for its manifestation.

To determine whether the development of thermotolerance is necessarily related to the synthesis of a set of specific proteins, such as heat-shock proteins, we examined the effects of heat (41−46°C), hypoxia, arsenite, cadmium, and ethanol on protein synthesis in Chinese hamster (HA−1) fibroblasts. In parallel experiments, we determined the kinetics of the induction of thermotolerance using cell survival as an end-point. Because it has been reported that heat induces resistance to adriamycin (Li and Hahn 1978), we also examined the possibility that such tolerance can be induced by other heat-shock protein inducers.

INDUCTION OF THERMOTOLERANCE AND HEAT-SHOCK PROTEIN SYNTHESIS AFTER INITIAL TREATMENT AT 46°C OR 41°C

To examine induction of thermotolerance, we chose two nonlethal heating conditions: (1) 46°C for 6 minutes, a heat shock that causes cell protein synthesis to decrease more than 95% following the heat treatment and (2) 41°C, at which total cell protein synthesis is not inhibited.

In the first set of experiments plateau-phase HA-1 cells (Li and Hahn 1980) were heated at 46°C for 6 minutes and incubated at 37°C for 0–24 hours before a second treatment at 45°C for 45 minutes, and then survival was assayed. Control cells were exposed to only one treatment at 45°C for 45 minutes, and the kinetics of thermotolerance development was measured by increases in survival above the control value (8×10^{-5}) (Fig. 1a). Thermotolerance was fully developed by 4–6 hours of incubation at 37°C.

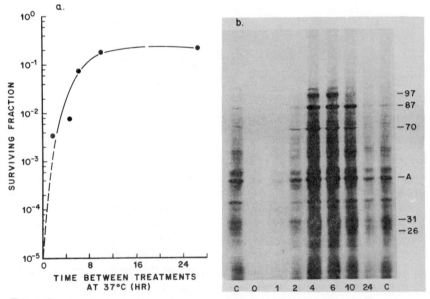

Figure 1
Induced thermotolerance and synthesis of heat-shock proteins in plateau-phase HA-1 cells exposed to 46°C for 6 minutes. After heat treatment, cells were incubated at 37°C for 0–24 hours before a second treatment at 45°C for 45 minutes or before labeling with [^{35}S]methionine. (a) Cell survival plotted as a function of time after the initial treatment. (b) Autoradiogram of an SDS-polyacrylamide slab gel of ^{35}S-labeled protein from 46°C heat-shocked cells. C, unheated control; 0-24, time (hr) of incubation before labeling; A, actin (M_r 43,000). $M_r \times 10^{-3}$ of proteins are indicated.

In parallel experiments, the effects of 46°C treatment on protein synthesis were examined. After 6 minutes heating at 46°C, cells were incubated at 37°C for 0–24 hours before labeling with [^{35}S]methionine for 1 hour at 37°C. Protein synthesis was drastically inhibited by the 46°C heat treatment, but recovered gradually during the 24-hour incubation period at 37°C, as seen by a greater incorporation of label at later time periods (Fig. 1b). When the proteins synthesized after heat shock were compared with those synthesized by unheated cells, the synthesis of certain proteins was greatly enhanced. At 6 hours after heating, three proteins with apparent molecular weights (M_r) of 70,000, 87,000, and 97,000 were synthesized in greater amounts than in control cells. Other, less prominent proteins (M_r = 59,000, 31,000, and 26,000) were also synthesized. The kinetics of synthesis of individual polypeptides was qualitatively different. The rate of the synthesis of M_r 70,000 and 97,000 proteins reached a maximum value 4–6 hours after heat shock and decreased to the control value by 8–10 hours. In contrast, synthesis of the M_r 87,000 protein was still appreciably enhanced by 10 hours after heating and returned to control rate by 24 hours.

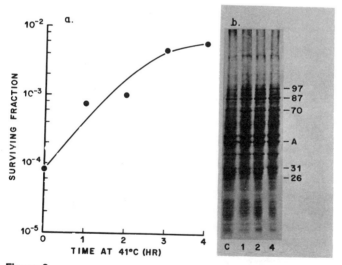

Figure 2
Induced thermotolerance and synthesis of heat-shock proteins in plateau-phase HA-1 cells exposed to 41°C for 1–4 hours. After heat shock, cells were immediately challenged by a second treatment at 45°C for 45 minutes or labeled with [^{35}S]methionine. (a) Cell survival plotted as a function of the length of the initial treatment. (b) Autoradiogram of an SDS-polyacrylamide slab gel of ^{35}S-labeled protein from 41°C heat-shocked cells. C, unheated control; 1–4, time (hr) of exposure to 41°C before labeling.

In the second set of experiments, plateau-phase HA-1 cells were exposed to 41°C for 1−4 hours. One group of cells was immediately challenged by a second treatment at 45°C for 45 minutes, and cell survival was assayed. The rest of the cells were pulse-labeled immediately with [35S]methionine for 1 hour at 37°C and protein synthesis was examined (Fig. 2). Thermotolerance was near its maximum by the end of each 41°C treatment. Incubation at 37°C between the two heat challenges did not result in an appreciable increase in survival (data not shown). Incubation at 41°C did not suppress total protein synthesis, but the synthesis of two proteins of M_r 70,000 and 87,000 was greatly enhanced in heated cells compared with unheated controls.

PERSISTENCE OF THERMOTOLERANCE AND OF HEAT-SHOCK PROTEINS

Monolayers of cells were first exposed to 45°C for 20 minutes and then incubated at 37°C for 6 hours. The 6-hour incubation was chosen because thermotolerance was then close to is maximum value. Cells were then labeled with [35S]methionine for 1 hour at 37°C, and incubated afterwards at 37°C for 0−36 hours in the absence of [35S]methionine. The rate of decay of thermotolerance was much slower than the rate of induction (Fig. 3). After 36 hours, cell survival decreased only from 30% to 10%. These survival results correlate well with the persistence of the heat-shock proteins synthesized between the sixth and seventh hour after heating. Up to 36 hours after pulse-labeling, no significant decay of the M_r 70,000 and 87,000 proteins was detected. In contrast, the M_r 97,000, 31,000, and 26,000 proteins decayed more rapidly.

INDUCTION OF THERMOTOLERANCE AND HEAT-SHOCK PROTEIN SYNTHESIS BY AGENTS OTHER THAN HEAT

In various biological systems, agents such as arsenite or cadmium induce proteins similar to those induced by heat (Ashburner and Bonner 1979; Johnston et al. 1980). It was our interest to test whether these agents also induced thermotolerance and synthesis of heat-shock proteins. In these experiments, we compared exponentially growing HA-1 cells exposed to 45°C for 15 minutes with cells treated with sodium arsenite or cadmium (Fig. 4). After an initial exposure to either heat shock, sodium arsenite, or cadmium chloride, the cells acquired a tolerance to the subsequent heat challenge, (45°C, 45 min), as evidenced by the increase in survival values. Six hours after either treatment, proteins of M_r 70,000, 87,000, and 97,000 were synthesized in greater amounts

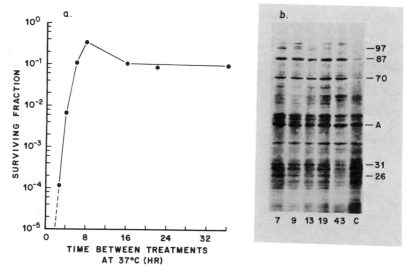

Figure 3
Persistence of thermotolerance and persistence of heat-shock proteins in plateau-phase HA-1 cells. After heat shock (45°C for 20 min), cells were incubated at 37°C for 0–36 hours before a second heat treatment (45°C for 45 min) and then the survival was assayed. To study the persistence of heat-shock proteins, cells were incubated after heat shock at 37°C for 6 hours, labeled with [^{35}S]methionine for 1 hour, and incubated afterwards at 37°C in the absence of [^{35}S]methionine for an additional 0–36 hours. (a) Cell survival plotted as a function of time after the initial heat shock. (b) Autoradiogram of an SDS-polyacrylamide slab gel of ^{35}S-labeled proteins from heat-shocked cells. 7–43, total time (hr) of incubation at 37°C after heat shock; C, unheated control.

than in untreated cells. Conversely, ethanol, an agent known to induce thermotolerance (Li and Hahn 1978) was also examined for induction of heat-shock protein synthesis and the same results were observed (Li and Werb 1982). In addition, recovery from hypoxia induced both the synthesis of heat-shock proteins (Li and Werb 1982; G.C. Li and D.C. Shrieve, in prep.) and transient thermotolerance (Fig. 4).

INDUCTION OF ADRIAMYCIN RESISTANCE BY HEAT, ETHANOL, OR SODIUM ARSENITE

As mentioned earlier, it has been reported previously that heat induces adriamycin resistance (Li and Hahn 1978). To test whether there is correlation between heat-shock protein synthesis and adriamycin tolerance, we examined the effects of heat, ethanol, or sodium arsenite on cellular survival following exposure to adriamycin (Fig. 5). Monolayers of

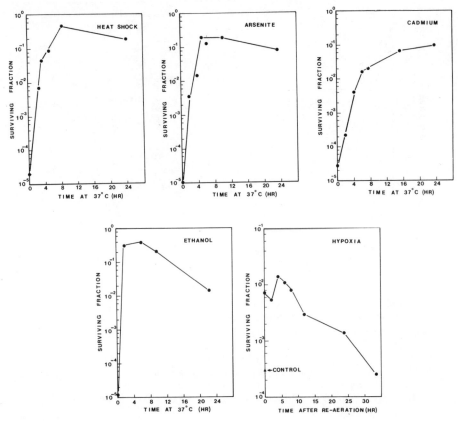

Figure 4
Kinetics of induction of thermotolerance by agents other than heat. Monolayers of exponentially growing HA-1 cells were initially exposed to 45°C heat shock for 15 minutes, or to 100 μM sodium arsenite for 1 hour at 37°C, or to 100 μM cadmium chloride for 1 hour at 37°C, or to 6% ethanol for 1 hour at 37°C, or to hypoxia for 16 hours at 37°C. Cells were then incubated at 37°C for 0–24 hours before a 45°C heat treatment. In each case, the second treatment was 45°C for 45 minutes. The surviving fractions are plotted as a function of the duration of 37°C incubation between the first and second treatment. Control cells were exposed to only one treatment at 45°C for 45 minutes. The kinetics of thermotolerance development was measured by increase in survival above the control value (3×10^{-4}).

cells were exposed to 45°C for 15 minutes, or to 6% ethanol for 1 hour at 37°C, or to 250 μM sodium arsenite for 1 hour at 37°C. After the initial treatment, some of the cells were immediately exposed to graded doses of adriamycin. The others were returned to 37°C for 6 hours before similar second challenges. After the initial exposure to heat, arsenite, or

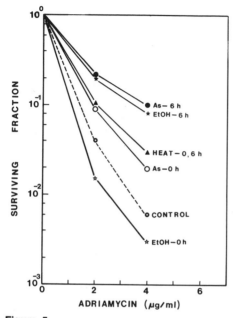

Figure 5

Survival of Chinese hamster HA-1 cells as a function of graded concentrations of adriamycin given as the second challenge. Monolayers of exponentially growing HA-1 cells were initially exposed to 45°C for 15 minutes (▲,▲), or to 6% ethanol for 1 hour at 37°C (☆,★), or to 250 μM sodium arsenite for 1 hour at 37°C (○,●). After the initial treatment, cells were immediately exposed to graded doses of adriamycin or incubated at 37°C for 6 hours before the adriamycin challenge. Lengths of time at 37°C incubation between the two challenges are as indicated. Survival values of cells exposed to adriamycin only are also shown (◒). These are included to demonstrate that it is resistance to adriamycin that develops, rather than the disappearance of a heat-induced sensitization (Hahn and Strande 1976).

ethanol, the cells clearly acquired a tolerance to adriamycin. The qualitative patterns of heat- or arsenite- or ethanol-induced adriamycin resistance were quite similar.

DISCUSSION

Our data suggest that heat-shock proteins may play a pivotal role in development of thermotolerance. Transcription is strongly affected by the heat-shock response (Ashburner and Bonner 1979); it seems likely that heat-shock proteins, perhaps through their association with the nuclear matrix (Levinger and Varshavsky 1981), may protect the genome from the adverse effects of heat and other environmental stresses. Several lines of evidence in HA-1 cells implicate heat-shock proteins in the development of thermotolerance.

1. Heat treatment, in the temperature range of 41−46°C, enhanced the synthesis of heat-shock proteins and induced a transient thermotolerance; this occurred under conditions in which the initial heat treatment either did not suppress total protein synthesis or drastically inhibited it.

2. When thermotolerance was fully developed, the rate of synthesis of most heat-shock proteins returned to control values.

3. There was good correlation between the persistence of some heat-shock proteins and the persistance of thermotolerance.

4. Agents known to induce thermotolerance induced synthesis of heat-shock proteins, and agents known to induce synthesis of heat-shock proteins induced thermotolerance.

In addition, the delay of onset of thermotolerance correlated well with the delay in the induction or enhanced synthesis of heat-shock proteins (Li and Werb 1982). It seems reasonable to hypothesize that heat-shock proteins may play a role in providing cells with an additional measure of heat resistance. However, because we do not know the precise functions of heat-shock proteins, it is also possible that the effect of heat shock on RNA or protein synthesis may simply reflect the state of cells after heating.

The synthesis of each protein increased and decreased at specific times after the initial heat treatment, depending on the duration and temperature of the initial treatment; e.g., the rate of synthesis of the M_r 97,000 protein tended to be greatly enhanced after cells were exposed to higher temperatures and for longer times. When the proteins induced by heat shock, ethanol, sodium arsenite, or hypoxia were compared, the most prominent polypeptide induced by all agents was the M_r 70,000 protein. The rate of synthesis of the M_r 87,000 protein was enhanced by sodium arsenite, 41−46°C heat shock, and ethanol, but less so by hypoxia. The fact that synthesis of the M_r 97,000 protein was significantly enhanced only at the higher temperature, longer heating time, or higher arsenite concentration implies that its synthesis is related to the severity of the external stresses. Thus, the proteins that correlated less well with development of thermotolerance may well have functions distinct from that of protecting heat-sensitive targets. Alternatively, they may represent a mosaic of events of which the overall effect is thermotolerance. Correlation studies using end-points other than survival might be powerful tools for the identification of the roles of individual proteins.

The physiological mechanism(s) of heat-shock proteins in conferring thermotolerance is unknown. However, our observation that agents that induce heat-shock proteins also confer a transient cellular resistnace to adriamycin, an agent known to bind to the anionic lipid domain in membrane (Bearer and Friend 1981), is intriguing. These data suggest that various cellular responses to stress may share some common mechanisms, for example, some of the protective effect may be at the level of cellular membrane. Our hypothesis is strengthened by the fact that heat induces resistance to ethanol (Li et al. 1980), an agent that also acts on membranes, but heat does not confer resistance to arsenite or cadmium (unpubl.).

Hyperthermia is currently being used to treat human malignancy on an experimental basis (Hahn 1982). The practical importance of understanding the function of heat-shock proteins in the induction of thermotolerance is emphasized by recent studies showing that thermotolerance can be induced in certain tumors and normal tissues (Law et al. 1979), and that heat-shock proteins can be induced in normal tissues (Currie and White 1981) and mouse tumors (unpubl.). Thus, our studies may be of clinical relevance in measuring the thermal sensitivity of target tissues.

ACKNOWLEDGMENTS

This work was supported by grants from the U.S. Public Heath Service (CA 31397 and CA 20529) and by the U.S. Department of Energy.

REFERENCES

Bearer, E.L. and P.S. Friend. 1981. Maintenance of lipid domains in the guinea pig sperm membrane. *J. Cell Biol.* **91:** 266 (Abstr.).

Currie, R.W. and F.P. White. 1981. Trauma-induced protein in rat tissues: A physiological role for a "heat-shock" protein? *Science* **214:** 72–73.

Hahn, G.M. 1982. Heat treatment of cancer. In *Cancer—principles and practice of oncology* (ed. V.T. De Vita, S. Hellman, and S.A. Rosenberg.), pp. 1811–1821. J.B. Lippincott Co., Philadelphia.

Hahn G.M. and D.P. Strande. 1976. Cytotoxic effects of hyperthermia and adriamycin on Chinese hamster cells. *J. Natl. Cancer Inst.* **57:** 1063–1067.

Henle, K.J. and L.A. Dethlefsen. 1978. Heat fractionation and thermotolerance: A review. *Cancer Res.* **38:** 1843–1851.

Law, M.P., P.G. Coultas, and S.B. Field. 1979. Induced thermal resistance in the mouse ear. *Br. J. Radiol.* **52:** 308–314.

Levinger, L. and A. Varshavsky. 1981. Heat shock proteins of *Drosophila* are associated with nuclease-resistant, high-salt-resistant nuclear structures. *J. Cell Biol.* **90:** 793–796.

Li, G.C. and G.M. Hahn. 1978. Ethanol-induced tolerance to heat and to adriamycin. *Nature* **274:** 699–700.

————. 1980. A proposed operational model of thermotolerance based on effects of nutrients and the initial treatment temperature. *Cancer Res.* **40:** 4501–4508.

Li, G.C. and Z. Werb. 1982. Correlation between synthesis of heat shock proteins and development of thermotolerance in Chinese hamster fibroblasts. *Proc. Natl. Acad. Sci.* **79:** 3218–3272.

Li, G.C., N.S. Petersen, and H.K. Mitchell. 1982. Induced thermal tolerance and heat shock protein synthesis in Chinese hamster ovary cells. *Int. J. Radiat. Oncol. Biol. Physics* **8:** 63–67.

Li, G.C., E.C. Shiu, and G.M. Hahn. 1980. Similarities in cellular inactivation by hyperthermia or by ethanol. *Radiat. Res.* **82:** 257–268.

Coexpression of Thermotolerance and Heat-shock Proteins in Mammalian Cells

John R. Subjeck and
James J. Sciandra
Division of Radiation Biology
Roswell Park Memorial Institute
Buffalo, New York 14263

Clinical hyperthermia has been used alone or as an adjunctive treatment in cancer therapy. Several factors have been shown to influence the ability of heated cells to form colonies including pH, cell-cycle redistribution, vascular effects, and thermal history (for review see Nussbaum 1982). The latter point has been of significant interest since a hyperthermia pretreatment can afford a significant protective effect to a later treatment. This induced thermotolerance has been well studied in mammalian cells (Henle and Dethlefsen 1978).

An analogous protective phenomenon has been studied in *Drosophila* and has been connected to the expression of heat-shock proteins by Mitchell and co-workers (Mitchell et al. 1979; Petersen and Mitchell 1981). A similar relationship has been suggested for yeast (McAlister and Finkelstein 1980) and slime mold (Loomis and Wheeler 1980), and recent work suggests that such a model may explain the thermotolerance phenomenon of mammalian cells (Subjeck et al. 1982a, b; Landry et al. 1982; Li et al. 1982). We have previously examined the relationship between rate (Subjeck et al. 1982b) and accumulation (Subjeck et al. 1982a) of heat-shock proteins and the kinetics of thermotolerance induction in Chinese hamster ovary (CHO) cells and consider in further detail here the possibility that a direct relationship between induced heat-shock protein levels and thermotolerance exists.

ONE- AND TWO-DIMENSIONAL GEL ANALYSIS

Figure 1 shows one-dimensional gels of control and heat-shocked CHO cells. The control cells received a 60-minute, [³⁵S]methionine pulse at 37°C (lane A) while the heat-shocked cells received a similar 60-minute pulse beginning 7.5 hours following a 12-minute, 45°C heat shock (lane B). Lanes A and B show results following Coomassie Brilliant Blue staining, while lanes C and D are autoradiographs of lanes A and B. Several heat-shock proteins are visible including major bands at 68 kD (hsp68), 89 kD (hsp89), and 110 kD (hsp110).

Two-dimensional gel electrophoresis (Fig. 2) was performed according to O'Farrell as modified and described (Hubbard and Lazarides 1979). Autoradiographs of control (Fig. 2a) and heat-shocked (Fig. 2b) CHO

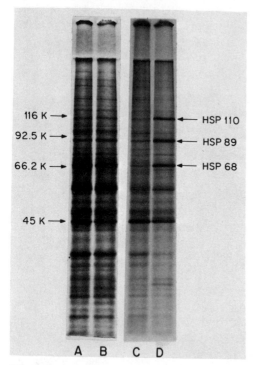

Figure 1
SDS polyacrylamide gel electrophoresis of control and heat-shocked CHO cells. A 60-minute [³⁵S]methionine pulse was applied to the controls and to the shocked cells (7.5–8.5 hr postshock). Lane A, protein pattern of cells cultured at 37°C, stained with Coomassie Brilliant Blue; lane B, protein pattern of CHO cells heat-shocked for 12 minutes at 45°C, stained with Coomassie Brilliant Blue; lane C, autoradiography of A; lane D, autoradiography of B. The three heat-shock proteins studied are designated and the others are visible. Left margin, molecular-weight standards: β-galactosidase (116,250), phosphorylase B (92,500), bovine serum albumin (66,200), ovalbumin (45,000). Equal amounts of protein were loaded using the Bio Rad protein determination assay.

cells are shown. Radiolabel was applied as indicated in Figure 1. Certain heat-shock proteins are visible, including hsp68 (1), hsp89 (3), and hsp110 (4). hsp89 appears as a molecular-weight doublet in Figure 2, both components being heat inducible. Other heat-induced peptides are visible including one at 66 kD (adjacent to hsp68) and one at 76 kD (2), as well as some lower-molecular-weight proteins (5).

A 97–100-kD heat-inducible protein has been observed in other mammalian cell lines. By two-dimensional analysis (Fig. 2), a protein which resembles this protein has been observed (arrow) but it does not appear to be heat inducible by the pretreatment conditions discussed here (12 min/45°C). If established, this identity would suggest that this protein is either not involved in thermotolerance or that it shares a function common to another protein in this system. A protein of this molecular weight is, however, strongly induced in this system by release from anoxia as is the 76-kD heat-shock protein, while hsp68 and hsp89 are only weakly induced by an anaerobic shock (J.J. Sciandra and J.R. Subjeck, unpubl.). Different types of stress may result in differential inductions of specific heat-shock proteins and provide useful information regarding their function.

CELL SURVIVAL STUDIES AND HEAT-SHOCK PROTEIN LEVELS

We have shown previously that the rate of protein synthesis of hsp68, hsp89, and hsp110 and actin is dramatically reduced by a 12-minute/ 45°C heat shock as measured by a 60-minute [^{35}S]methionine pulse delivered at increasing times at 37°C after the heat shock (Subjeck et al. 1982b). While actin recovered to near normal rates of synthesis in 8–10 hours, the three heat-shock proteins briefly achieved an induced synthesis rate of 2.5–3 times the normal level, following which these rates rapidly returned to near control values. In this study, the ability of CHO cells to form colonies, as measured following the application of a second severe heating period (thermotolerance), dramatically increased during the period of rapid heat-shock protein synthesis. The acquisition of thermal resistance also appeared to level off at a time approximately coincident with the repression of heat-shock protein synthesis. While the development of thermotolerance appears to correlate with the rate of heat-shock protein synthesis (as measured by a pulse label), the magnitude of thermotolerance expressed would be expected to relate to the total amount of heat-shock protein that accumulates in the cell (or in the integral of the rate analysis). To examine this further, we applied the [^{35}S]methionine label at the beginning of heat shock and allowed the cells to incubate in the presence of the label between heat shock and increasing times to estimate the total incorporation over the time interval in question (Subjeck et al. 1982a). In this way a curve of the total

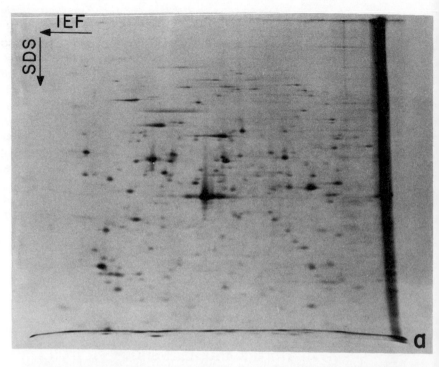

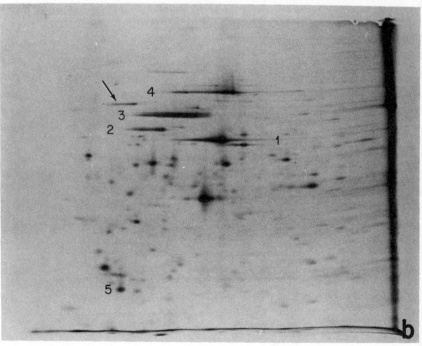

accumulation or synthesis of heat-shock proteins between heat shock to increasing times can be plotted and compared directly to a plot of survival (expressed thermotolerance) over the same interval. Heat-shock protein accumulation and thermotolerance curves could then be directly superimposed for comparative purposes. In such a study we have shown that hsp68, hsp89, and hsp110 accumulate after heat shock in a manner that is temporally coincident with the magnitude of thermotolerance expression. Such an examination yields two dependent variables at each time point after heat shock: surviving fraction and total heat-shock protein synethsis. If the examined heat-shock proteins were responsible for thermotolerance, a relationship between these dependent variables should then exist. Such an analysis is presented in Figure 3. In this figure, the survival of cells to a second 27-minute/45°C challenge (at various times following the 12-min/45°C pretreatment) is plotted against total label uptake (normalized to control synthesis levels over the same periods). This semi-log representation suggests that over this interval an exponential or dose-effect relationship between survival and each of the three major heat-shock proteins may exist. It is noteworthy in view of this result that induced heat-shock proteins only are represented in such an analysis and basal (preheat-shock) heat-shock protein levels would not be included. However, the relationship expressed in Figure 3 should be interpreted cautiously and additional studies at different post-(challenge) and pretreatment conditions are necessary to establish such a connection.

CONCLUSIONS

Available evidence strongly recommends that expression of heat-shock proteins and thermotolerance share some connection. This should prove to be useful since many of the radiobiological details of thermotolerance expression in mammalian cells, subject to a variety of induction and postincubation conditions, have been worked out. Specifically, maximal induction should be distinguished from maximal expression and care should be taken to ensure that measured parameters are independent of cellular lethality. If in addition, heat-shock proteins are responsible for thermotolerance, then the tolerance phenomenon may provide a tool to examine the natural function of heat-shock proteins, since their natural function and their role in thermotolerance should be related.

Figure 2
Isoelectric focusing/SDS-polyacrylamide gel electrophoresis of control (*a*) and heat-shocked (*b*) CHO cells. Figures are autoradiographs of labeled gels as defined in Fig. 1. hsp68 (1), hsp76 (2), hsp89 (3), and hsp110 (4). Other heat-shock proteins are visible. Equal amounts of protein were loaded.

410

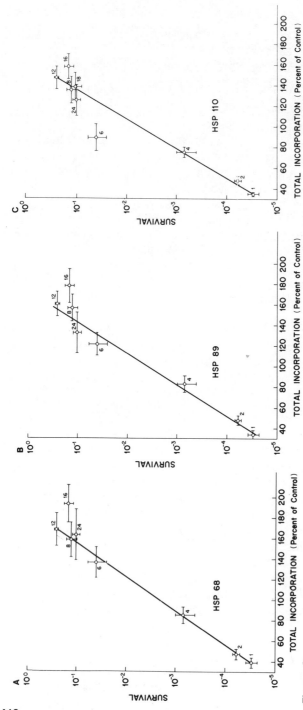

Figure 3

CHO cells are heat-shocked (12 min/45°C). *Y* axis: The acquisition of survival resistance as measured by a second 27-minute/45°C challenge is indicated at time (hours indicated) postshock. Error bar: S.E.M. from three to six independent experiments. *X axis:* The total incorporation of [³⁵S]methionine into hsp68 (*A*), hsp89 (*B*), and hsp110 (*C*) from heat shock (12 min/45°C) to the same time points is plotted. The percent total incorporation to time *t* is determined by normalizing to the total control incorporation (at 37°C) for the same time interval at each point. Incorporation determined by cutting the protein from the gel and employing a scintillation spectrophotometer (Subjeck et al. 1982a). S.E.M. plotted for ³⁵S incorporation is determined from four or five experiments and an average of seven gels per point. Data from each gel is included independently. Correlation coefficients: hsp68, .981; hsp89, .971; hsp110, .929.

REFERENCES

Henle, K.J. and L.A. Dethlefsen. Heat fractionation and thermotolerance: A review. *Cancer Res.* **38:** 1843–1851.

Hubbard, B.D. and E. Lazarides. 1979. Copurification of actin and desmin from chicken smooth muscle and their copolymerization in-vitro to intermediate filaments. *J. Cell Biol.* **80:** 166–182.

Landry, J., P. Chretien, D. Bernier, L.M. Nicole, N. Marceau, and R.M. Tanguay. 1982. Thermotolerance and heat shock proteins induced by hyperthermia in rat liver cells. *Int. J. Radiat. Oncol. Biol. Phys.* **8:** 59–62.

Li, G.C., N.S. Petersen, and H.K. Mitchell. 1982. Induced thermal tolerance and heat shock protein synthesis in Chinese hamster ovary cells. *Int. J. Radiat. Oncol. Biol. Phys.* **8:** 63–67.

Loomis, W.F. and S. Wheeler. 1980. Heat shock response of *Dictyostelium*. *Dev. Biol.* **79:** 399–408.

McAlister, L. and D.B. Finkelstein. 1980. Heat shock proteins and thermal resistance in yeast. *Biochem. Biophys. Res. Comm* **93:** 891–824.

Mitchell, H.K., G. Moller, N.S. Petersen, and L. Lipps-Sormiento. 1979. Specific protection from phenocopy induction by heat shock. *Dev. Genet.* **1:** 181–192.

Nussbaum, G. (ed.). 1982. *Physical aspects of hyperthermia,* vol. 8. American Association of Physics in Medicine Monograph. In press

Petersen, N.S. and H.K. Mitchell. 1981. Recovery of protein synthesis after heat shock: Prior heat treatment affects the ability of cells to translate mRNA. *Proc. Natl. Acad. Sci.* **78:** 1708–1711.

Subjeck, J.R., J.J. Sciandra, and R.J. Johnson. 1982a. Heat shock proteins and thermotolerance: A comparison of induction kinetics. *Br. J. Radiol.* **55:** 579–584.

Subjeck, J.R., J.J. Sciandra, C.F. Chao, and R.J. Johnson. 1982b. Heat shock proteins and biological response to hyperthermia. *Br. J. Cancer* **45** (Suppl. 5): 127–131.

Cell-cycle Phase Response and the Induction of Thermotolerance and Heat-shock Proteins

Robert R. Klevecz, Gary A. King, and Carolyn H. Buzin
City of Hope Research Institute
Department of Cell Biology
Duarte, California 91010

Our interest in heat-shock proteins stems from earlier studies on the effects of heat shock and hyperthermia on clonogenicity, thermotolerance induction, and cellular timekeeping processes in cultured mammalian cells (King et al. 1980; Klevecz et al. 1980). Hamster V79 cells display a marked increase in their tolerance to thermal damage following a regime of heat conditioning (Henle and Dethlefsen 1978). The observable effect of thermotolerance induction is a transient increase in the resistance of cells to a potentially lethal heat shock due to prior continuous or acute heat treatment. Cells treated continuously at temperatures between 41.5°C and 42.5°C are initially killed at a relatively high rate and with first-order kinetics, but after several hours the slope of the survival curve decreases (Dewey et al. 1977). Thermotolerance is transient and recycling experiments have largely ruled out the prior existence of a heat-resistant subpopulation, though temporal heterogeneity and cell-cycle redistribution may well be factors in the kinetics of induction.

Induction of thermotolerance reveals kinetic behavior that is intriguing when viewed in the context of cellular oscillators and oscillatory timekeeping mechanisms. Maximum thermotolerance development occurs in V79 cells under heat pretreatment conditions which are just sufficient to cause a full oscillator reset. It is consistent to suggest that heat-shock proteins, which have been implicated in the acquisition of thermotolerance, are preferentially translated, phase-specific proteins and may be found at a particular phase in synchronized cells or whenever a random synchronous or quiescent culture is driven to the appropriate region of phase space by any perturbation.

PHASE RESPONSE

When mitotically selected V79 cells are given heat shocks at temperatures between 42°C and 45°C through one synchronous cell cycle, a discontinuity is observed in the family of phase-response curves generated. In Figure 1, the phase responses of cells to 10-minute pulses at 42°C, 43°C, and 45°C at 0.5-hour intervals through the cell cycle are shown together with a simulated phase-response curve predicted by hard resetting of a limit cycle oscillator. Cells pulsed at 43°C or 45°C for 10 minutes in the first hour after mitotic selection are only slightly advanced or delayed in the subsequent mitosis relative to the paired untreated control. There follows a pattern of increasing delays up to 4 hours when a rather abrupt shift in response occurs, giving a second minimum in delay at 5 hours. Pulses given after 5 hours in the cycle again show increasing delays up to the initiation of the first postselection mitotic wave. The two phase-response curves are nearly identical, while 10-minute pulses at 42°C through the cell cycle produce only a slight advancing or delaying effect on cell division. This sharp transition in phase delay between 42°C and 43°C is of particular interest because of the fact that 42.5°C appears in the literature as a transition point in the induction of thermotolerance by acute versus continuous hyperthermia (Sapareto et al. 1978).

PHASE RESET AND THERMOTOLERANCE

Phase reset and cell-cyle redistribution are coupled to the development of thermotolerance. Induction of maximum thermotolerance occurs under pretreatment conditions just sufficient to cause full-phase reset. When mitotically selected V79 cells are treated at 3.75 hours after mitosis with various durations of heating at 42°C, 43°C, and 45°C, division delay increases continuously with increasing duration of heating until a maximum continuous delay of 4 hours is achieved (Fig. 2A). Beyond that point, discrete or quantized increments in delay occur with increasing exposure time.

Similarly, when random V79 cells are preheated for increasing durations at 42°C, 43°C, and 45°C and then assayed 2, 4, or 8 hours later for thermotolerance development, maximum thermotolerance is induced by that duration of heating at each temperature which is just sufficient to

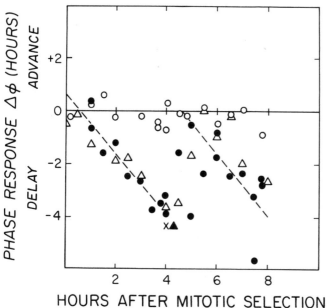

● 43°C 10min
Δ 45°C 10min
o 42°C 10min
x 42°C 40min
▲ 45°C 10min ⟶ 37°C 5min ⟶ 45°C 10min
--- limit cycle simulation

Figure 1

Phase response to conditioning heat pulses. Synchronized V79 cells were heat-shocked by complete immersion of 25 cm² plastic flasks in a Lauda model K2/R water bath. Water-bath temperatures were maintained within .05°C of desired temperature during the heating interval and were calibrated against a NBS thermometer. Means of first mitotic wave following synchronization were compared for each pair of heat-shock and control cultures as a function of time in the cycle at which the heat shock was begun. Heat shocks of 45°C, 10 minutes (Δ), 43°C, 10 minutes (●), 42°C, 10 minutes (o), and 42°C, 40 minutes (x) were employed. Positive values of Δφ indicate that heated cultures divided sooner than controls; negative values, later than controls. Methods of cell growth, synchronization, time-lapse video microscopy, and analysis of generation time and phase response have all been described in detail elsewhere (King et al. 1980; Klevecz et al. 1980). Phase response predicted by oscillator model (Klevecz et al. 1980) assuming destruction of 90% oscillator variables (------).

cause a full 4-hour reset (Fig. 2B and arrows on Fig. 2A). Additional heat pretreatment beyond the minimum necessary to produce a full reset results in diminishing thermotolerance.

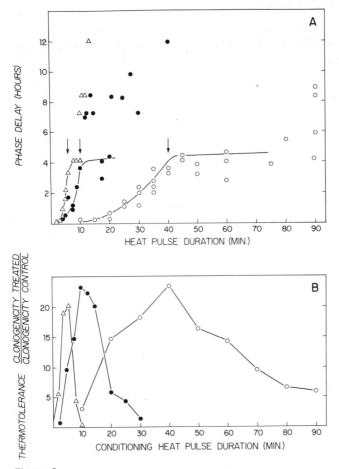

Figure 2

(A) Division delay following 45°C (△), 43°C (●), and 42°C (○) heat pulses for intervals from 2 to 90 minutes applied at 3.75 hours into the V79 cell cycle. Delays in the first synchronous division following heat shock are shown relative to paired controls as in Fig. 1 except that here delay is plotted as a positive value. Arrows indicate heat pulse duration necessary for maximum thermotolerance in Fig. 2B. (B) Thermotolerance development following 45°C (△), 43°C (●), or 42°C (○) heat pulses for durations of 2–90 minutes. Random V79 cells, 24 hours after subculture, were trypsinized and counted, and appropriate numbers of cells were seeded in 25 cm^2 flasks containing 10 ml of media which had been gassed for 24 hours previously in an atmosphere of 95% air and 5% CO_2. Inoculated flasks were then sealed and cells were allowed to attach by incubation at 37°C for 4 hours prior to heat treatment. Following pretreatment, cells were incubated for 4 hours at 37°C then reheated at 45°C for 24 minutes. Flasks were then returned to 37°C and assayed for clonogenicity 7–10 days later. Thermotolerance is expressed as the mean number of colonies per pretreated flasks divided by the mean number of colonies per unpretreated controls. A measurable enhancement of survival (two- to fourfold) appears within 15 minutes of conditioning heat treatment and plateaus by 4 hours.

KINETICS OF PHASE RESET

By perturbing the cells with two heat pulses, we have attempted to determine whether the results are consistent with a resetting of the putative timekeeping oscillator to a new phase or whether the cells are simply arrested at the cycle phase in which the heat pulses were applied. From Figures 1 and 2A, it can be seen that treating synchronized cells at 3.75 hours after mitotic selection with 43°C, 10-minute heat pulses causes a 4-hour delay and treating with 43°C, 20-minute heat pulses causes an ~ 8-hour delay. If the 20-minute heat treatment is given as two 10-minute heat pulses separated by an incubation at 37°C, the cells respond to the second pulse as though the first pulse had reset them to an oscillator phase roughly equivalent to a point 1 hour after mitosis. This redistribution occurs instantaneously, or nearly so, for if the two pulses are separated by as little as 10 minutes at 37°C, no delay beyond the 4-hour delay produced by a single pulse is manifest. According to our model the cells which were at 3.75 hours in the cell cycle have been reset to an oscillator phase equivalent to 0−1 hour in the cell cycle and respond to the second heat pulse as cells in this cycle phase respond to single heat pulses; that is, they show no further delay. As the time at 37°C between the two pulses is increased to 4 hours, the delay produced by the two pulses increases beyond 4 hours, suggesting the cells have again traversed into a sensitive phase and respond as would cells in the phase equivalent to 2−4 hours after mitosis. Similar reset kinetics are seen in thermotolerance development. If random V79 cells are pretreated with two 43°C, 10-minute heat pulses separated by 10 minutes at 37°C and then assayed 4 hours later for thermotolerance, subsequent thermotolerance development is identical to that produced by a single 43°C, 10-minute heat pretreatment and greater than that produced by a 43°C, 20-minute pretreatment.

SUMMARY/CONCLUSIONS

In summary, the induction of thermotolerance behaves as though it were dependent upon the cell reaching a particular region in oscillator phase space. It is consistent to suggest that heat-shock and stress proteins that have been implicated in the acquisition of thermotolerance (Li et al. 1982), are nothing more than preferentially translated, phase-specific proteins and for that reason are not unique to heat-shocked cells. Rather, they will be found at particular phases in synchronized cells or whenever a random or synchronous culture is driven to the appropriate phase space region by any perturbation. This is supported by the observation that heat-shock proteins appear in response to such phase-shifting agents as ethanol, anoxia, and a variety of chemical and mechanical stresses (Ashburner and Bonner 1974; Guttman et al. 1980).

We are currently assaying heat-shock protein synthesis and membrane fluidity as a function of cell-cycle phase and perturbation history.

REFERENCES

Ashburner, M. and J. Bonner. 1974. The induction of gene activity in *Drosophila* by heat shock. *Cell* **17:** 241–254.

Dewey, W.D., L.E. Hopwood, S.A. Sapareto, and L.I. Gerweck. 1977. Cellular responses to combinations of hyperthermia and radiation. *Radiology* **123:** 463–474.

Guttman, S.D., C.V.C. Glover, C.P. Allis, and M.A. Gorovsky. 1980. Heat shock, deciliation and release from anoxia induce the same set of polypeptides in starved *T. pyriformis*. *Cell* **22:** 299–307.

Henle, K.J. and L.A. Dethlefsen. 1978. Heat fractionation and thermotolerance: A review. *Cancer Res.* **38:** 1843–1851.

King, G.A., J.O. Archambeau, and R.R. Klevecz. 1980. Survival and phase response following ionizing radiation and hyperthermia in synchronous V79 and EMT-6 cells. *Radiat. Res.* **84:** 290–300.

Klevecz, R.R., G.A. King, and R.M. Shymko. 1980. Mapping the mitotic clock by phase perturbation. *J. Supramolec. Struct.* **14:** 329–342.

Li, G.C., N.S. Petersen, and H.K. Mitchell. 1982. Induced thermal tolerance and heat shock protein synthesis in Chinese hamster ovary cells. *Int. J. Radiat. Oncol. Biol. Phys.* **8:** 63–67.

Sapareto, S.A., L.E. Hopwood, W.C. Dewey, M.R. Raju, and J.W. Gray. 1978. Effects of hyperthermia on survival and progression of chinese hamster ovary cells. *Cancer* **38:** 393–400.

Summary

Alfred Tissières
Départment de Biologie Moléculaire
Université de Genève
Genève, Switzerland

The main feature of the heat-shock or stress response is the vigorous activation of a small number of specific hsp genes, previously either silent or active at low levels. New mRNAs are actively transcribed from these genes and translated into heat-shock proteins. At the same time, the translation of the messengers made prior to the stress and still present in the cells is curtailed. What is known about the hsp genes? What controls their transcription under stress or heat-shock conditions? What are the mechanisms involved in the regulation of these genes and their products? What is known about heat-shock proteins, about their function, and about the physiological response to stress? These are some of the main questions discussed in this volume. A typical characteristic heat-shock or stress response has been observed in all organisms investigated so far and there are indications that some heat-shock proteins have been conserved in species as distant as *Drosophila*, yeast, and man. Moreover a recent finding shows that there is homology between a bacterial hsp gene and the corresponding genes in yeast and *Drosophila*. These observations suggest that the response to heat shock or to stress is universal.

HEAT-SHOCK-ACTIVATED GENES

What is the organization of these genes, can particular signals essential for heat-shock induction be recognized in the nucleotide sequences flanking these genes, and can we learn something about the mechanism of coordinate induction from their study?

The major heat-shock genes in *Drosophila melanogaster,* coding for the heat-shock polypeptides hsp83 (or 84), 70, 68, 27, 26, 23, and 22 have all been cloned, mapped in detail, and sequenced (Lis et al. 1978; Livak et al. 1978; Schedl et al. 1978; Artavanis-Tsakonas et al. 1979; Craig et al. 1979; Holmgren et al. 1979; Corces et al. 1980; Craig and McCarthy 1980; Holmgren et al. 1981; Voellmy et al. 1981; Ingolia and Craig 1982b). Their organization is briefly described in the articles by Craig et al. and Török et al. (both this volume). They fall into three groups. The hsp83 in the first group is present as a single copy at chromosomal locus 63BC (Holmgren et al. 1979). The second group of hsp genes is more complex and consists in most strains of *Drosophila melanogaster* of five copies of hsp70, two at locus 87A and three at 87C, with considerable homologies on their coding and 5′-sequence regions (Craig et al. 1979; Holmgren et al. 1979; Ish-Horowicz et al. 1979) and the hsp68 gene with a single copy at locus 95D, partially homologous with hsp70 (Holmgren et al. 1979). Evolutionary implications have been drawn from the study of hsp70 in different *Drosophila* species (Leigh Brown and Ish-Horowicz 1981) and from their flanking sequences (Török et al., this volume). The third group is that of the small heat-shock genes coding for hsp27, 26, 23, and 22. They are all encoded at locus 67B, as a single copy each, on a DNA segment of about 11 kb (Corces et al. 1980; Wadsworth et al. 1980; Voellmy et al. 1981).

It has been shown from hybridization experiments that there is partial homology between the four small hsp genes (Corces et al. 1980). This was confirmed recently by DNA sequencing: the four small hsp genes are homologous over about one-half of the coding sequence to the 3′ ends, indicating that their functions may be related (Ingolia and Craig 1982b; R. Southgate, pers. comm.). These results also suggest duplication from an ancestral gene.

Several observations support the view that some of the heat-shock genes are also expressed at some time during normal development and thus their products are not uniquely needed after heat shock or some other form of stress. For instance, Sirotkin and Davidson (1982) and Sirotkin (this volume) have reported that hsp26 and hsp22, besides being induced by heat shock, are also expressed in the course of development. A recent finding of Ingolia and Craig (1982a) (see also Craig et al., this volume) has similar implications. They show that a number of genes, which they call cognates, are homologous to the hsp70, but present at different loci on the chromosomes and not induced by heat shock. They are normally expressed during development. It will of course be interesting to compare the sequences surrounding these genes with those of hsp70, and also to study their transcription following transformation or transfection using systems similar to those discussed below.

Heat-shock proteins from species as distant as yeast, *Drosophila,* chicken, and man appear to share structural homologies (Schlesinger et al., this volume), and Craig et al. (this volume) report homologies in hsp70 gene sequences in yeast, and even, remarkably, bacteria and *Drosophila.* Further comparison with genes from other species such as chicken and man will be interesting and it is probable that this information will soon become available.

In order to study transcription of *Drosophila* heat-shock genes, hsp70 from *Drosophila* was introduced into mouse cells by cotransformation with the herpes thymidine kinase gene (Corces et al. 1981 and this volume), into monkey COS cells by transfection with vectors containing an SV40 origin of replication (Mirault et al.; Pelham and Bienz; both this volume), or injected into *Xenopus* oocytes (Voellmy and Rungger 1982; Pelham and Bienz; Voellmy and Rungger; both this volume). In all four instances, hsp70 was found to be transcribed only upon raising the normal temperature of growth of the recipient cells by a few degrees, indicating that the mechanism of heat induction must be remarkably similar in these different cell types. Corces et al. have constructed a hybrid gene containing 1.3 kb of the 5′ sequence of the *Drosophila* hsp70 linked to the coding region of a human growth hormone gene. After cotransformation, the hybrid gene was expressed in mouse cells only after raising the temperature, indicating that the sequences in the 1.3 kb of *Drosophila* DNA were sufficient to place the growth hormone gene under heat-shock control. Using a series of deletions upstream of hsp70, they concluded that a region of 249 bp extending from −47 to +207 was sufficient to determine the heat-shock control of transcription. By constructing hsp70 mutant genes with fragments of different sizes upstream of the gene in Znc, the sequence of about 350 bp directly upstream of the coding sequence, and introducing these genes into COS cells by means of vectors from SV40, Mirault et al. and Pelham and Bienz (both this volume) were able to show that a stretch of less than 70 bp directly upstream of the gene was sufficient for the heat-shock response to take place. The crucial sequence appears to lie between −70 and −40.

It should be noted that in the COS cells the foreign genes are replicated to a high copy number, clearly an unusual situation. Indeed, assuming that some protein factor(s) was needed for normal control mechanisms to take place, this factor(s) would probably become limiting under these conditions.

A short imperfect inverted repeat, present 11 nucleotides upstream from the TATA box at position −44 to −56 in front of all hsp70 genes, as as first noted by Ingolia et al. (1980) and Holmgren et al. (1981), is discussed by Mirault et al. and Pelham and Bienz (both this volume). By comparing the regions upstream of several hsp70 genes, the latter authors have deduced an apparent consensus sequence, pointing out that it might be implicated in control. Obviously, attempts should also be made to use homologous systems. To this end, the P element, a transposon of *Drosophila,* has been cloned and used as a vector to bring about transformation (G.M. Rubin and A. Spradling, pers. comm.), and this looks very promising.

Are all hsp genes really coordinately regulated? Strictly speaking this does not seem to be the case, as there are particular situations where some of the heat-shock genes are expressed while others are not. Thus, the question of coordinate induction of hsp genes may be rather complex.

CHROMATIN STRUCTURE

Is chromatin structure altered in regions of active transcription? If so, what are the changes involved and what brings them about? Could this be a primary effect of gene regulation? In order to approach these problems, use has been made of nucleases—DNase I or micrococcal nuclease—to probe the accessibility of DNA to these enzymes in particular regions of chromatin. For this purpose, the heat-shock system in *Drosophila* is very suitable. The genes can easily be turned on by raising the temperature of the cells a few degrees, or by some other means. The chromatin can be studied while the genes are fully active or during the period of recovery after heat shock, when they become gradually switched off. The extent of nuclease digestion can be assayed making use of probes of known sequences from within the genes and from the surrounding regions using cloned DNA. The work of Keene and Elgin, Wu, Levy and Noll, and Eisenberg and Lucchesi (all this volume) deals with these questions. Very fine mapping of DNase I-hypersensitive and - sensitive sites are reported, at, or close to, the 5′ ends of the small heat-shock genes and hsp70 and hsp83. These sites, which are thought to be free of nucleosomes, could be the entry sites for RNA polymerase and possible regulatory factors.

Levinger and Varshavsky (1982 and this volume) have sought to determine whether variant nucleosomes are present, depending upon the functional state of the DNA. They report that nucleosomes from transcribed *copia* and hsp70 contain a large proportion of ubiquitin-H2A (uH2A) semihistone. This is in contrast to nontranscribed regions, as AT-rich satellite DNA, which lacks uH2A but contains a high amount of nonhistone protein D1.

REGULATION

Until a few years ago, it was thought that the heat-shock response, as we know it in *Drosophila*, was confined to higher eukaryotes. Then we learned that yeast responded in a way strikingly similar to that of more complex organisms (Miller et al. 1979) and now we hear that the response to heat in bacteria is also similar, as reported in the elegant work of Yamamori et al. and Neidhart et al. (both this volume). Though it is difficult at this time to evaluate to what extent this heat response is the same as in higher cells, what seems particularly interesting is that in bacteria a group of over a dozen genes, strongly activated by heat, are under the control of one gene called *hin* or *htpR*. Four of the proteins induced by heat have been identified so far and in the words of Neidhardt et al. "they are a strange group," having to do with phage morphogenesis and also essential for bacterial growth. What makes this work so attractive is the use of genetics to approach some crucial questions, as to the mechanism of induction and the function of heat-shock proteins. The possibility of using genetics to solve these problems could also be worked out with yeast or other microorganisms, such as *Dictyostelium* (see Loomis et al., this volume).

Search for proteins binding to specific DNA sequences, thus regulating the expression of particular genes, has seemed very difficult in eukaryotes. Jack and Gehring (this volume) have used cloned DNA segments from heat-shock genes of *Drosophila* as probes in a filter binding assay. They find that a partially purified protein fraction binds to sequences upstream and in some instances downstream of the heat-shock genes. So far they have failed to detect such binding sites on sequences of genes other than hsp genes, in the several cloned DNAs that were investigated.

Lindquist et al. (this volume) have made a careful analysis of the induction of heat-shock proteins and the recovery of normal protein synthesis following heat shock in *Drosophila*. As a consequence of their data, they propose that heat-shock proteins autoregulate their own synthesis.

How are heat-shock mRNAs preferentially translated in heat-shocked cells, whilst the translation of messengers made before heat shock and still present at the higher temperature is curtailed? How are heat-shock messengers recognized? In what way do they differ from other messengers? Do they bind to ribosomes and form the initiation complex more effectively? Are the initiation factors involved, or should the answer be found at the level of elongation? Initially we thought that the messengers made at normal temperatures were in some way sequestered into free ribonucleoproteins (RNP) following heat shock. This does not appear to be the case, as these messengers are in some instances found in polysomes. It should be noted that the extent of inhibition of normal protein synthesis after heat shock differs quite markedly in different species: thus, in yeast, chicken, and *Xenopus,* the inhibition is sometimes barely detectable.

This field has attracted much research, leading to sometimes different but not necessarily contradictory conclusions, as the answer may not be simple. The experiments of Ballinger and Pardue, working with *Drosophila,* and those of Thomas and Mathews, with HeLa cells, (both this volume) point to a specific block of elongation after heat shock with messengers made at normal growth temperature. This might, at least partly, be responsible for the observed inhibition of protein synthesis. Krüger and Benecke (this volume) failed to find a soluble factor responsible for this effect. Instead, they found that ribosomes from nonheat-shocked cells were able to restore the ability to translate normal messengers. Ernst et al. (this volume) have used reticulocyte and HeLa cell lysates, and in both systems they find that the phosphorylation of the initiation factor eIF-2, which takes place at the higher temperature, inhibits initiation. Hickey and Weber (this volume) conclude from their investigation in HeLa cell extracts that differences in the efficiency of initiation between heat-shock and normal messengers might be sufficient to explain the preferential translation of heat-shock messengers. In *Xenopus* oocytes, the heat-shock response appears to act entirely at the translational level: hsp70 mRNA is synthesized during oogenesis and it accumulates in an inactive, masked state until heat shock induces its translation (Bienz, this volume).

It was found recently (Glover; Sanders et al.; both this volume) that S6, the only ribosomal protein which is normally phosphorylated, was rapidly dephosphorylated after heat shock. This is particularly interesting, as the extent of this phosphorylation has been correlated to the efficiency of translation (Wool 1979).

HEAT-SHOCK PROTEINS

What is the function of heat-shock proteins, where are they located in the cell, are they present as monomers or as aggregates, do they interact with other macromolecules or with particular cellular components, are they chemically modified, have they been conserved through evolution, and therefore do they show homologies in quite distant species? Schlesinger et al. (this volume) have studied some of these problems. They have purified heat-shock proteins from chicken cells and observed that hsp89, 70, and 24 fractionated as oligomers or in a complex with other proteins. They found that the antibodies raised against chicken hsp70 cross-reacted with heat-shock proteins of similar subunit molecular weight in widely divergent species from yeast to man. A second antibody raised against hsp89 from chicken also cross-reacted with the corresponding heat-shock proteins from flies to humans. These results, together with those of Craig et al. (this volume) indicate a remarkable conservation of heat-shock proteins and hsp genes throughout evolution. Using immunofluorescent staining, Schlesinger et al. (this volume) were able to conclude that in chicken cells hsp70 and hsp24 probably interact with the cytoskeleton. Further involvement of the cytoskeleton in the heat-shock response is reported by Biessmann et al. (this volume). After raising the temperature of *Drosophila* tissue culture cells to 37°C, two proteins of 46K and 40K normally found in the cytoskeleton bound to the microsomal fraction appear to aggregate around nuclei. The 46K protein is related to vimentin, a component of the cytosketetal filament in verteberates. The vimentin cytoskeleton disintegrates after heat shock and appears to aggregate around the nucleus. The effect of heat shock on the cytoskeleton and its interaction with heat-shock proteins is important and deserves to be further investigated. Other effects such as the decrease in the membrane-associated Na^+/K^+ ATPase activity might be equally important (Burdon, this volume).

Evidence has been presented to indicate that the transforming protein of RSV, pp60^{src}, interacts in a complex with two other cellular proteins, pp50 and a heat-shock protein, hsp90 (Brugge et al. 1981; Oppermann et al. 1981). It should be noted that hsp90 from higher vertebrates, probably equivalent to hsp84 from *Drosophila* (R. Voellmy, pers. comm.), appears to be easily detectable in all cells and is only increased by two- to fivefold after stress. Yonemoto and Brugge (this volume) present here a study of this protein complex and of the kinetics of its association. The biological significance of this phenomenon is not known.

There is considerable interest in the localization of heat-shock proteins in cells, as it may give hints as to their function. Since the initial paper of Mitchell and Lipps (1975) showing that in salivary gland cells of *Drosophila* all heat-shock proteins were found in nuclei after heat shock, with the exception of hsp84, several laboratories have dealt with this subject. Similar observations have been made in *Chironomus* (Vincent and Tanguay 1979) and *Drosophila* (Arrigo et al. 1980; Velasquez et al. 1980; Arrigo and Ahmad-Zadeh 1981). Generally hsp84–90 appears to be mainly localized in the cytoplasm, whilst hsp70, 68, 27, 26, 23, and 22 are found in high proportion in the nucleus. When the cells are returned to the normal temperature after heat shock, heat-shock proteins move to the cytoplasm. While hsp68, 70, and 84 are mostly soluble in the extracts, the small heat-shock proteins are present in RNP aggregates sedimenting at 20–30S (Arrigo et al. 1980; Arrigo and Ahmad-Zadeh 1981). This seems important and therefore these aggregates should be thoroughly characterized. The observation of Ingolia and Craig (1982b) that the small heat-shock proteins from *Drosophila* share surprising homologies with the α-crystallin may eventually give a clue to their function. By means of a monoclonal antibody, Lin et al. (this volume) have shown that hsp100 in mammalian cells is localized in or close to the Golgi apparatus. The possibility that hsp100 may be involved in the metabolic reactions of the Golgi is discussed.

Much more biochemical work is needed on heat-shock proteins from different organisms, on their full characterization, and the preparation of monoclonal antibodies against them (see Welch et al., this volume). It is likely that valuable information, particularly on the function of heat-shock proteins perhaps in connection with other approaches, will then emerge.

PHYSIOLOGICAL RESPONSE

The papers discussed in this section touch upon most aspects of the stress response, genes, regulation, and proteins. When early embryos of *Drosophila* are subjected to a severe heat shock (for instance 40–41°C for 30 min, when even heat-shock gene expression is curtailed), abnormalities of morphogenesis, called phenocopies, are induced in the adult fly. The cause of these developmental defects is logically sought in the perturbation of the normal programs of gene expression taking place at a critical time. Thus, a careful study of these programs and their disturbances in gene expression caused by heat shock should eventually indicate what are the critical steps in development and in the induction of the abnormalities (Mitchell and Petersen; Petersen and Mitchell; both this volume). Are only a few genes, or possibly one, involved, or is it much more complex, with perhaps a whole sequence of events? These dramatic defects in development caused by heat can be prevented by a lower temperature (35°C) pretreatment which induces heat-shock gene expression but does not shut off normal protein synthesis. A first mild heat-shock pretreatment as mentioned above has been found to enhance survival greatly following a more severe heat shock. Thus, the first mild heat shock that induces the synthesis of heat-shock proteins appears to protect the organism from an injury which would have otherwise been fatal. It produces thermoprotection, leading to transient enhanced resistance to heat, or thermotolerance. This thermotolerance has been observed with yeast (McAlister and Finkelstein 1980; Finkelstein and Strausberg, this volume), *Dictyostelium* (Loomis and Wheeler, this volume), plant cells (Key et al.; Altschuler and Mascarenhas; both this volume), *Drosophila* (Mitchell et al. 1979; Petersen and Mitchell 1981), and HeLa (Gerner and Schneider 1975) and Chinese hamster cells (Li and Hahn 1978; Li et al.; Subjeck and Sciandra; both this volume). A correlation between the cellular level of heat-shock proteins and the degree of thermotolerance has been noted, indicating that the role of heat-shock proteins, or of some of them, might possibly be to afford protection to the cell against excessive injury. Loomis and Wheeler (this volume) have isolated a mutant from *Dictyostelium* which failed to show thermal resistance following a mild heat shock. While hsp70 was made, although at a reduced rate, no small heat-shock proteins could be detected in this strain, indicating that the small heat-shock proteins (or one of them) might be responsible for thermoprotection.

An increase in the body temperature of 3° brought about by the drug LSD, an elevation of ambient temperature, or the injection of a bacterial pyrogen all induce a marked increase in the synthesis of an hsp74 protein in the brain, heart, liver, and kidney of the rabbit (Brown and Cosgrove, this volume). Thus, in the intact mammalian organs, only one of the heat-shock-induced proteins, hsp74, was detected.

After exposing tissue to the trauma of slice preparation, hyperthermia, or the ligation of major arteries, hsp71 became the major protein synthesized in all organs of mice or rats (White and Currie, this volume). This protein was found by peptide mapping to be related to a protein of 73K which is a major constituent of all cells. Several cardiovascular cell types, including aortic smooth muscle, myocardial, and cardial mesenchymal cells, can all be stimulated to synthesize heat-shock proteins, mainly an hsp71 and also in some cases a 78K glucose-regulated protein (Hightower and White, this volume). It is difficult to correlate heat-shock proteins in the 70K region, given the slighly different molecular weights by different authors on the basis of their electrophoretic mobility. Clearly, a full characterization of heat-shock proteins in the 70K region discussed above by peptide mapping or two-dimensional gel electrophoresis is needed.

Virus infection with adeno (J.R. Nevins, pers. comm.), herpes simplex (C.M. Preston and E. Notarianni, pers. comm.), polyoma, and SV40 (E.W. Khandjian and H. Türler, pers. comm.) leads to enhanced synthesis of proteins, some of which have been identified as heat-shock proteins. It appears that this might result from some early viral functions.

The work on heat shock in mammals suggests that heat-shock proteins must be important in humans. They must be made in fever, inflammatory processes, and different kinds of stress and this should have practical consequences.

Half a century ago, Tillett and Francis (1930) in the laboratory of O.T. Avery at the hospital of the Rockefeller Institute discovered in the sera of patients with various infectious and inflammatory diseases a particular fraction which was subsequently identified as C-reactive protein by Abernethy and Avery (1941). They established the appearance of this protein, or protein family, in the serum as a response to infection, inflammation, and tissue damage. They also introduced the term "acute phase protein" to designate this fraction. A great deal of work has been done in recent years on C-reactive protein(s). They appear to be synthesized de novo in the liver in large amounts following cell injury. These proteins have been conserved in distant species, such as man, fish (Baldo and Fletcher 1973), and even horseshoe crabs (Robey and Liu 1981). We would therefore like to know whether the C-reactive protein fraction is in any way related to heat-shock proteins.[1]

CONCLUDING REMARKS

There is little doubt that the interest in the heat-shock or stress response will increase in the years to come. The study of the genes involved poses a number of general problems of nucleotide sequence organization and control. Recently, in the case of hsp70 genes, a crucial, rather short sequence necessary to place these genes under heat-shock control has been investigated. Whether similar or other "heat-shock signals" exist in front of other heat-shock genes in *Drosophila* and other species, and to which extent these signals differ from other control elements on DNA, remains to be known. Moreover, this work should eventually relate to development as it turns out that some of the heat-shock genes in *Drosophila* are also developmentally regulated. This means that the function of some of these genes is not uniquely required following stress, but that it is also needed during normal development.

The response to stress which is supposed to protect the cell against excessive injury appears to be universal. It has been conserved through evolution in very distant species and its functions therefore are likely to be of fundamental importance. It is probable that several different functions are involved, at least for the heat-shock proteins of different size classes. The search for these functions remains a difficult problem to solve. At least two lines of work however may prove fruitful in this respect: (1) the use of mutants with microorganisms and (2) the study of the interaction of the heat-shock proteins with other cellular components.

A considerable increase in the work on higher cells including man can be expected in the future, as the stress response must occur as a result of a variety of agressions to the organisms, such as infectious diseases, acute inflammation, some types of poisoning, and massive accidental damage to tissues.

---

[1]A great deal of work is presently done on the "acute phase response" and the proteins specifically made under these circumstances. See Kushner et al. (1982).

REFERENCES

Abernethy, T.J. and O.T. Avery. 1941. Occurrences during acute infections of a protein not normally present in the blood. *J. Exp. Med.* **73:** 173–182.

Arrigo, A.P. and C. Ahmad-Zadeh. 1981. Immunofluorescence localization of the small heat shock protein (hsp23) in salivary gland cells of *Drosophila melanogaster. Mol. Gen. Genet.* **184:** 73–79.

Arrigo, A.P., S. Fakan, and A. Tissières. 1980. Localization of the heat-shock induced proteins in *Drosophila melanogaster* tissue culture cells. *Dev. Biol.* **78:** 86–103.

Artavanis-Tsakonas, S., P. Schedl, M.-E. Mirault, and L. Moran. 1979. Genes for the 70,000 dalton heat shock protein in two cloned *D. melanogaster* DNA segments. *Cell* **17:** 9–18.

Baldo, B.A. and T.C. Fletcher. 1973. C-reactive protein-like precipitin in plaice. *Nature* **246:** 145–146.

Barnett, T., M. Altschuler, C.N. McDaniel, and J.-P. Mascarenhas. 1980. Heat shock induced proteins in plant cells. *Dev. Genet.* **1:** 331–340.

Brugge, J.S., E. Erikson, and R.L. Erikson. 1981. The specific interaction of the Rous sarcoma virus transforming protein, pp60^{src}, with two cellular proteins. *Cell* **25:** 263–272.

Corces, V., A. Pellicer, R. Axel, and M. Meselson. 1981. Integration, transcription and control of a *Drosophila* heat shock gene in mouse cells. *Proc. Natl. Acad. Sci.* **78:** 7038–7042.

Corces, V., R. Holmgren, R. Freund, R. Morimoto, and M. Meselson. 1980. Four heat shock proteins of *Drosophila melanogaster* coded within a 12-kilobase region in chromosome subdivision 67B. *Proc. Natl. Acad. Sci.* **77:** 5390–5393.

Craig, E.A. and B.J. McCarthy. 1980. Four *Drosophila* heat shock genes at 67B: characterisation of recombinant plasmids. *Nucleic Acid Res.* **8:** 4441–4457.

Craig, E.A., B.J. McCarthy, and S.C. Wadsworth. 1979. Sequence organisation of two recombinant plasmids containing genes for the major heat shock induced protein of *D. melanogaster. Cell* **16:** 575–588.

Gerner, E.W. and M.J. Schneider. 1975. Induced thermal resistance in HeLa cells. *Nature* **256:** 500–502.

Holmgren, R., V. Corces, R. Morimoto, R. Blackman, and M. Meselson. 1981. Sequence homologies in the 5′ regions of four *Drosophila* heat shock genes. *Proc. Natl. Acad. Sci.* **78:** 3775–3778.

Holmgren, R., K. Livak, R. Morimoto, R. Freund, and M. Meselson. 1979. Studies of cloned sequences from four *Drosophila* heat shock loci. *Cell* **18:** 1359–1370.

Ingolia, T.D. and E.A. Craig. 1982a. *Drosophila* gene related to the major heat-shock-induced gene is transcribed at normal temperatures and not induced by heat shock. *Proc. Natl. Acad. Sci.* **79:** 525–529.

————. 1982b. Four small *Drosophila* heat shock proteins are related to each other and to mammalian α-crystallin. *Proc. Natl. Acad. Sci.* **79:** 2360–2364.

Ingolia, T.D., E.A. Craig, and B.J. McCarthy. 1980. Sequence of three copies of the gene for the major *Drosophila* heat shock induced protein and their flanking regions. *Cell* **21:** 669–679.

Ish-Horowicz, D., S.M. Pinchin, P. Schedl, S. Artavanis-Tsakonas, and M.E. Mirault. 1979. Genetic and molecular analysis of the 87A7 and 87C1 heat-inducible loci in *D. melanogaster. Cell* **18:** 1351–1358.

Kushner, I., J.E. Volanakis, and H. Gewwiz, eds. 1982. C-Reactive protein and the plasma protein response to tissue injury. *Ann. N.Y. Acad. Sci.* **389.**

Leigh Brown, A.J. and D. Ish-Horowicz. 1981. Evolution of the 87A and 87C heat shock loci in *Drosophila. Nature* **290:** 677–682.

Levinger, L. and A. Varshavsky. 1982. Selective arrangement of ubiquinated and D1 protein-containing nucleosomes within the *Drosophila* genome. *Cell* **28:** 375–385.

Li, G.C. and G.M. Hahn. 1978. Ethanol-induced tolerance to heat and to adriamycin. *Nature* **274:** 699–701.

Lis, J.T., L. Prestidge, and D.S. Hogness. 1978. A novel arrangement of tandemly repeated genes at a major heat shock site in *D. melanogaster. Cell* **14:** 901-919.

Livak, K., R. Freund, M. Schweber, P. Wensink, and M. Meselson. 1978. Sequence organisation and transcription at two heat shock loci in *Drosophila. Proc. Natl. Acad. Sci.* **75:** 5613–5617.

McAlister, L. and D.B. Finkelstein. 1980. Heat shock proteins and thermal resistance in yeast. *Biochem. Biophys. Res. Commun.* **93:** 819–824.

Miller, M.J., N.H. Xuong, and E.P. Geiduschek. 1979. A response of protein synthesis to temperature shift in the yeast *Saccharomyces cerevisiae. Proc. Natl. Acad. Sci.* **76:** 5222–5225.

Mitchell, H.K. and L.S. Lipps. 1975. Rapidly labelled proteins on the salivary gland chromosomes of *Drosophila melanogaster. Biochem. Genet.* **13:** 585–602.

Mitchell, H.K., G. Moller, N.S. Petersen, and L. Lipps-Sarmiento. 1979. Specific protection from phenocopy induction by heat shock. *Dev. Genet.* **1:** 181–192.

Oppermann, H., W. Levinson, and J.M. Bishop. 1981. A cellular protein that associates with the transforming protein of Rous sarcoma virus is also a heat shock protein. *Proc. Natl. Acad. Sci.* **78:** 1067–1071.

Petersen, N.S. and H.K. Mitchell. 1981. Recovery of protein synthesis after heat shock: Prior heat-treatment affects the ability of cells to translate mRNA. *Proc. Natl. Acad. Sci.* **78:** 1708–1711.

Robey, R.A. and T.-Y. Liu. 1981. Limulin: A C-reactive protein from *Limulus polyphemus. J. Biol. Chem.* **256:** 969–975.

Schedl, P., S. Artavanis-Tsakonas, R. Steward, W. Gehring, M.E. Mirault, M. Goldschmidt-Clermont, L. Moran, and A. Tissières. 1978. Two hybrid plasmids with *D. melanogaster* DNA sequences complementary to mRNA coding for the major heat shock protein. *Cell* **14:** 921–929.

Sirotkin, K. and N. Davidson. 1982. Developmentally regulated transcription from *Drosophila melanogaster* chromosomal site 67B. *Dev. Biol.* **89:** 196–210.

Tillett, W.S. and T. Francis. 1930. Serological reactions in pneumonia with a non protein somatic fraction of pneumococcus. *J. Exp. Med.* 561–571.

Velasquez, J.M., B.J. DiDomenico, and S. Lindquist. 1980. Intracellular localization of heat shock proteins in *Drosophila. Cell* **20:** 679–689.

Vincent, M. and R.M. Tanguay. 1979. Heat shock induced proteins present in the cell nucleus of *Chiromonus tentans* salivary gland. *Nature* **281:** 501–503.

Voellmy, R. and D. Rungger. 1982. Transcription of a *Drosophila* heat shock gene is heat induced in *Xenopus* oocytes. *Proc. Natl. Acad. Sci.* **79:** 1776–1780.

Voellmy, R., M. Goldschmidt-Clermont, R. Southgate, A. Tissières, R. Levis, and W. Gehring. 1981. A DNA segment isolated from chromosomal site 67B in *D. melanogaster* contains four closely linked heat shock genes. *Cell* **23:** 261-270.

Wadsworth, S.C., E.A. Craig, and B.J. McCarthy. 1980. Genes for three *Drosophila* heat shock induced proteins at a single locus. *Proc. Natl. Acad. Sci.* **77:** 2134–2137.

Wool, I.G. 1979. The structure and function of eukaryotic ribosomes. *Ann. Rev. Biochem.* **48:** 719–754.

POSTERS

G. ALIPERTI and M.J. SCHLESINGER, Dept. of Microbiology and Immunology, Washington University Medical School, St. Louis, Missouri: Recovery of normal chicken embryo fibroblast mRNAs after heat shock is inhibited by actinomycin D and cycloheximide.

M. ALTSCHULER and J.P. MASCARENHAS, Dept. of Biological Sciences, State University of New York, Albany: The synthesis of heat-shock and normal proteins at high temperatures in plants.

B.G. ATKINSON,[1] M. SOMERVILLE,[1] and D.B. WALDEN,[2] [1]Dept. of Zoology and [2]Dept. of Plant Sciences, University of Western Ontario, London, Canada: The effect of heat shock and other stresses on the gene expression of differentiating myogenic cells.

E. BERGER,[1] R. IRELAND,[1] K. SIROTKIN,[2] M.A. YUND,[3] and D. OSTERBUR,[3] [1]Dept. of Biology, Dartmouth College, Hanover, New Hampshire; [2]Dept. of Microbiology, University of Tennessee, Knoxville; [3]Dept. of Genetics, University of California, Berkeley: The regulation of small hsp gene activity is under dual control.

N. BOURNIAS-VARDIABASIS and C.H. BUZIN, Division of Cytogenetics and Cytology, City of Hope Medical Center, Duarte, California: Mild heat pretreatment protects *Drosophila* embryonic cells from inhibition of differentiation caused by hyperthermia.

T. BRADY and M. SORLEY, Dept. of Biological Sciences, Texas Tech University, Lubbock: Chromosomal binding of pyridoxine (vitamin B_6) at the tyrosine aminotransferase (TAT) gene site during heat shock and pyridoxine-induced gene activation.

P.A. BROMLEY and R. VOELLMY, Dept. of Molecular Biology, University of Geneva, Switzerland: The effect of a reversible heat shock on the expression of RSV and vesicular stomatitis virus in infected cells.

P.A. BROMLEY and R. VOELLMY, Dept. of Molecular Biology, University of Geneva, Switzerland: Massive heat-shock polypeptide synthesis in late chicken embryos—A convenient system for the study of protein synthesis in highly differentiated organisms.

R. Caizzi, C. Caggese, and F. Ritossa, Institute of Genetics, Bari, Italy: Modifications of the 70K heat shock protein in *Drosophila.*

R. Camato, L. Nicole, and R.M. Tanguay, Dept. of Medicine, Université Laval, Ste. Foy, Canada: Histone gene expression and histone modifications during heat shock in *Drosophila* cells.

A. Dangli,[1] C. Grond,[2] R. Kabisch,[1] and E.K.F. Bautz,[1] [1]Molekulare Genetik, Universität Heidelberg, Federal Republic of Germany; [2]Dept. of Genetics, University of Nijmegen, The Netherlands: Heat-shock puff 93 D in *D. melanogaster*—Ultrastructural and immunocytochemical aspects of organization and homology to other species.

K. Dybvig, G. Aliperti, C.D. Clark, D. Liteanu, and M.J. Schlesinger, Dept. of Microbiology and Immunology, Washington University Medical School, St. Louis, Missouri: Cloning of the gene coding for the 24,000 D heat-shock protein (hsp24) from chicken embryo fibroblasts.

G. Graziosi, F. Micali, A. Di Marcotullio, A. Savoini, F. de Cristini, and R. Marzari, Instituto di Zoologia e Anatomia Comparata, Università di Trieste, Italy: Heat shock of the *Drosophila* egg—Mortality, cellular multiplication, DNA replication, and hsp synthesis.

J.J. Heikkila, L. Gedamu, K. Iatrou, and G.A. Schultz, University Biochemistry Group, University of Calgary, Canada: Expression of a set of fish genes following heat or metal ion exposure.

H. Iida and I. Yahara, Tokyo Metropolitan Institute of Medical Science, Japan: Durable synthesis of heat-shock proteins in yeast cells and chicken embryonic cells during entering the resting state.

J.K.C. Knowles and K. Hemminki, Dept. of Medical Chemistry and Dept. of Genetics, University of Helsinki, Finland: Induction of heat-shock puffs in *D. hydei* salivary gland cells by microinjection of a chromatin fraction.

S. Kurtz, L. Petko, and S. Lindquist, Dept. of Biology, University of Chicago, Illinois: Heat-induced genes of *S. cerevisiae.*

K.W. Lanks, Dept. of Pathology, State University of New York, Downstate Medical Center, Brooklyn: Heat shock induces synthesis of the 85K glucose-regulated protein.

W. Levinson, G. Li, H. Opperman, D. Johnston, and J. Jackson, Dept. of Microbiology, University of California, San Francisco: Chemical induction of heat-shock proteins.

R.E.J. Mitchel and D.P. Morrison, Dept. of Radiation Biology, Atomic Energy of Canada Limited, Chalk River, Ontario: Heat-shock induction of radiation resistance and recombinational repair ability in *S. cerevisiae.*

J. Mohler and M.L. Pardue, Dept. of Biology, Massachusetts Institute of Technology, Cambridge: Genetic analysis of the region containing the 93 D heat-shock locus.

R. Morimoto and M. Meselson, Harvard Biochemical Laboratories, Cambridge, Massachusetts: The hyperthermal protective role of the mammalian 68K stress-induced protein.

R. Morimoto, J. Schaffer, and M. Meselson, Harvard Biochemical Laboratories, Cambridge, Massachusetts: In vitro transcription of *Drosophila* actin and heat-shock genes in HeLa cell-free extracts.

J.R. Nevins, Rockefeller University, New York: Induction of synthesis of the HeLa cell 70kD heat-shock protein by the adenovirus E1A gene product.

T.S. Nowak, Jr., W.D. Lust, and J.V. Passonneau, NINCDS, National Institutes of Health, Bethesda, Maryland: Metabolic correlates of amphetamine-induced hyperthermia in mouse brain.

F.P.A.M.N. PETERS, C.J. GROND, P.A. SONDERMEIJER, and N.H. LUBSEN, Dept. of Genetics, University of Nijmegen, The Netherlands: Chromosomal arrangement of heat-shock locus 2-48*B* of *Drosophila hydei.*

J. PLESSET, J.J. FOY, and C.S. MCLAUGHLIN, Dept. of Biological Chemistry, University of California, Irvine: Heat shock in *S. cerevisiae*—Quantitation of transcriptional and translational effects.

O. PONGS, W. KOERWER, and A. POETING, Lehrstuhl für Biochemie, Ruhr-Universität Bochum, Federal Republic of Germany: Two small heat-shock proteins are ecdysterone inducible proteins in salivary glands of *D. melanogaster* larvae.

C.M. PRESTON and E. NOTARIANNI, Institute of Virology, Glasgow, Scotland: HSV immediate-early polypeptides induce heat-shock proteins.

I. RUBIN and H. SWIFT, Dept. of Pathology, University of Chicago, Illinois: Effect of growth conditions and treatment patterns on the heat-shock response of chick embryo cells.

F. SCHÖFFL and J.L. KEY, Dept. of Botany, University of Georgia, Athens: Soybean heat shock—Molecular studies on mRNAs for a group of heat-shock proteins.

J.C. SILVER and D. PEKKALA, Dept. of Microbiology, University of Toronto, Canada: Effect of heat shock on synthesis and phosphorylation of nuclear and cytoplasmic proteins in the fungus *Achlya.*

R.M. SINIBALDI and P.W. MORRIS, Dept. of Biological Chemistry, University of Illinois Medical Center, Chicago: *D. melanogaster* Kc cell heat-shock proteins and the nucleoskeleton.

T.P. SNUTCH and D.L. BAILLIE, Dept. of Biological Sciences, Simon Fraser University, Burnaby, Canada: Heat-shock induction in *C. elegans.*

R.M. TANGUAY and M. VINCENT, Dept. of Medicine, University Laval, Ste. Foy, Canada: Intracellular distribution and modification of cellular and heat-shock proteins during heat shock in *Drosophila.*

G.P. THOMAS, Cold Spring Harbor Laboratory, New York: Isolation and preliminary characterization of recombinant phage containing the gene for the human 90K stress protein.

G.P. THOMAS and M.B. MATHEWS, Cold Spring Harbor Laboratory: Alterations in gene expression of HeLa cells exposed to amino acid analogs.

J. VELAZQUEZ and S.L. LINDQUIST, Dept. of Biology, University of Chicago, Illinois: Studies with monoclonal antibodies against hsp 70.

R. VOELLMY,[1] P. BROMLEY,[2] and H.P. KOCHER,[3] [1]Dept. of Biochemistry, University of Miami School of Medicine, Florida; [2]Battelle, Geneva Research Center, Switzerland; [3]Dept. of Medical Biochemistry, University of Geneva Medical Center, Switzerland: Structural similarities between heat-shock proteins from different eukaryotes.

R. VOELLMY[1] and D. RUNGGER,[2] [1]Dept. of Biochemistry, University of Miami School of Medicine, Florida; [2]Dept. of Animal Biology, University of Geneva, Switzerland: Heat-induced transcription of *Drosophila* heat-shock genes in *Xenopus* oocytes.

S.C. WADSWORTH, Cell Biology Group, Worcester Foundation for Experimental Biology, Shrewsbury, Massachusetts: A family of related proteins is encoded by the major *Drosophila* heat-shock gene family.

D.B. WALDEN,[1] B.G. ATKINSON,[2] C.L. BASZCZYNSKI,[1] J. BOOTHE,[1] and M.W. KLINCK,[1] [1]Dept. of Plant Sciences; [2]Dept. of Zoology, University of Western Ontario, London, Canada: Temperature shift response in *Zea mays.*

Index